STALKING THE SHADOWS
My Ex, His Obsession, and the Digital Chains

STALKING THE SHADOWS
My Ex, His Obsession, and the Digital Chains

by
Dr. Rachel Levitch

Copyright © 2026 by With Healthy Families First Broadcast and Media
WHFF.tv and WHFF.Radio
All rights reserved. No part of this book may be reproduced, scanned,
or distributed in any printed or electronic form without permission.
First Edition: January, 2026
Printed in the United States of America
ISBN: [ISBN number with hyphens]

Section One: Stalking the Shadows — My Experience of Narcissistic Abuse (Autobiographical)

Introduction: Personal Story and Contextual Background

This section invites the reader into the lived experience of narcissistic abuse, beginning with my personal story. It explores the chronology of events, the erosion of boundaries, and the profound emotional and psychological impact. These detailed accounts lay the foundation for understanding the patterns and mechanisms explored in Section Two.

Chapter 1: Autobiography: Stalking the Shadows
My Ex, His Obsession, and the Digital Chains

Chapter 2: Autobiography: Understanding the Character of My Ex
Narcissism, Obsession, and the Cycle of Control

Chapter 3: Autobiography: Psychological Profile
The Illness Behind the Obsession

Chapter 4: Autobiography: The Link Between Narcissistic Injury and Cyberstalking
The Psychological Impact on the Observed

Chapter 5: Autobiography: *The Unyielding Presence of Cyberstalking and Harassment*
Unwanted "Check-Ins" and "Pop-Ups"

Chapter 6: Autobiography: *Narcissistic Abuse In The Digital Age*

How Technology Empowers Stalkers and

Manipulators

Chapter 7: Autobiography: *The Fragmented Self*
    Emotional and Physical Effects

Chapter 8: Autobiography: *The Breaking Point*
    The Narcissistic Injury and Psychological Breakdown

Chapter 9: Autobiography: *The Never-Ending Grip of Narcissistic Abuse*
    Identity Theft: A Weapon of Control

Chapter 10: Autobiography: The Unseen Prisoner of Narcissistic Abuse Breaking Point
    The Silent Suffering: Recognizing Narcissistic Abuse

Section Two: Understanding Narcissistic Dynamics and Cyberstalking (Autobiographical with Analysis)

Introduction: From Personal Experience to Broader Insights
This section shifts from the narrative to a more analytical perspective, exploring the psychological mechanisms underlying narcissistic behaviors. The personal story continues, but now with a more authoritative voice and references to build understanding.

Chapter 1: The Silent Suffering
    Recognizing Narcissistic Abuse

Chapter 2: *Foundation First*
    The Link Between Rejection and Narcissistic Rage

Chapter 3: Linking Narcissistic Rage and Cyberstalking
    The Covert Manipulation Behind Every Message

Chapter 4: Manipulation Tactics: The Smear Campaign
    Manipulating Perception Through Lies and Social Influence

Chapter 5: Manipulation Tactics: Narcissists' Use of Victim Responses to Bolster Their Self-Image
    The Long-Term Effects: Social Isolation and the Need for Rebuilding Trust

Chapter 6: Manipulation Tactics: The Psychological Game and Delusion of Superiority
    A Deeper Look Into Narcissistic Behavior

Chapter 7: Digital Abuse: The Cyberstalking and Invasion of Privacy
Narcissistic Delusion in the Digital Age

Chapter 8: Digital Abuse: The Sadistic Nature of Narcissism
Invasion of Privacy and Control

Chapter 9: Digital Abuse: Narcissistic Superiority and Manipulation During Mental Health Struggles
How Narcissists Exploit Others' Vulnerabilities

Chapter 10: The Digital Veil
Hiding Behind Technology to Perpetuate Abuse

Continue the Journey in

"Behind the Digital Chains: Cyber Abuse, Narcissism, and the Path to Psychological Freedom"

The story doesn't end here. In the next volume, Dr. Rachel Levitch takes readers deeper into the long-term consequences of narcissistic abuse and the ways technology can be weaponized to control, harass, and manipulate. Sections Three and Four explore how narcissistic behaviors extend beyond personal relationships into digital environments, legal battles, and the fractured dynamics of family life. Through case studies, autobiographical examples, and psychological analysis, readers will discover how obsession, entitlement, and the digital tools of surveillance create invisible prisons that persist even after physical separation.

This continuation is not just about understanding the abuse—it is about reclaiming power, clarity, and psychological freedom. From confronting cyberstalking and identity manipulation to healing from the emotional toll of Complex PTSD, this next book offers both insight and hope for anyone seeking to navigate, survive, and ultimately break free from the lasting grip of narcissistic abuse.

DR. RACHEL LEVITCH

About the Author

Dr. Rachel Levitch is a multidisciplinary professional recognized as a cybersecurity expert, privacy engineer, author, and philanthropist. She is the CEO and Founder of the **Cognitive Institute of Dallas, Charles Edda & Charles Bouley** and, **With Healthy Families First Broadcast and Media [WHFF.TV and WHFF.RADIO]** where she leads initiatives supporting families navigating cognitive health challenges, child welfare systems, and digital literacy.

With advanced credentials including a **PhD in Performance Improvement in Technology**, and certifications such as **CRISC, CISO, and CMQE**, Dr. Levitch combines technical expertise with a deep understanding of human behavior to address the intersection of technology, privacy, and psychological well-being. Her work spans cybersecurity, privacy engineering, and the design of systems that safeguard individuals in the digital age.

Dr. Levitch is also a media producer and author. Her writing and documentaries explore narcissistic abuse recovery, parenting in the technology age, and navigating the hidden psychological effects of digital manipulation. Through her professional and philanthropic efforts, she empowers individuals and communities to reclaim autonomy, protect privacy, and build resilience in an increasingly connected world.

Dr. Rachel Levitch

For more about Dr. Levitch and her work, readers can visit:

- Official Website – Cognitive Institute of Dallas: https://cognitiveinstituteofdallas.org/dr-rachel-levitch-community-service-and-commitment-to-outreach/

- Media & Advocacy Projects – WHFF.TV Staff Profile: https://whff.tv/staff-dr-rachel-levitch.html

- Professional Expertise & Consulting – Kolabtree Profile: https://www.kolabtree.com/find-an-expert/rachel-l.23656

- Media/Film Work – IMDb Profile: https://www.imdb.com/name/nm16162859/

- Thought Leadership on Technostress & AI – CX Quest Article: https://cxquest.com/dr-rachel-levitch-on-invisible-labor-technostress-and-the-ai-economy/

Disclaimer

Disclaimer: Personal Experience and Educational Purpose

The experiences shared in this book are based on my personal recollections and perspective. When I refer to terms like "ex-husband," "ex-partner," and "ex," these are used to represent the same individual from my past. However, this narrative is not intended to serve as a factual or definitive legal account but rather as a personal reflection on the emotional and psychological effects of narcissistic abuse, cyberstalking, and harassment.

Some details have been altered or fictionalized for privacy and safety reasons. Where relevant, specific names, locations, and identifying features have been changed to protect the identities of those involved, particularly in cases where legal prosecution has not occurred or where such details are not public knowledge.

Cyberstalking and harassment are illegal behaviors under current law. The actions described in this book are shared with the intention of offering educational insight and teachable moments to help others understand the dangers of manipulative behavior, how it can escalate, and its legal and emotional impact. This book aims to provide resources and tools to those who may find themselves in similar situations and encourages people to seek legal and psychological support.

Dr. Rachel Levitch

While my experiences have led me to the conclusions shared here, I recognize that these are personal reflections and may not represent the actions or intentions of others involved. This is not intended to accuse or charge anyone with wrongdoing but rather to offer insight into the emotional, psychological, and legal implications of living under such behaviors.

For more on disclaimer visit:

https://rachellevitch.io

Disclaimer Power

A Word on Exposing the Narcissist's Power: "F Him"

Let me make this clear: exposing the truth is a direct, unapologetic act of deflation toward any narcissist who has tried to control or harm you. When you unmask them, they lose the only power they have—the ability to manipulate through <u>secrecy.</u> It's empowering, it's freeing, and it's one of the most powerful acts of self-protection.

In my case, revealing the truth of my experiences, calling out the manipulation, and showing the narcissist for who he truly is, doesn't just provide me with a sense of justice, but it takes back my power. Every time I speak my truth, every time I share what was hidden, every time I expose the toxic behaviors and the game they were trying to play, I am taking control of my narrative.

And to him—my ex, the narcissist, the manipulator—F you. For all the damage, the lies, the control, the manipulation, the harassment. You no longer hold power over me. This is my story, my journey, and you are no longer part of it.

This is my declaration of independence. You are not worth my silence anymore. This is me stepping into my truth. And to anyone who feels like they're being controlled, manipulated, or gaslit—speak your truth. There's power in it. They thrive on your silence, on your fear. Don't let them take that from you. Expose them.

Preface

Writing this book hasn't been easy. It's a raw and difficult journey to relive the past, especially when the echoes of ongoing investigations and the anonymity of my abuser still hover over every word. In the fall of 2025, I experienced a major illegal hack—one that continues to be investigated—and it's this breach of my personal and professional life that ultimately pushed me to share my story.

This is not just about an abusive relationship. It's about the quiet, covert ways narcissism can infiltrate your life, often disguised behind technology and manipulated narratives. The line between safety and fear becomes blurred when someone like this is bent on controlling you, and for me, this control came through digital harassment, obsessive stalking, and psychological manipulation.

The impact has been far-reaching: my finances, my safe spaces, my business, and even my family have all been affected. But through it all, there's one thing that remains untouched—my mental health, my gift of life, and my unwavering belief that I am not the sick one here. I am not the manipulator. I am not the covert narcissist. I live in the freedom of knowing I am not broken, and I am not to blame for what was done to me.

In this book, I invite you into the reality of what it's like to face a narcissist in their truest form. I want to show you how this kind of emotional and psychological abuse can often go unnoticed or misunderstood, especially when masked by technology and rage. I hope that by sharing my experience, others who are living in similar situations will see the signs sooner, protect themselves more effectively, and find the strength to reclaim their autonomy.

Though my story has been one of pain, it has also been one of immense resilience. It's a story about healing, reclaiming control, and finally understanding that you are not the problem. The narcissist is. I hope that through these pages, you will find the strength to break free and live in the freedom that I now cherish every day.

My hope.

I hope this book prepares you for the moment when fear arrives uninvited—when you realize your privacy has been breached, your future has been peered into, and the places you once felt safest no longer feel secure. After a serious hack, the fear isn't abstract; it's visceral. It's the sudden awareness that someone has crossed invisible but sacred boundaries, that your personal life, your work, your identity, and your sense of safety have been intruded upon without consent. That kind of violation changes how you see the world.

I also hope this book helps you understand why seeking help can feel so isolating. Local law enforcement often does their best, but many agencies are not equipped with the digital tools, training, or jurisdiction needed to investigate sophisticated cyber intrusion or covert stalking. This gap can leave victims feeling dismissed, misunderstood, or forced to explain harm that doesn't leave visible marks. That disconnect can deepen the fear and amplify the sense of being alone.

Most of all, I hope this book exposes the unique danger of the covert narcissist—the kind who hides behind anonymity, technology, and plausible deniability. This form of narcissism thrives in the shadows, fueled by rage, control, and the belief that remaining unseen means remaining untouchable. It is calculated, persistent, and deeply destabilizing. Naming it matters. Understanding it matters.

If you recognize yourself anywhere in these pages, know this: the fear you feel is not weakness, and the violation you experienced is real. You are not imagining it. You are not "too sensitive." And you are not the one who is broken. My hope is that this book gives language to what has been happening to you, clarity where there has been confusion, and the reassurance that freedom—real freedom—begins with the truth.

Dr. Rachel Levitch

Federal Resources

Federal Bureau of Investigation (FBI) - Internet Crime Complaint Center (IC3)

URL: https://www.ic3.gov

Description: The FBI's IC3 provides a platform for individuals to report internet crimes, including cyberstalking and online harassment. If you believe you are a victim of digital abuse, the IC3 is a critical resource for filing a complaint and seeking federal investigation.

Office for Victims of Crime (OVC)

URL: https://ovc.ojp.gov

Description: The OVC is part of the U.S. Department of Justice and provides information, funding, and resources for crime victims. They offer a wide range of support services, including information on victim compensation and safety planning.

National Domestic Violence Hotline (NDVH)

URL: https://www.thehotline.org

Description: While focused on domestic violence, the NDVH provides support for those experiencing emotional, psychological, or digital abuse. They offer confidential support, safety planning, and resources for those in danger.

U.S. Department of Justice - Stalking Resource Center

URL: https://www.ncjrs.gov/pdffiles1/ojp/225325.pdf

Description: The Stalking Resource Center offers resources, research, and training to victims of stalking, law enforcement, and others involved in the legal process. They provide educational materials to better understand the dynamics of stalking, including cyberstalking.

Cybersecurity and Infrastructure Security Agency (CISA)

URL: https://www.cisa.gov

Description: CISA provides resources for protecting personal information and preventing online attacks. If you believe you are the victim of a cybercrime or hacking, their website offers guidance and a variety of resources to help protect your digital life.

Dr. Rachel Levitch

Nonprofit Resources

Cyber Civil Rights Initiative (CCRI)

URL: https://www.cybercivilrights.org

Description: CCRI is dedicated to helping victims of non-consensual pornography, cyberstalking, and other forms of online harassment. They provide free legal aid and support, along with educational resources about cyber abuse.

National Center for Victims of Crime (NCVC)

URL: https://victimsofcrime.org

Description: NCVC offers support for all types of crime victims, including those who have experienced emotional, financial, or digital abuse. They provide a variety of resources, including a victim assistance helpline.

The National Cyber Security Alliance (NCSA)

URL: https://staysafeonline.org

Description: NCSA promotes online safety and security. They provide resources for individuals on how to protect their information and manage digital threats, including cyberstalking and online harassment.

LoveisRespect

URL: https://www.loveisrespect.org

Description: LoveisRespect offers confidential support and resources for young people experiencing relationship abuse, including digital abuse and stalking. They provide text, chat, and phone support.

Safe Horizon

URL: https://www.safehorizon.org

Description: Safe Horizon is a nonprofit that supports victims of domestic violence, sexual assault, child abuse, and other violent crimes. They offer a range of services, including emotional support, legal advocacy, and help with safety planning.

State Resources

It's also crucial to contact local state resources to address legal matters, file complaints, or connect with additional services. Most states have specific victim assistance programs, local hotlines, or law enforcement agencies that handle cases of harassment and stalking.

State-specific resources: Visit your state's government website or search for state-level domestic violence organizations for additional support. Many states also have specialized cybercrime task forces.

Section One Introduction

Stalking the Shadows — My Experience of Narcissistic Abuse

Section One invites the reader into the lived experience of narcissistic abuse, beginning with my personal story. This is not a theoretical exploration; it is the narrative of what actually happened, the chronology of events, and the profound impact on my sense of self. Here, the reader will see the patterns, feel the uncertainty, and witness the erosion of boundaries from a first-person perspective.

The experiences recounted in this section highlight the insidious ways in which narcissistic personalities operate: through control, manipulation, and obsession. These behaviors are often subtle, appearing as concern or attention, but beneath the surface lie patterns of coercion and psychological dominance. Technology amplifies these dynamics, creating new avenues for intrusion, monitoring, and influence.

This section also explores the emotional consequences of such abuse—the erosion of trust in one's own perceptions, the hypervigilance, and the constant negotiation between fear and self-preservation. Readers will encounter not only the actions of the narcissist but also the responses they provoke: confusion, doubt, and the struggle to maintain a coherent sense of self.

By beginning with the personal narrative, the goal is to provide a foundation of empathy and understanding. Readers will not only comprehend the chronology of events but also grasp the emotional and psychological landscape in which these behaviors unfold. The detailed accounts create the groundwork for Section Two, which analyzes these behaviors and their underlying psychological mechanisms.

Through this narrative lens, the reader can experience the story as it unfolded, allowing them to later understand the patterns and implications with clarity. Section One is both a memoir and a case study, showing the human cost of narcissistic abuse and preparing the reader for the deeper psychological exploration to come.

Chapter 1

Stalking the Shadows: My Ex, His Obsession, and the Digital Chains

In my life, there is a shadow that never truly fades, a presence that haunts me daily despite the physical distance that separates us. This shadow is my ex-husband—his obsession with controlling my life and keeping me tethered to him has been a constant, insidious force for over a decade. Though we've been separated for years, his actions have proven that for some people, separation is not an end, but merely the beginning of a more relentless kind of abuse.

This story is not just about a broken relationship or a toxic ex who cannot let go. It is about something much darker: a profile of an individual whose psychological disorder—Narcissistic Personality Disorder (NPD)—has shaped his every action toward me, turning my life into a nightmare of harassment, cyberstalking, and manipulation. This is the story of how I've been forced to survive under the weight of someone who believes my freedom and independence are his to control, and how his obsession has kept me locked in a digital prison for years.

The Breakdown of Our Relationship: A Dangerous Pattern Begins

We were once a family, but even from the start, his need for control was evident. At first, it was subtle—suggestions disguised as "advice," little comments about how I should dress, whom I should spend time with. It seemed like harmless concern at first. But over time, these behaviors escalated. The controlling nature of his personality started to take shape as emotional manipulation—guilt-tripping me for wanting time alone, for spending money on myself, for having interests outside of our relationship. What seemed like possessiveness turned into constant surveillance. He checked my phone, monitored my emails, and questioned me about every interaction.

The relationship was built on a foundation of coercion and subtle threats, but it all came crashing down when I left. I wanted freedom. I needed to regain my sense of self, but to him, my departure was a personal attack—an injury to his pride that he could not bear. That's when the stalking began.

The Escalation: Stalking, Cyber Harassment, and Identity Theft

Leaving him was not the end of our story—it was just the beginning of a more dangerous chapter. For someone with Narcissistic Personality Disorder, rejection isn't just a hurt ego; it's an existential crisis. My ex could not accept that I had walked away. I was no longer his property, his extension. I was now an object to be possessed at all costs. His obsession became an obsession not only with my physical presence but with my digital life.

What began as emotional manipulation quickly escalated into criminal behavior. He hacked into my personal accounts, accessed my emails, and stole my passwords. He even went so far as to tamper with my financial records, using my Social Security Number to take control of my accounts. His intrusion didn't stop at my finances—it seeped into every area of my life. My social media, my private files, my personal conversations—all became a battleground for him to monitor, control, and manipulate. Even when I physically moved across the country, he remained an ever-present shadow in my life. The digital chains he forged were invisible, but their grip was undeniable.

Psychological Impact: Complex PTSD, Anxiety, and Rebuilding a Life

The emotional and psychological toll of this constant harassment was profound. It wasn't just about stolen information or violated privacy—it was about the never-ending anxiety of not knowing where he would strike next. The trauma from living under his constant surveillance created a psychological fracture in me. I developed Complex PTSD—a form of post-traumatic stress disorder born from long-term exposure to trauma and abuse.

Every ping from my phone, every unexpected email, was a trigger. His harassment rewired my brain, leaving me on edge, unable to trust myself or anyone around me. The constant gaslighting, the feeling of being hunted in my own home, was suffocating. I could no longer distinguish between the real world and the digital nightmare he had woven around me.

The financial damage was just as severe. My ex's control over my identity, my credit, my financial future made it impossible for me to start over. I couldn't find stable housing, couldn't get credit, and found myself fighting to reclaim every piece of freedom he had stolen from me.

And then there were the children. He used them as pawns in his war against me. By manipulating their emotions and our shared responsibilities, he kept me tethered to him in ways that were as cruel as they were effective. My relationship with my children was deeply affected by his behavior, as they grew up surrounded by his emotional instability and manipulation.

Narcissistic Personality Disorder: The Root of the Abuse

At the heart of all this lies Narcissistic Personality Disorder—the disorder that explains why he has never been able to let go, and why he has continued to torment me even after everything we've been through. Narcissists believe they are entitled to the people around them—they view others not as individuals but as extensions of themselves, to be controlled, manipulated, and used to feed their fragile ego.

When I left him, he didn't just lose a partner—he lost control, and for someone with NPD, that's unforgivable. His need to dominate, to punish me for rejecting him, turned into a digital war. His obsession with my downfall became his mission. His rage manifested in relentless cyberstalking, hacking, and a continual campaign to undermine my life. This is the story of how that obsession became my prison.

Moving Forward: Protecting Myself and My Children

Despite everything, I am not just a victim—I am a survivor. But survival doesn't mean an easy road or a clean slate. The damage inflicted on me, and my family, has been immense. The scars are not just psychological but deeply ingrained in every aspect of our lives. Rebuilding myself and reclaiming my sense of peace has been a long and grueling process. Healing is slow—not just for me, but for my children too.

For years, I fought for the freedom to live a life that was no longer controlled by my ex. But my fight didn't stop at my own well-being. The toll this has taken on my children is a burden I carry every day. He has control of them. He has manipulated their emotions, kept them isolated from me, and—most painfully—has used them as pawns in his twisted power play. Though my oldest is now an adult, the damage he has done to the relationship with my children still feels fresh, and the ache of that loss is deep.

The psychological manipulation doesn't end when the relationship does. The damage done by someone who feels entitled to your life, your identity, and your family doesn't vanish overnight. His constant hacking, his control over communication, and his emotional manipulation have kept me cut off from my children. Despite being an 18-year-old young adult, my oldest has been taught to fear me, distrust me, and see me as the enemy. They've all been subjected to the same gaslighting tactics I endured—convincing them that I was the one who was unstable, the one who was dangerous.

The fact that I am not physically near them—due to his interference—does not make it easier. I have learned that distance doesn't always mean freedom, especially when it comes to emotional control. He doesn't need to be in the same room with us to continue his torment. He has weaponized technology to ensure that I can't even have a conversation with my children without his interference. They've been turned against me, not through their own actions, but by his relentless manipulation.

Dr. Rachel Levitch

This book is not just a cathartic release. It is a call to action. A plea for help, a demand for justice. It is my attempt to break the silence that has been imposed on me and my children. I am seeking justice, not just for the years of suffering I have endured, but for the ongoing psychological warfare he continues to wage against us. This is not just about sharing my personal story—it is about exposing the dark reality of cyberstalking, identity theft, and the way someone with a personality disorder can manipulate not just one life, but an entire family.

The stakes are high. Every day, I feel the weight of my children's absence in my life. I long to rebuild a relationship with them, but the digital chains he has forged around them are suffocating. I am forced to wait, watch, and hope that one day, the truth will reach them—that they will see the reality of who I am and not the distorted image he's created. But I can't wait forever. They deserve to be free of his control just as much as I do.

My request is simple: I seek legal measures to protect my family from this ongoing torment. I seek the ability to be free—to finally sever the connection he continues to wield over us. I need the law to recognize the severity of this situation—the damage he has caused, not just to me, but to my children, who have been robbed of the mother they once knew and loved.

I am taking this step because I can't continue to live in fear that he will keep destroying what's left of my family. I am speaking out because I will not allow him to keep taking. Every step forward I've made, he has tried to pull me back—dragging me back into his world of manipulation, control, and lies. I cannot allow that anymore.

This book is not only a reflection of the trauma we've endured, but a testament to the resilience it takes to break free from the grip of someone like him. The journey toward healing is a constant battle, and I am still fighting every day. But I write this not only for myself but for my children, who need to understand that their mother never stopped fighting for them, and never stopped loving them, even when she had to fight from afar.

I want this to be a message of hope for others trapped in similar circumstances, people who feel as though they have no voice. You do not have to stay silent. You do not have to live in fear. There is power in speaking the truth, even when it's terrifying. My truth is out now, and it is my way of reclaiming the life that was stolen from me.

As I continue to fight for my family, I hope my story can be a beacon for others. This is not just my battle—it's a fight for freedom. A fight for justice. And a fight to reclaim what's rightfully mine: my life, my identity, and the love of my children.

Psychological Profile of My Ex

Key Traits to Consider:

Obsessive and Narcissistic Tendencies

Narcissism: My ex exhibits a profound need for control that hasn't diminished with time. Even after our relationship ended, he refused to accept that I could live my life independently. His obsession with maintaining power over me is relentless, and this constant need for control has resulted in years of harassment. He cannot let go of past grievances and is determined to punish me for leaving, perceiving it as an attack on his fragile ego.

Dr. Rachel Levitch

Obsessive: His fixation on punishing me for leaving has become a consuming obsession. Over the years, he's turned this fixation into a form of digital stalking and emotional manipulation, continuously trying to infiltrate every corner of my life. His refusal to move on or accept boundaries has led him to escalate his actions, proving that his obsession knows no limits.

Pattern of Gaslighting and Manipulation

From the moment we separated, his manipulation didn't stop. He gaslit me at every turn—making me doubt my own reality. His actions ranged from playing mind games to manipulating our financial and legal matters, hacking my personal information, and undermining my stability. He constantly twisted the narrative, trying to make me believe that I was the one in the wrong, while he continued to exert control over me.

His manipulation also extended into our finances and emotional well-being. By blocking housing, tampering with documents, and using the children against me, he ensured that I couldn't move on or reclaim my freedom.

Possible Traits of Borderline Personality Disorder (BPD) or Antisocial Personality Disorder (ASPD)

BPD: The extreme emotional reactions I witnessed from him, particularly when he felt abandoned, suggest traits of borderline personality disorder. He displayed black-and-white thinking, where anything less than total submission to his will was seen as an unforgivable betrayal. His revenge fantasies often seemed to fuel his actions.

ASPD: His behavior also aligns with traits of antisocial personality disorder. His disregard for my rights and his constant manipulation for personal gain indicate a lack of remorse for the harm he caused. His actions, which have been criminal in nature, show a willingness to hurt others without guilt. This is especially evident in how he used deceit and control to undermine my finances and disrupt my life.

Inability to Co-Parent or Act Rationally in Shared Parental Matters

One of the most damaging aspects of my ex's behavior has been his use of the children as emotional leverage. His inability to act rationally in shared parental responsibilities is clear—he manipulates our kids to maintain control over me. He's used them to create guilt and confusion, keeping me isolated from them, even though they have their own lives and voices.

His inability to put the children's best interests first has caused emotional damage, not just to me but to them. The manipulation is ongoing, and the consequences of his actions are becoming more and more apparent, particularly as they grow older and I am kept out of their lives.

Projection and Victim BlamingPerhaps one of the most insidious tactics my ex uses is projection. He justifies his harassing actions and emotional abuse by claiming to be the victim of my leaving. He believes that I "disrespected" him by asserting my independence, and in his mind, everything he does is justified because of what he perceives as my "wrongdoing."

This victim blaming allows him to evade any responsibility for his behavior. Instead of accepting accountability, he continues to view himself as the aggrieved party, which is why he feels justified in making my life a continuous battleground.

Behavior Indicators to Watch: Retaliatory Actions

My ex has demonstrated that he is willing to retaliate in any way possible. His need for vengeance doesn't subside with time. He's likely to continue pursuing legal threats, cyber harassment, and using every tool available to him to disrupt my life. The more I try to move forward, the more desperate and dangerous his behavior becomes. His actions, at times, seem to be guided by the sole intention of punishing me for taking control of my own life.

Lack of Empathy

He has no empathy for the emotional, physical, or psychological toll his actions have on me. His behavior is entirely self-serving, and he refuses to acknowledge the harm he has caused. This lack of empathy is not just hurtful; it makes him dangerous. His ability to inflict harm without remorse is one of the most terrifying aspects of his behavior.

Need for Control

My ex's constant need for control is evident in everything he does. Whether it's hacking into my accounts, trying to manipulate the children for his own gain, or using financial leverage to destabilize my life, he can't stand seeing me make choices that do not include him. Every victory I have—whether personal, financial, or emotional—feels like a loss for him, and he is determined to reverse any progress I make.

Inability to Move On

One of the most toxic things about his behavior is his fixation on the past. He cannot accept that we are no longer together, and this obsession prevents him from moving forward. This lack of closure is not just emotional—it manifests as a continuous need to control and punish me for leaving him. His inability to move on means I will never truly be free unless I enforce strict boundaries.

How This Affects My Safety:

Escalation of Behavior

I've learned the hard way that my ex's behavior will likely escalate over time. As he feels his grip on me loosen, he'll only grow more desperate to reassert control. This could mean more hacking, more threats, and a worsening of his stalking behavior. His need to punish me for leaving will likely continue until legal measures are taken to stop him.

Difficulty Breaking the Cycle

I'm trapped in a pattern of abuse that won't break easily. Until my ex feels like he's regained control or "won," he will continue his manipulation and harassment. This cycle of abuse is exhausting, and breaking free from it isn't just about physical separation—it's about securing my legal and emotional freedom.

Dr. Rachel Levitch

Psychological Impact

The emotional toll of this constant manipulation and cyberstalking has left deep psychological scars. My experience with complex PTSD, anxiety, and hypervigilance has made it incredibly hard to trust others or even feel safe in my own home. I fear his intrusion every day, and the damage to my mental health continues to unfold as I try to rebuild my life.

Protective Strategies:

Legal Protection

I've taken legal action, including filing motions and requests for a protective order. It's essential that the court recognizes this as a pattern of abuse, not just a series of isolated incidents. The legal system needs to protect me from further harm and stop him from using our children and my personal information against me.

Increased Security

To protect myself, I've taken steps to increase security: implementing encrypted communication, using new devices, and working with cybersecurity experts to safeguard my online presence. I also make sure to document every instance of his attempts to infiltrate my life, whether digital or physical.

Therapeutic and Psychological Support

Healing from this kind of abuse is a journey. I've sought therapy for myself and for my children to address the trauma we've experienced. It's important to have the support of professionals—whether a forensic psychologist to document abuse for court, or trauma therapists to help us work through the complex PTSD and anxiety.

The Journey Ahead: Reclaiming My Life, One Step at a Time

I am not just dealing with someone who is obsessive, narcissistic, and capable of profound emotional manipulation—I am facing a persistent and dangerous force that seeks to undermine my every attempt at independence. His behavior goes far beyond mere invasions of privacy; it is a deliberate, calculated effort to assert control over my life, my children, and my future.

The manipulation, the harassment, the constant surveillance—it's not just about maintaining a relationship. It's about domination. His need to feel powerful and in control of me and my decisions has driven him to criminal acts. Identity theft, cyberstalking, emotional abuse—these are all tools in his arsenal, meant to hold me captive, even when we are no longer together.

A recent example illustrates the depth of his control tactics: After I moved to Connecticut, hoping for a fresh start and physical distance from his influence, his reach didn't stop. I learned that despite everything—despite my name change, new identity, and efforts to protect my privacy—he had still been receiving updates from the Social Security Administration, allowing him access to my new personal information. When I reached out to SSA to address this, I realized that he had maintained control of certain aspects of my financial and legal life even from miles away in Texas.

And this wasn't just a fluke. As I've worked through the fallout of this realization, it's become clear that his ability to manipulate systems extends far beyond mere surveillance. My ex uses these personal vulnerabilities to remake the narrative in his favor, ensuring that I cannot escape the emotional grip he's placed on me—even from afar. Every time I take a step forward, he finds a way to pull me back into his orbit.

I know that without strong legal and personal boundaries, he will continue his relentless campaign to regain control. His behavior won't simply fade away. It is as entrenched in him as his ego. His narcissistic rage will only intensify as he tries to make me feel small, insignificant, and powerless once more. That's why I fight for my legal protection—to ensure that my rights, my privacy, and my children's well-being are not held hostage by his manipulation.

One of the more insidious ways he tried to undermine my safety came when I was working in a highly toxic environment. Although I had hoped for a supportive space where I could rebuild my career and confidence, that hope quickly faded as I began to feel isolated and targeted—not just by my colleagues, but by those in leadership positions as well. My professional boundaries, which I had requested to be respected for both my personal well-being and my success, became a point of contention. When I secured my first major successful deal, something I had worked hard for, instead of celebrating it as a win, some of my coworkers tried to undermine it—spreading gossip, downplaying my success, and taking credit for my efforts. These actions weren't just disrespectful; they were damaging to my career and self-esteem.

I voiced my concerns to the management, explaining that the lack of respect for my personal boundaries and the subtle discrimination I was facing were creating a toxic environment. However, instead of addressing these issues, they took the opposite approach, essentially dismissing my concerns and telling me to handle it on my own. I was left with the painful realization that even the places where I sought support and professional growth could become tools of oppression. They weren't just neglecting their responsibility—they were enabling the harassment. The sense of isolation I once tried to escape in that space quickly turned into an environment where emotional and professional abuse became the norm.

But this is not just a fight for survival. It is a fight for freedom. And it is not a fight I face alone. While his behavior may isolate me physically, I am not isolated emotionally. There is a growing network of people who understand what I am going through. I lean on professionals, friends, and advocates who offer their support in ways that allow me to regain my strength. It's a long, slow, often painful process, but it's a process of healing.

I have been victimized, yes, but I am not a victim. I am a survivor. Each step I take toward legal action, each boundary I set, each moment of self-reclamation brings me closer to a life outside his reach. This is a journey—not just to break free from his control but to find the version of myself that exists beyond the fear and trauma he has inflicted. It is about healing from the scars he's left but also about rediscovering the strength I have within to rebuild my life on my own terms.

This fight is not just for my own freedom; it is also for my children, for their futures, and for the lives we can build when we are finally free from his toxic influence. I will continue to protect them, even if it means fighting with every tool at my disposal—legal, emotional, and psychological. I will show them that there is a way to break the cycle of abuse, a way to step into freedom and self-determination.

There are days when the weight of this battle feels overwhelming. There are moments when the constant vigilance takes its toll on my mental and emotional well-being. But I know that survival is not just about enduring—it's about fighting back. It's about using the strength I have to create a life that is mine, a life where I am no longer bound by his influence.

I will not be silenced. I will not be controlled. And I will not be afraid. This journey is ongoing, and though it is filled with obstacles, it is also filled with hope. Because with every step forward, I am reclaiming my power, and with every challenge I overcome, I am proving to myself and the world that I am not just surviving—I am thriving.

This is the story of my journey: the fight to reclaim my life, my peace, and my future from the hands of someone who has tried to steal them away. And though the battle may seem endless, I am reminded that every story of abuse has an end—and it's my story to write.

Mini Part 1: Mental and Emotional Resilience: Building Inner Strength Amidst the Storm

The journey of rebuilding my life after leaving my abuser has not been a simple one. It's been a constant battle against the waves of trauma, the narcissistic manipulation, and the overwhelming feeling of being hunted in the digital world. It is not just about surviving the external threats of cyberstalking, harassment, and surveillance. It's about surviving the internal warfare—the constant strain on my mental and emotional health.

I've had to learn to outsmart the damage done by years of emotional manipulation. To understand that I can no longer trust the thoughts that have been ingrained in me for so long—the thoughts that told me I wasn't good enough, that I couldn't succeed without his control, that I was nothing without his approval. Every day, I have to remind myself that the narrative I lived for years—the one where he was always right, where his needs trumped mine, where my thoughts and feelings didn't matter—is not my truth anymore.

Reframing the Trauma

It would be easy to look at everything I've been through and assume that the mental toll would be unbearable—that the emotional scars would always remain raw. But I've had to make the choice every day to reframe my trauma, to take the pieces of myself that he tried to destroy and start putting them back together, one small step at a time.

This process isn't linear. There are days when the weight of everything—his constant surveillance, the haunting sense of his presence in my life, the never-ending legal battles—feels unbearable. On those days, I remind myself that survival isn't just about enduring, it's about thriving despite the odds. I can survive the consequences of his actions, but I refuse to let them define my future.

In therapy, I've learned that mental resilience isn't about ignoring the pain or pretending it doesn't exist. It's about acknowledging the pain while learning to use it to fuel my growth. Every day, I battle my self-doubt and the feelings of powerlessness that his years of manipulation embedded in me. But I've learned to counter them with something stronger: the realization that I am capable. I am worthy. And most importantly, I am free.

Understanding Narcissistic Abuse and Its Psychological Impact

One of the hardest things I've had to confront is the truth about narcissistic abuse. It's insidious. It doesn't just damage you physically—it rewires your mind. The gaslighting alone was enough to make me question my own reality. The constant barrage of emotional neglect, betrayal, and lies left me wondering who I really was outside of his manipulation.

When you live with someone who is narcissistic, you become conditioned to believe that you exist only in relation to them. Your thoughts, your emotions, and your desires are secondary to theirs. Over time, your sense of self starts to erode. Your boundaries become blurred, and the world around you becomes warped.

I spent years in this fog, never fully trusting myself. The trauma shattered my identity in so many ways, and it took me a long time to realize that I didn't need his validation to define who I was. For the first time in my life, I am learning to trust my own voice and my own judgment.

The Road to Recovery

The road to recovery isn't smooth. It's riddled with moments of doubt, fear, and pain. But it's also filled with moments of clarity—small victories that remind me of how far I've come.

I have started to rebuild my sense of self by setting clear boundaries and learning to say "no" to anything that compromises my well-being. Saying "no" wasn't easy. For years, I was trained to people-please, to accommodate everyone's needs at the expense of my own. But now, I see the power in standing firm. Saying "no" isn't an act of rejection; it's an act of self-preservation.

I also try to focus on the small wins—the moments when I catch myself thinking or acting in ways that reflect the new person I'm becoming. When I set a boundary with someone, I celebrate that. When I make a decision that puts my well-being first, I acknowledge it. These small acts of self-love and self-respect are the foundation of my mental resilience.

Each time I feel the grip of anxiety—whether it's triggered by his attempts to contact me or the constant feeling of being surveilled—I take a moment to breathe. To remind myself that I'm safe in this moment, that I have control over how I respond. I am learning to control my reactions, to manage my anxiety, and to create space between my emotions and the situation at hand.

It's not always easy, and there are days when the weight of it all feels unbearable. But I know that healing isn't a destination. It's a journey that will take time, patience, and most importantly, self-compassion. Every step forward—no matter how small—matters.

Reclaiming My Strength: Rebuilding a Life of Empowerment

The journey I am on is a journey of reclamation. It is the process of healing from the trauma inflicted upon me and, in many ways, it's a personal revolution—a conscious decision to rewrite the story of my life. What was once an experience rooted in disempowerment is now becoming a path toward self-empowerment. And though the scars may never fully fade, the pain no longer defines who I am, nor does it dictate how I move forward.

Mental resilience, when rooted in the aftermath of narcissistic abuse and emotional manipulation, isn't merely about gritting your teeth and getting through the day. It's about developing a framework of mental fortitude—a psychological recalibration that enables you to navigate the cognitive dissonance, the gaslighting, and the pervasive sense of being controlled. It's about recognizing and dismantling the patterns of thought that have been programmed into your psyche over years of manipulation. It is the art of reframing your narrative and reclaiming your agency.

Reframing Trauma: The First Step in Healing

Trauma is a complex beast. It doesn't just live in your mind; it often manifests in your body, in your emotions, and even in your behaviors. For so long, I internalized my experiences as truths—truths about my worth, my capabilities, my rights. For years, the narrative of my life was dictated by someone else, shaped by their ability to control my thoughts, my beliefs, and my very sense of self.

But I have learned that trauma is not the end of the story, it's a chapter—albeit a painful one—that must be rewritten. Cognitive Behavioral Therapy (CBT), coupled with trauma-informed care, has been instrumental in reshaping my internal landscape. Through these methods, I have learned to identify and challenge distorted thinking patterns. I have learned to see myself not as a victim of my past, but as someone who is in the process of reclaiming control over the narrative of my life.

This process begins with recognizing the trauma's impact—understanding how it altered my perceptions of myself and the world around me. Cognitive restructuring has become a cornerstone of my recovery. When I find myself spiraling into self-doubt or feeling triggered by his presence or by the lingering fear of his surveillance, I remind myself: these thoughts are not truths. They are echoes of an old, broken narrative that I am actively dismantling. The key is in practicing self-awareness—identifying when those thoughts are beginning to take hold and redirecting them in a way that is more aligned with my truth and my future.

This is not a passive process. It's a conscious, intentional act of deprogramming. The abusive patterns that have been ingrained in me must be replaced by healthier, more adaptive thoughts and behaviors. This requires a willingness to confront discomfort—to sit with the vulnerability of acknowledging just how deep the manipulation runs. But in doing so, I am rebuilding my identity from the ground up, piece by piece.

Boundary Setting: The Cornerstone of Psychological Defense

Setting boundaries is perhaps the most powerful tool in my emotional toolbox. When you have lived in an environment where boundaries were either disrespected or nonexistent, learning to establish and uphold them becomes a radical act of self-preservation. In fact, this act of establishing firm, non-negotiable boundaries is a direct challenge to the narcissistic abuse I endured—an assertion that my needs matter.

Boundaries are not just physical; they are also emotional, cognitive, and spiritual. They encompass the way I interact with people, how I prioritize my emotional health, and how I manage my personal space—both in the literal and figurative sense. Emotional boundaries are often the most difficult to maintain, particularly when the person violating them is someone you once loved or trusted. For years, I allowed myself to subordinate my emotional needs to his whims and demands. But that was never a true relationship. That was control masked as love.

I've learned that boundaries are not about creating distance for the sake of punishment; they are about creating a protected space within which I can thrive. This has meant setting clear limits on the ways in which I communicate with him—no more allowing his emotional manipulation to sway my decisions, no more engaging in his constant gaslighting. When I say "no," it's not just a word; it's a firm declaration that my thoughts, feelings, and actions are my own.

Setting boundaries is also about recognizing when a boundary has been crossed and taking immediate action to enforce it. It is about regulating myself in moments of emotional distress—learning to self-soothe when I feel triggered, and not allowing him to pull me back into his orbit. Every time I successfully enforce a boundary, I am reclaiming my sense of control. And with that, I reclaim my sense of self-worth.

Healing Through Resilience: The Intersection of Strength and Vulnerability

True resilience doesn't mean the absence of vulnerability. If anything, the ability to be vulnerable is a hallmark of true emotional strength. Vulnerability is not weakness; it is the willingness to show up—to acknowledge the parts of me that still hurt, to let myself feel without becoming overwhelmed by the emotions.

In my therapy sessions, I've come to understand that emotional resilience is about embracing the full spectrum of human experience—both the joy and the pain. Emotional regulation isn't about suppressing emotions; it's about learning to navigate them in a way that honors both my feelings and my well-being. When I experience moments of anger, sadness, or fear, I do not reject them; I welcome them as part of the healing process. I learn to validate my emotions, but I also refuse to let them dictate my actions. I choose how I respond.

The process of healing is about integrating the painful parts of my past into my current self, not denying them. Post-Traumatic Growth (PTG) theory teaches us that trauma does not only break us—it can also make us stronger, wiser, and more resilient. I believe that the pain I've endured has given me a depth of understanding that I did not have before. I've developed the emotional intelligence to recognize when I am being manipulated or controlled, and I've learned the mental flexibility to adapt to difficult circumstances without losing my center.

There are still moments where the past seems to reach out and remind me of my vulnerability—when I feel the pang of grief, when the fear of surveillance makes my heart race, when the memories of emotional abuse threaten to pull me under. But each time I confront these feelings, I get a little stronger. I begin to understand that the pain is not permanent—it is simply part of the ongoing process of healing.

And with every moment of self-reclamation, I move further away from the person I once was—a person who believed she had no voice, no control, no power. The person I am becoming is someone who not only survives, but thrives. Someone who faces fear, but no longer bows to it.

A Life Reclaimed

The resilience I am cultivating is not an abstract concept. It is a tangible force that drives my everyday decisions, how I engage with others, and how I navigate the emotional landscape of my life. It is about living with intentionality—choosing to act in ways that honor my true self and align with my core values.

It is through mental resilience and emotional fortitude that I am learning to rebuild. With each step I take toward healing, I am not just breaking free from the trauma of my past—I am forging a new path forward. One where boundaries are respected, where self-care is prioritized, and where I am free to write the next chapter of my life on my own terms.

This is my journey—a journey of resilience, reclamation, and transformation.

Mini Part 2: Self-Care—The Pillars of Reclamation

Self-care is often misunderstood as a luxury, something nice to do when life is easy and manageable. But for someone healing from emotional trauma and psychological abuse, self-care is not an indulgence—it is a necessity. It's the framework that supports mental and emotional resilience, the practice that allows me to rebuild after years of depletion and manipulation.

For me, self-care is not just about spa days or moments of solitude. It's about creating a sustainable, intentional lifestyle that prioritizes my well-being, from the inside out. It is about regaining control over my body, my emotions, and my mental state—a reclamation of the space that was once violated by trauma.

1. The Physical Body—Rebuilding from the Inside

The impact of emotional abuse and cyberstalking is invisible on the surface but profoundly felt in the body. Stress, anxiety, and fear often manifest as physical tension—tight shoulders, headaches, fatigue, sleeplessness. After years of neglecting my own physical health in the wake of constant emotional turmoil, I realized that true healing must encompass the whole person. Self-care begins with the body, and it's essential to treat it with the same compassion that I give my mind and heart.

One of the first steps in my healing process was to restore physical balance. That started with sleep. Trauma has a way of disrupting natural rhythms. My body would often wake up in the middle of the night, alert to every sound, expecting danger. This hypervigilance interfered with my ability to rest, leaving me exhausted and unable to think clearly during the day. Sleep hygiene became a non-negotiable part of my self-care routine.

I began with simple changes: no screens before bed, setting a consistent bedtime, and creating a quiet, dark space for sleep. I also explored relaxation techniques like progressive muscle relaxation (PMR) and deep breathing exercises to help regulate my body's fight-or-flight response. Over time, these practices helped my nervous system recalibrate and allowed me to experience deeper, more restorative sleep.

Nutrition was another area where I needed to make significant changes. When you're in survival mode, healthy eating often takes a backseat. But nourishing my body with whole foods—especially those rich in omega-3 fatty acids, magnesium, and vitamins—became a key part of my self-care plan. These nutrients play a critical role in mood regulation and stress management, and I felt the effects almost immediately. I had more energy, my moods became more stable, and I was better equipped to handle the emotional ups and downs.

Exercise, too, became a crucial pillar of my healing process. It wasn't just about weight loss or physical appearance. It was about creating a channel for releasing pent-up emotions and stress. Movement, especially yoga and walking, allowed me to reconnect with my body in a way that was grounded and compassionate. Yoga provided a meditative practice that not only strengthened my body but also brought me back to my breath, reducing anxiety and helping me to center myself in the present moment.

2. Mental Health—Cultivating Inner Peace

While self-care for the body is essential, it is the mental health practices that truly sustain my long-term healing. For someone who has endured years of narcissistic abuse and cyberstalking, mental well-being becomes a battlefield. The constant gaslighting, the invasive thoughts, the emotional instability—these are the tools of manipulation that tear at your mind. But even in the aftermath, reclaiming mental peace is possible.

Therapy has been foundational in my self-care practice. Not just any therapy, but trauma-informed therapy that specifically addresses the needs of survivors of abuse. Cognitive Behavioral Therapy (CBT) has helped me identify and confront negative thought patterns—thoughts that once crippled my ability to function and see myself as worthy of peace. The therapist I work with has guided me in reframing the false beliefs instilled by years of emotional manipulation. One of the most empowering realizations was understanding that I am not crazy—I am not overreacting. What I've experienced is real and it has impacted me on a profound level.

But therapy alone isn't enough. Healing also requires self-compassion—the ability to hold yourself with the same love and understanding that you would offer a dear friend in pain. One of the most powerful self-care tools I've embraced is self-compassion exercises. These are small but transformative practices in which I actively give myself permission to feel pain, to be imperfect, and to heal at my own pace. This is crucial when you've been conditioned to believe that your needs don't matter or that your pain is insignificant.

Mindfulness practices have also been a core part of my mental self-care. When my thoughts spiral or when flashbacks to the abuse emerge, I use grounding techniques to bring myself back to the present. One practice I use is the 5-4-3-2-1 grounding technique, which helps me connect with my immediate surroundings and interrupts the flood of anxious or intrusive thoughts. Being present in my body, feeling the solid ground beneath me, and focusing on the details around me helps anchor my mind, bringing me back to the here and now.

Journaling has also been a vital practice for me. Writing has become a safe space to express emotions, articulate my thoughts, and reflect on my healing process. It's not about creating perfect prose, but about using words to externalize feelings and make sense of my inner world. Journaling also serves as a powerful tool for tracking progress—whether it's noting moments of strength, personal victories, or identifying areas of continued struggle. The practice of writing offers clarity in moments of confusion and provides a safe outlet for processing complex emotions.

3. Emotional Health—Creating Safe Emotional Boundaries

Perhaps the most difficult, yet essential, part of self-care has been learning to protect my emotional energy. It's easy to become entangled in the emotions of others—especially when you've been manipulated for so long. Learning to establish and enforce emotional boundaries was a process I had to commit to each and every day.

Emotional boundaries are not just about saying "no" to other people—they are about recognizing what is my responsibility and what is not. For example, I no longer take responsibility for my ex's emotions or reactions. His anger, his need for control, his emotional manipulation—all of these are no longer my burden to bear. In fact, part of my healing has been learning to detach emotionally from his behavior. I do not need to engage with it, nor do I need to internalize it. His toxicity is his to carry, not mine.

Creating these boundaries was especially difficult when it came to managing my feelings around guilt and shame—feelings that were often weaponized by my abuser to control me. Boundaries are not only external; they are internal. I've had to redefine the relationship with myself—to forgive myself for staying in the abusive dynamic for as long as I did and to release any shame about my survival.

Self-care also means giving myself permission to feel—without judgment. I allow myself to experience joy and sadness, pride and frustration, all in equal measure. Emotions are not something to be feared or suppressed; they are a natural and necessary part of the healing process.

4. The Role of Support Systems—Not Walking Alone

One of the most powerful forms of self-care is recognizing that healing is not a solitary pursuit. I lean on those who understand the complexities of trauma recovery—whether it's therapists, support groups, or close friends who offer their empathy, encouragement, and guidance. A healthy support system is essential in sustaining emotional resilience, as isolation only amplifies the weight of trauma.

Surrounding myself with people who validate my experiences has been crucial in maintaining psychological safety. Having others who understand that my need for boundaries isn't about being difficult, but about protecting my mental health, has been incredibly empowering. The support I receive is not about rescuing me from my challenges, but about providing encouragement, offering perspective, and reminding me that I am not alone in this fight.

The Art of Living Well Again

Self-care is not a one-time act—it is a lifestyle that must be continuously cultivated. It requires patience, consistency, and compassion toward myself. In learning to prioritize my needs—both mentally and physically—I am rebuilding a life that honors my worth. This is my act of resistance, my reclamation of self.

With each step I take in nurturing my body, mind, and soul, I am forging a path toward a future where I am free from the control of those who have sought to undermine me. I am not just surviving. I am learning to live well again, on my own terms.

Mini Part 3: Navigating Triggers—Recognizing the Past and Reclaiming the Present

Navigating Triggers in the Aftermath of Abuse: Emotional Survival in a Hostile Environment

The aftermath of abuse is not just about healing physical wounds or moving away from the abuser. It's about reclaiming a sense of self in a world that constantly tries to define you by the trauma you've experienced. For many survivors of abuse—whether it's emotional, financial, digital, or physical—the journey doesn't end once the immediate threat is removed. The emotional landscape remains scarred, the memories still raw, and the triggers that were planted during the abuse continue to seep into daily life, disrupting even the smallest moments of peace.

Triggers don't simply appear as faint echoes of the past; they can be violent flashbacks, unexpected emotional crashes, or surging waves of panic that transport you back to the worst moments of your trauma. They may seem like random, fleeting events at first, but for someone who has been subjected to ongoing manipulation or control, these triggers are much more than brief moments—they are windows into a time when you had no control over your life, when every thought, decision, and action was dictated by someone else's needs and demands.

In the aftermath of a toxic relationship, the emotional labor required to heal is intense. It's not just about managing the immediate fallout but about rebuilding the core of who you are: your identity, your confidence, and your belief in your own worth. Every day spent away from the abuser is a step toward regaining that lost sense of self. But these triggers can set you back—not just in your mind but in your body. The nervous system, particularly when it's been subjected to long periods of stress or manipulation, becomes hypersensitive. A glance, a sound, a situation, or even a well-meaning interaction can send you spiraling back into a space where you feel invisible, small, and powerless.

The Work Environment as a Reflection of Toxic Abuse Dynamics

When we think about work environments, they often serve as microcosms of the larger world—places where both support and toxicity coexist. In the context of abuse recovery, the workplace or any other "support" system can become a landmine of triggers, manipulation, and subtle attacks that serve to undermine your progress. This is especially true when your relationship with the people around you is one that was supposed to offer support but instead leaves you feeling isolated and unseen.

Imagine entering a space where you're constantly reminded that your needs don't matter. Where your requests for basic accommodations—whether for emotional well-being, safety, or personal space—are either dismissed or met with hostility. In an environment like this, you're not simply trying to function day-to-day; you are emotionally surviving. Every interaction feels like a negotiation of power, every word spoken or left unsaid becomes a subtle challenge to your sense of self-worth. And the more you try to assert yourself, the more you risk being painted as difficult, high-maintenance, or too emotional.

The emotional survival required in this kind of environment is exhausting. Just like a survivor of abuse might find themselves perpetually watching their back, always alert for the next emotional attack, someone in a toxic work environment can feel the same way. Every conversation feels loaded with potential conflict. Every day is a test of whether you will be able to maintain your emotional integrity or whether the next small violation will break you. In both cases—whether in an abusive relationship or a toxic work environment—the underlying dynamic is the same: power and control.

How Triggers Operate in Toxic Spaces

Triggers in a toxic environment are not just random occurrences. They are carefully timed manipulations, often designed to keep you off-balance. The difference between personal relationships and workplace dynamics in this regard is that at work, you're expected to "perform" regardless of your emotional state, even when the environment actively works to erode your sense of stability.

Take, for example, a moment where a basic personal request—such as needing to have a private space to eat, given dietary restrictions or religious observances—becomes an issue of contention. What might seem like a simple need is often manipulated, twisted, and undermined in ways that trigger deeper fears of rejection or disrespect. The very act of asserting your need for privacy, space, or respect becomes a source of resentment from those around you, as if you are being unreasonable for wanting something as basic as recognition and accommodation.

This process of emotional undermining is not always direct. It can appear in the form of microaggressions—small, seemingly insignificant comments or behaviors that, over time, wear down your confidence and sense of autonomy. For example, people might sabotage your space by introducing unwanted items into your personal food storage area. They might ignore clear boundaries, such as when you ask that a shared space be respected because it aligns with your specific religious or health needs. But the real emotional toll of these actions lies in the intentional disregard for your personal identity and safety.

When you are repeatedly told—through words or actions—that your boundaries don't matter, you're left to question your own worth. In a work environment, just as in a personal relationship, these moments chip away at your foundation. It becomes exhausting to be the only one standing firm in your needs, constantly defending your right to exist as you are, to be recognized for who you are, and to be treated with the same basic respect that others receive.

Breaking Free from the Cycle: A New Kind of Strength

It's clear that emotional survival in a toxic environment requires not only resilience but also strategic self-protection. For those who've lived through abuse, these dynamics feel all too familiar.

The emotional turbulence caused by triggers, the feeling of constantly being pushed to the edges of what feels safe, and the desensitization to boundary violations can make it difficult to see a way out. But there is a way forward.

It begins with recognizing that healing from emotional wounds does not happen in a vacuum. You cannot heal in spaces that are designed to undermine your progress. In the face of ongoing triggers, it's critical to reclaim control over how you respond. This means becoming aware of the emotional patterns that play out when you are triggered and learning how to intercept those patterns before they overwhelm you. It also means setting firm boundaries, and ensuring that those around you understand that your needs must be respected. Your well-being should always be your first priority, and anyone who consistently violates that should no longer be given room in your emotional space.

Ultimately, this is about transforming your relationship with your triggers. Rather than allowing them to control you, you must acknowledge them, process their origins, and then move beyond them. It's a slow, often painful process, but it is a process of reclaiming your agency—the agency that was stripped away during the abuse—and using that power to craft a new narrative, one that's defined by strength, sovereignty, and self-determination.

1. The Role of Triggers in the Workplace Environment

When I reflect on my experience within what I can only describe as a toxic support environment, I can't help but draw parallels to a hostile workplace. On paper, the shelter or any support system is meant to be a space where you receive the help you need to heal, grow, and reclaim your life. But just as in a dysfunctional workplace, an environment that should feel safe and supportive can, over time, morph into a zone of emotional turmoil and manipulation. This is especially true when the people who are supposed to be your advocates—those who you trust to have your back—begin to subtly undermine your needs and goals. The dynamics that unfold are not only emotionally draining but deeply triggering, often leading you to question your worth and your place in the system that was meant to support you.

What starts as a seemingly well-intentioned support network can, over time, turn into a space where gaslighting and emotional neglect take center stage. Instead of feeling validated in your experiences, you are met with comments like, "Why do you need our help? Can't you do this on your own?" or "I know you're not complaining already." These dismissals are often couched in the guise of concern, but in reality, they are subtle ways of shifting blame onto you. They make you feel as though your legitimate needs are somehow burdensome or unworthy of attention. The more you try to assert your boundaries and make it clear that your situation requires ongoing support, the more these manipulative comments chip away at your confidence.

This kind of emotional erosion creates a pressure to constantly prove your independence—a pressure that's not only unreasonable but deeply isolating. It is a constant reminder that your value is being measured by your ability to "handle things on your own," a narrative that ignores the very real trauma and limitations you are facing. The irony, of course, is that when you stop meeting with them for updates or when you choose to keep some of your concerns to yourself, you're punished for not being more engaged, for not being more "compliant." The energy shifts, and now you're made to feel as though you've done something wrong. In reality, it's their failure to provide the support you need that is the root cause of your decision to distance yourself.

Gaslighting: The Underlying Tactic in a Toxic Work Environment

Gaslighting is a term often used to describe emotional manipulation that causes the victim to doubt their own perceptions and reality. In my case, as in many toxic environments, the use of gaslighting was subtle at first. It started with small comments—questions that were framed as concerned but were actually undermining. These were followed by instances where my personal needs were ignored, my boundaries disregarded, and my identity disrespected.

The gaslighting began slowly, making me second-guess my own instincts and needs. At first, when I expressed concerns about my dietary restrictions or privacy, I was met with empathy or promises of help. But soon, that turned into a quiet form of mockery and disregard. When I asked for accommodations for my religious dietary needs, for example, I was met with comments like, "You can't really expect us to cater to you," or "Why do you need special treatment? We all have to deal with less." It wasn't just about denying my request for a safe and appropriate eating space—it was about shaming me for having needs at all. It was as if my basic requirements for safety and respect were now framed as inconveniences to others, and somehow, my survival was becoming their burden.

The more I tried to assert my boundaries—whether around food, space, or emotional needs—the more I was met with responses that made me feel small and unreasonable. This is the core of gaslighting: not just denying someone's experience but actively making them question whether their needs are valid or if they are simply too difficult or dramatic. Over time, I began to feel isolated, not just from those around me, but from my own instincts. I had become so used to dismissing my own needs that I started questioning whether I had the right to even ask for help. The gaslighting left me trapped in a constant cycle of self-doubt.

The Erosion of Personal Boundaries in the Name of "Support"

Another form of emotional manipulation came in the violation of boundaries under the guise of "help." For example, my request to have a designated area where I could store my food—food that aligned with my religious beliefs—was met with what seemed like a positive response. The staff cleaned and set up areas for me, even going so far as to put up signs to ensure that no one would interfere with my designated space. At first, this felt like a victory. But the environment quickly shifted. People began tampering with my space, leaving items that were in direct conflict with my dietary needs and beliefs. It was as if my boundaries were being deliberately tested—not just by the residents, but by the very people who were supposed to be protecting those boundaries.

This act wasn't a simple mistake or a misunderstanding. It was an intentional act of disregard. The boundaries I tried to establish—small as they were—were seen as obstacles to others, not as requests for respect. They were consistently ignored, manipulated, and violated in ways that triggered deep feelings of vulnerability and powerlessness. Every instance of tampering with my food area, every time I was forced to confront someone else's violation of my space, was a vivid reminder of the boundaries that were never respected during my abusive relationship. It was a painful, ongoing reminder that even in spaces where I should have felt safe, my right to exist as I am was constantly being challenged.

Emotional Survival in a Toxic Work Environment

All of these dynamics—gaslighting, boundary violations, emotional neglect—take an incredible toll on your mental and emotional well-being. Navigating these spaces requires emotional survival skills that no one should have to cultivate in a space that is supposed to be supportive. What starts as an attempt to rebuild your life and find safety turns into a constant battle for your sense of self-worth. It's as if you're forced to fight not only for your survival but for your right to be respected as a person with needs, desires, and boundaries that matter.

In this toxic work environment, I wasn't just fighting to get my basic needs met—I was fighting to maintain my identity and my sanity in the face of emotional manipulation. I had to navigate the treacherous waters of gaslighting and boundary violations without losing sight of my ultimate goal: to survive, to heal, and to reclaim the life that was taken from me.

Breaking the Cycle of Emotional Erosion

The real challenge in this kind of environment is breaking the cycle. When you've been gaslighted, manipulated, and emotionally abused, it becomes easy to believe that you don't deserve better or that your needs are just too much to ask for. But the first step in reclaiming your power is recognizing the manipulation for what it is. It's about drawing a firm line between what you are willing to tolerate and what you absolutely will not accept. Over time, I learned that my boundaries were not negotiable, and that I didn't have to justify my needs to anyone. I also learned that my emotional survival was not a burden, and that seeking help was not a weakness.

It's a difficult journey, but each small act of self-assertion—each moment where you stand firm in your need for respect and dignity—becomes a victory. And slowly, you begin to break free from the manipulative grip of the environment around you, reclaiming control over your emotions, your boundaries, and your life.

2. Recognizing Triggers: Emotional and Physical Responses

What I did not expect was how emotionally charged these interactions would become. Triggers in the workplace aren't just about emotional responses to rude comments or dismissive behavior. Triggers are deeply embodied. They live in the body as well as the mind. When my personal space was violated—whether it was someone cooking in areas that I had specifically requested to be left alone or tampering with items in my designated storage space—I would feel an immediate physical reaction.

My body would tense, my chest would tighten, and my thoughts would spiral. The feeling of being violated all over again—like I wasn't entitled to personal space, like I didn't have the right to set boundaries—was overwhelming. These feelings weren't just about a specific incident; they were a direct line back to years of emotional manipulation, when I was constantly told that my needs didn't matter.

The physical response I experienced was a flight-or-fight reaction—a legacy of living in environments where my safety and well-being were compromised. But what I've come to understand is that triggers aren't personal. They are not about the people who are triggering me; they are about my trauma response to situations that feel too familiar, too unsafe. It's not the violation that causes the reaction; it's the recollection of what has been done to me in the past that triggers such a strong emotional and physical response.

3. Setting Boundaries in a Toxic Environment

The biggest challenge in a toxic work environment is knowing when and how to set boundaries that protect you—not just from external harm but from internal chaos as well. My boundaries were constantly being tested by those around me who didn't understand the gravity of my personal trauma. And it wasn't just the staff—it was the other residents too. In fact, I had to learn quickly that assertive communication was essential, but so was emotional detachment.

In a space that is meant to support recovery and growth, emotional self-defense became necessary for survival. I had to stop expecting people to "get it" or to suddenly respect my boundaries out of kindness or goodwill. Instead, I had to learn to enforce my needs clearly, firmly, and without apology.

For instance, when it became clear that I was not being heard, I asserted my boundaries by requesting that my personal space be respected. When my dietary needs were repeatedly violated, I no longer hesitated to speak up and remind others that my dietary restrictions were non-negotiable. But the emotional toll of constantly advocating for myself was exhausting. The more I insisted on my needs being respected, the more I could feel the resentment building in the environment—especially as the focus shifted to me being difficult or high-maintenance.

However, self-care in these moments wasn't just about asserting my needs—it was about letting go of the fear that others would reject me for setting boundaries. I had to remind myself that my well-being and peace of mind were far more important than trying to maintain the peace with people who didn't understand what I had been through.

4. Emotional Regulation in High-Stress Situations

In moments when my triggers were overwhelming—when I felt my emotional boundaries crumbling under the weight of disrespect—I had to turn inward to my emotional regulation techniques. Deep breathing and grounding exercises became a way to self-soothe. I'd take a moment to focus on my breath, grounding myself in the present moment, away from the people around me and their toxic energy.

In these situations, mindfulness became my anchor. By focusing on my physical sensations—how my feet felt planted on the floor, how the air felt against my skin—I was able to detach from the emotionally charged environment and center myself again. The more I practiced emotional regulation, the more I realized that I could stay grounded in my truth without getting sucked into the drama or emotional chaos.

It's not about controlling the behavior of others. It's about controlling my response—understanding that I can't always stop people from violating my boundaries, but I can control how I respond to it. I no longer allowed myself to become reactive. Instead, I chose to respond from a place of strength, knowing that my emotional health would never again be for sale to anyone who sought to manipulate or harm me.

The Path Forward: Recognizing the Power of Boundaries and Triggers

When I reflect on my journey to reclaim my life and professional identity, it's clear that boundaries aren't just about drawing a line between myself and others—they're about asserting my worth in spaces where the goal often feels to diminish me. For so long, I let others define my value, dictate my narrative, and control my emotional responses. In environments like a toxic workplace, I was consistently surrounded by individuals who lacked respect for my personal space, my needs, and my values. It wasn't just about physical space or the demands of the job—it was about my right to exist and operate on my own terms, to uphold my beliefs, and to be treated with dignity and professionalism.

For instance, when my kosher dietary needs were disregarded, it wasn't simply about food—it was about people refusing to acknowledge my identity. The tampering with food areas, the constant violations, and the mocking comments that followed made me feel disrespected and invalidated. In the midst of that environment, I had to learn that my needs were valid—my boundaries were valid, even if others didn't understand or respect them. Saying "no" wasn't just about food; it was about protecting the essence of who I was.

I also had to face something deeper—the pressure to conform to a system that wasn't designed to meet my needs, but rather to control my behavior. There were moments when I felt as though my need for privacy, my desire to be left alone in my own space, was seen as an inconvenience. I found myself trying to fit into a mold of what they thought I should be: more "compliant," more "easy to manage," more willing to sacrifice my needs for the group's comfort. It wasn't overt—it didn't feel like a direct attack—but the subtle insinuations, like "Why do you need our help?" or "Can't you just deal with it yourself?" made me feel like I was bothering people by existing as I am. They wanted me to fit into the box they had for me, but I couldn't—and I shouldn't.

Those moments were triggering, but they also became opportunities to recognize and assert my boundaries. The key to my recovery has been learning that boundaries are not negotiable—they are a vital part of my self-worth. It's not just about keeping people away; it's about saying, "I have the right to take up space in this world." Those boundaries had to be set again and again, like a muscle that had to be rebuilt after years of being ignored or violated. Saying "no" became my act of self-respect, and each time I did it, I got stronger.

But that strength didn't come overnight. I had to confront the triggers—those flashbacks, those memories of feeling overlooked, unappreciated, and unimportant. In the workplace, I found myself facing my own vulnerabilities in ways I hadn't anticipated. The shift in the team's attitude—from "we're here to support you" to "why are you making things harder?"—triggered the same old feelings of being dismissed and undervalued. The constant undermining of my professional boundaries, my requests for personal space, and my attempts to navigate a stressful work environment felt like a replay of the toxic dynamics I had experienced in past relationships—particularly with my ex.

But rather than retreating, I began to see these triggers as signals—reminders that I had to fight even harder for the respect I knew I deserved. Every time my professional boundaries were tested, I responded with compassion, not just toward my colleagues but toward myself. I reminded myself that I had already come so far—I wasn't the same person I was when I first entered the workplace, unsure and hesitant. I had gained the strength to push back against disrespect.

With each challenge, I built emotional resilience. It wasn't easy—at times, it felt like a battle—but I started to view these confrontations as opportunities to affirm who I was, not just who others thought I should be. I began to reclaim my self-worth—and it wasn't tied to how others treated me. It was about what I would accept and what I wouldn't.

For example, when my supervisor would conduct surprise "check-ins" on my progress or workspace, I initially felt violated and unsure of how to respond. But instead of shrinking back, I reinforced my professional boundaries—reminding them that I had every right to privacy and respect in my role. Each time I stood up for myself, it was like adding another layer to the foundation of my self-resilience. I wasn't just getting by—I was actively protecting myself and reinforcing my sense of value in the workplace.

The truth is, triggers don't just disappear. They're a natural part of the healing process—like old wounds that resurface before they truly heal. But I've learned that I don't need to let those triggers dictate how I live my life. They are not a reflection of my weakness—they are just reminders that there is more work to do. And every time I face a trigger, I have the chance to respond differently, to choose growth over fear.

The work is ongoing, but I can already see how far I've come. Each time I say no to manipulation or disrespect, I am reminding myself of my strength. Each time I set a boundary that others may try to violate, I am reclaiming control over my life, my narrative, and my future. I am no longer the person who cowers in the face of rejection or abuse. I am learning to honor myself, to trust my instincts, and to respond to the world around me with the same compassion and respect that I am learning to give myself.

Reclaiming Power: The Path to True Freedom

This is not just a fight for safety. It is a fight for self-empowerment. And with each step forward, I am creating a future where I am free to be my authentic self, free to live without fear, free to thrive without the weight of the past holding me back.

For so long, I lived in a shadow of fear, uncertainty, and manipulation. I felt like I was constantly looking over my shoulder, not just physically but emotionally and psychologically. Every step I took felt like it was under the watchful gaze of someone who was determined to keep me in a state of helplessness. It wasn't just the physical threats or the emotional manipulation—it was the mental toll of never being able to fully trust my surroundings or the people in them. I was constantly second-guessing myself, questioning my instincts, and wondering whether the next step would lead to more harm.

Dr. Rachel Levitch

But that's not my story anymore.

Every day, I make choices that reshape my reality—choices that no longer come from a place of fear or submission, but from a place of strength and clarity. This is the essence of empowerment: not just knowing I have the right to exist as I am, but actively embracing it. Every time I assert a boundary—whether it's about privacy, safety, or respect—I am choosing myself. Every time I stand up to manipulation, I am reclaiming my power. This isn't just about surviving the past, it's about using my experiences as fuel to propel me forward.

In the past, when I was in a toxic work environment, I found myself trapped in a space that didn't just threaten my well-being—it challenged my identity. The constant microaggressions, the violations of my personal and professional boundaries, the subtle gaslighting that made me feel like I was asking too much just to be treated with basic respect—it all contributed to a mental space where I began to doubt my own worth. But I began to see each challenge as an opportunity to reaffirm my values and assert my rights. Each time I pushed back, I felt a little bit stronger. And with each step I took, I realized that no one else can define me unless I let them.

The journey to self-empowerment is not without its challenges. It's not easy to shake off years of trauma, manipulation, and emotional abuse. But with every day that passes, I feel the weight of the past becoming lighter, the grip of fear loosening its hold on my life. The fear that once dictated my actions no longer has the power it once did. I no longer cower under the fear of being misunderstood or attacked. Instead, I'm choosing to move forward in truth, embracing my past without letting it control me.

This is freedom: the ability to move through the world as your true self, unapologetically and unafraid. It's a process, but it's also a realization: I am enough. I always have been. The abuse I endured doesn't define my future. The manipulation I experienced does not dictate my present. Every boundary I set, every decision I make in alignment with my values, is a reclamation of my power. And I am doing this not just for myself, but for the future I am building—one where my children will see me as a model of strength, resilience, and self-respect.

In learning to trust myself again, I am showing my children that it is possible to break free from the cycle of abuse. I am teaching them that surviving is not enough. You don't just survive abuse—you thrive in spite of it. You don't just cope with manipulation—you learn how to build a life that doesn't revolve around it. That's what I want for them, and that's what I am giving myself: a life where I can move forward without constantly looking over my shoulder. Where my authenticity is not negotiable, and my sense of self is unshakable.

This journey has taught me that true freedom is not passive. It's an active, intentional act of standing up for yourself, of recognizing your worth, and of making choices that honor who you are. It's the ability to set boundaries without guilt, to live without fear of judgment, and to embrace the full scope of your being—flaws, strengths, and all. It's about knowing that you are worthy of a life free from manipulation and control, a life where your dreams are not just possible—they are within reach.

Every step forward is a small victory. And while the journey is ongoing, I am no longer afraid to take it. I am no longer afraid of the obstacles that lie ahead, because I know that I am equipped to handle them. I have survived this far. And the woman I am becoming is someone who not only survives—but thrives.

With each day, I am rebuilding my life from the ground up. I am creating a space where I can be unapologetically myself, where my children can flourish without the shadows of fear and manipulation, and where my future is filled with the hope and promise of the life I deserve.

And this is just the beginning.

Chapter 2

Understanding the Characteristics of My Ex: Narcissism, Obsession, and the Cycle of Control

The behavior of my ex-husband is not just emotionally abusive; it is the result of a deeply ingrained need to control, manipulate, and dominate. His actions over the years have followed a clear pattern—a cycle of emotional manipulation and harassment that has escalated over time. But it wasn't always so obvious. The early stages of our relationship seemed relatively normal, with subtle hints of possessiveness and manipulation disguised as "caring" behavior. However, over time, these tendencies became more pronounced, revealing the narcissistic and obsessive traits that have governed his every action since.

Narcissism: A Desire for Control Masked as Love

At the core of my ex's behavior lies narcissism—a pattern of grandiosity, an excessive need for admiration, and a complete disregard for the feelings of others. Narcissists often see their partners not as equals but as extensions of themselves, tools to be used to enhance their own sense of importance. Early in the relationship, his actions seemed charming, even flattering. He made me feel like I was the center of his world. But, over time, this "love" began to feel suffocating, as if my personal autonomy was being slowly stripped away.

In hindsight, I see that everything he did was designed to feed his ego. From undermining my confidence to making me feel responsible for his moods, he sought to maintain an unequal power dynamic in our relationship. His love was not truly love—it was about possession. The more he could control my thoughts, actions, and emotions, the more validated he felt in his own mind.

Obsession: The Need to Control and Maintain Power

One of the most devastating aspects of dealing with a narcissistic personality is the obsession that comes with it. Once someone like my ex perceives a loss of control—whether it's through separation, conflict, or boundary-setting—their response is often a desperate, even dangerous, obsession with regaining control. This obsession manifests in a variety of ways, but for him, it was most clearly seen in his constant digital surveillance. He would show up at places I hadn't mentioned being, make bizarre comments about things I had done or seen, and leave cryptic messages in places that only someone deeply embedded in my life would know.

His cyberstalking wasn't just about tracking my movements—it was about disrupting my life in ways that felt intentional and destabilizing. He wanted to make me doubt my own sanity, to make me feel like I couldn't even trust my own reality. For example, when I would return to my apartment, I began noticing strange things—objects moved, documents misplaced, even my food tampered with in ways that made no sense. I now know these were deliberate attempts to make me question what was happening in my own home.

The use of technology as a tool of control is perhaps one of the most insidious ways that narcissistic, obsessive individuals harm their victims. It allows them to maintain a constant presence in your life, even when they are not physically around. It is an invisible form of harassment that is difficult to prove but incredibly effective in shattering a victim's sense of safety and reality.

Antisocial Traits: A Complete Lack of Empathy

Another disturbing feature of my ex's behavior is his antisocial tendencies—a lack of empathy and an ability to manipulate and disregard the suffering of others without remorse. This can be seen in his willingness to violate my privacy, not just through digital means, but also by engaging in behaviors that others would deem morally reprehensible. It wasn't enough for him to continue controlling me emotionally—he took it a step further by actively trying to sabotage my career, manipulate my financial situation, and even target my emotional well-being through covert means.

Antisocial individuals often lack empathy for others, and in his case, this meant he was able to justify the harm he was causing. He wasn't concerned with how his actions affected me or our children. His focus was entirely on himself—on his need to control, his need to have power, and his need to avoid being "exposed" for the damage he had done.

I remember the shock of discovering that he was aware of private things—like confidential conversations I'd had with my therapists or financial advisers. It was clear that he had been accessing my personal information without my consent. These moments were terrifying, not just because of the breach of privacy but because they proved that he was going to any lengths to stay connected to my life, even if it meant crossing ethical and legal lines.

The Narcissistic Cycle of Abuse: A Perpetual Loop

The cycle of abuse that my ex perpetuates is one I have come to recognize as both predictable and relentless. First, there is the idealization phase, where he appears loving and attentive, showering me with affection and praise. Then comes the devaluation phase, where he starts undermining me, controlling me, and asserting his dominance. Finally, when I inevitably push back or attempt to break free, there is the discard phase, where he lashes out, resorts to harassment (like cyberstalking or physical intimidation), and tries to convince me that I am the one who is "crazy" or "irrational."

The key to understanding this cycle is recognizing that the abuse isn't just about hurt feelings or conflict—it's about a complete power imbalance. It's about someone actively trying to make you feel small and insignificant so that they can feel big and important. They will stop at nothing to retain control, even if it means crossing every boundary, legal and moral.

Impact on the Victim: The Psychological Toll

The psychological impact of this kind of harassment is devastating. The constant emotional manipulation and surveillance leave you feeling isolated, doubtful, and anxious. You start questioning your own instincts, and this creates a sense of disorientation and self-doubt that is hard to overcome. The more a narcissistic person undermines your sense of reality, the more they reinforce their control over you. It becomes a slow, insidious form of psychological imprisonment, where the victim is trapped in a state of perpetual confusion, unable to break free.

The emotional toll goes beyond just the present moment. It lingers, often resurfacing in physical symptoms such as chronic stress, insomnia, or anxiety. The psychological scars left by narcissistic abuse are often invisible, but they are deep and lasting. This is why healing from narcissistic abuse takes time—and why it requires intentional, focused work on rebuilding one's sense of self and personal boundaries.

The Impact on the Victim: The Lingering Scars of Narcissistic and Obsessive Abuse

The toll that this type of narcissistic, obsessive, and manipulative behavior takes on a victim cannot be underestimated. It's not just about feeling hurt or betrayed in the moment—it's the long-term psychological damage that leaves deep scars. When you're repeatedly subjected to emotional abuse, surveillance, and constant manipulation, the trauma becomes ingrained in your sense of self and how you interact with the world around you.

For me, the effects of cyberstalking and gaslighting were subtle at first but quickly escalated. There were times when I questioned my own perceptions—was I being too sensitive? Was I reading too much into things? That was the goal: to make me doubt myself so thoroughly that I'd lose confidence in my ability to navigate reality. My thoughts and actions became so controlled by him that I no longer trusted my instincts, my decisions, or even my emotions.

One of the most destructive effects of being in an environment shaped by narcissistic abuse is psychological fragmentation. This occurs when the constant cycles of idealization, devaluation, and discard erode your sense of worth and your trust in others. You start to internalize the false narratives being pushed onto you. In my case, I would question whether I was truly worthy of respect, of safety, of autonomy. The gaslighting was relentless: I was made to feel that everything I said and did was somehow wrong or exaggerated. It wasn't just what he did—it was the subtle, consistent undermining of my entire existence.

This emotional chaos isn't just draining—it's designed to break down your boundaries and make you emotionally vulnerable. The constant surveillance, whether physical or digital, is not just a violation of privacy; it's a conscious effort to isolate you, to make you feel like you're always being watched, always under scrutiny. It strips away your sense of safety, your personal agency, and any hope of independence.

The fear of being watched is like a constant weight, pressing on your chest. You don't feel like you can ever be free. For example, when I would attempt to leave the workplace to go back to my apartment for short periods, I couldn't shake the feeling that someone was there—watching, waiting. The routine cars parked outside my apartment were just the surface of a much deeper fear: the fear that every space I entered was no longer my own. My privacy, my sanctuary, had been erased.

This invasive behavior wasn't just psychological—it was emotional terrorism, a way of creating a sense of chaos and powerlessness in my life. Over time, this wear and tear eroded my mental resilience, and I found myself constantly second-guessing my own actions, feeling the need to justify every step I took. I became hypervigilant, always scanning my environment for signs of intrusion or threat. The pressure to stay "on alert" for an unseen enemy was exhausting.

The Rebuilding Process: Emotional Resilience as a Path Forward

Yet, amid the wreckage, there was a spark of determination to rebuild. Understanding the trauma and emotional toll that comes with cyberstalking and narcissistic abuse was the first step toward healing. I had to confront the reality that the person I was before this trauma didn't exist in the same way anymore. The emotional landscape had changed, and I would need to redefine my sense of self, my safety, and my worth.

Healing from this kind of abuse requires radical self-compassion and recovery from the self-doubt instilled in me over years of manipulation. I had to come to terms with the fact that his actions were never about me—they were always about his need for control. And though I had been victimized, I did not need to remain a victim. Healing meant not just surviving the trauma, but learning to thrive in spite of it.

A crucial part of this journey was taking back control of my boundaries. Before I could begin healing, I had to learn that boundaries aren't just about physical space—they are about emotional sovereignty. In the context of narcissistic abuse, boundaries are about reclaiming your worth, about refusing to be treated as an afterthought, and about asserting your right to exist without fear of invasion. Once I started setting these boundaries, even when it was uncomfortable or difficult, I could feel myself strengthening.

When I started to disconnect from the abusive dynamics and reinstate my boundaries, it became a form of self-preservation. This meant not just cutting off contact, but also shifting the narrative in my mind. I could no longer afford to live in the shadow of someone else's expectations, control, and emotional manipulation. Every time I set a boundary, whether it was telling the workplace staff about my dietary needs or refusing to engage with my ex's attempts at contact, I was reminding myself that my needs matter and that I deserved respect.

The Battle for Privacy and Autonomy: A Test of Strength

The journey toward reclaiming my privacy and autonomy has been anything but straightforward. In many ways, it feels like the most profound test of my strength—one that I didn't ask for, but one I now have no choice but to face. The violations I've experienced, both physical and digital, have stripped me of much of my sense of safety. But reclaiming my privacy is not just about restoring the past—it's about creating a new future, one where my needs, my boundaries, and my autonomy are no longer compromised.

At the beginning of this journey, I had to confront the harsh reality that my sense of privacy had been completely undermined. My home—my personal space—had become a battlefield. It wasn't just the intrusions I experienced in my apartment, with people watching me from outside or the feeling that my every move was being monitored. It was the constant fear that no space was truly my own anymore, that even the most private parts of my life had been touched by someone who believed they had a right to control everything I did. The digital space was no better—my emails, my phone, my personal documents—all were fair game for someone who had no regard for my well-being or my right to privacy.

The first step in regaining control was acknowledging the violations. I had to admit to myself that my boundaries had been repeatedly disrespected—not just by my ex, but by systems, people, and environments that I had once trusted. This wasn't about self-blame; it was about coming to terms with the fact that my autonomy had been taken away from me, and I needed to act to get it back.

Taking back my privacy meant understanding that I could not allow these constant violations to continue without response. I had to assert my right to safety in ways that were both physical and emotional. Setting up security systems like cameras and more advanced Wi-Fi protections was my first move—simple steps that, on the surface, might seem like mere deterrents, but for me, they represented much more than that. They were statements of defiance against the invasive control that had shaped my life for so long. Every camera I installed was a small victory over the forces that had made me feel vulnerable—it was a visible commitment to myself that I wouldn't allow my life to be manipulated anymore.

But physical security measures, while essential, aren't the whole picture. I also had to protect myself emotionally—this was, and still is, where the most difficult work happens. How do you regain a sense of self-respect after years of being made to feel small, insignificant, and watched? How do you move forward when every part of you has been conditioned to expect betrayal, manipulation, and emotional abuse?

I found the answer in boundaries—clear, unapologetic, and protective. When I began setting boundaries, it wasn't just about saying "no" to others; it was about learning how to say "yes" to myself. At first, this felt unnatural—almost impossible. My boundaries were violated so often, so thoroughly, that I had a hard time even recognizing them anymore. But as I began to regain my sense of self, I started to see these boundaries as necessary tools of survival, not just abstract concepts. The more I said "no" to things that violated my privacy, the more I began to regain control of my narrative, of my life.

Even something as simple as declining interactions that felt intrusive—whether it was a request for personal details from someone at the workplace, or my ex trying to engage in manipulative small talk—was a reclaiming of power. Each refusal was a tiny victory. It didn't mean that the threat was over, but it meant that I had the strength to assert that I would no longer accept being treated as less than.

Rebuilding my autonomy meant taking control over the way I allowed others to affect my life. It meant detaching from the emotional chaos that my ex constantly tried to instigate. His actions were designed to emotionally bankrupt me, to push me into a state where I felt like I was running on fumes. For a long time, I felt like I was simply surviving—going through the motions, day by day. But reclaiming my privacy and autonomy was about something much more profound than that—it was about asserting that I mattered.

There were moments when I was overwhelmed—when the magnitude of what I was trying to do felt too big to face. The process of reinstalling security, of organizing my documents, of protecting myself emotionally from old wounds—it all felt like too much at times. There were days when I didn't want to fight anymore. But even on those days, I reminded myself that the process was not just for today, it was for tomorrow. It was for the life I was building, a life free from manipulation, from fear, from the daily struggle of wondering if I would be invaded again.

In many ways, each step forward is a personal revolution. It is me saying, "I choose myself." When I installed cameras, when I started changing my passwords, when I cut off communication with people who were toxic—each step was a defiant statement against the power that had tried to strip me of my identity. Autonomy is freedom, and I wasn't going to let anyone take that from me again. The more I reclaimed, the more empowered I became.

But it isn't just about physical space. It's about claiming your emotional space—refusing to let anyone manipulate your thoughts or beliefs about who you are. And it's about learning that privacy is not just a right; it's a reflection of your dignity. My ability to protect my own space, my own energy, and my own sense of self became an essential part of my healing process.

Reclaiming privacy is not a one-time event; it is an ongoing commitment to yourself. It is a constant test of strength—one that demands vigilance, but also compassion. It's about saying, "This is mine," and protecting it with everything you've got. And as I take each step forward, I feel the weight of the past lifting—slowly, steadily—until the person I was before these invasions of privacy no longer defines me. Instead, I become the person who chooses freedom, chooses safety, and chooses peace.

Dr. Rachel Levitch

One of the most defining moments in my journey to reclaiming my autonomy came during my time at the workplace. It wasn't just the physical space that felt unsafe—it was the constant undermining of my needs by people who were supposed to be my support system. I had expressed multiple times that, as a kosher individual, I needed specific accommodations for my food. I explained that violating my dietary rules wasn't just an inconvenience; it was a deep violation of my personal boundaries and beliefs. I needed the space to prepare my meals safely, and I needed the kitchen areas to remain untouched, to respect my religious practice.

Instead of receiving the support I needed, I found myself being targeted. The same advocates who had once said they would assist me in creating a safe space for my needs became increasingly dismissive of my concerns. They overrode my request to keep my dietary restrictions private and clear from others. I was repeatedly harassed by other residents—food items I didn't bring or approve of were placed in my designated space, with blatant disregard for the emotional and spiritual impact this had on me. At one point, pork was left on my shelf in an area I had specifically marked as kosher only. It wasn't just food; it was a symbolic violation of my very identity, something that spoke louder than just an oversight—it was intentional.

At first, I tried to rationalize it. Maybe the staff wasn't aware of how deeply this affected me. But after bringing it up repeatedly and receiving dismissive responses like "just clean it," I realized this wasn't an accident or misunderstanding. It was a pattern of emotional disregard. The situation escalated from one of carelessness to one of active neglect. I had asked for help—all I wanted was a safe, clean space to prepare my meals in accordance with my beliefs—but instead, I was made to feel as though I was being difficult, demanding, or too sensitive.

This wasn't just about food. It was about the lack of respect for my autonomy and my dignity. I felt the same helplessness that I'd experienced in past relationships where my needs had been minimized, ignored, or trampled over. It became a microcosm of the broader manipulation and control that I had been subjected to. The message I was getting was loud and clear: You don't matter enough to make an effort for.

This situation, small as it may have seemed on the surface, was another turning point. It was a stark reminder that even in spaces meant for healing, the absence of true respect for personal boundaries can be devastating. And it was a reminder that no one else was going to protect my autonomy but me.

Tying It Into the Larger Narrative:

As I navigated the emotional toll of this experience, I was forced to confront the deep vulnerability I had been left with by years of abuse. This wasn't just a physical violation of space—it was a psychological violation as well. And as much as I fought to stay strong, to assert my rights, I realized that no one else could fight this battle for me. The power to protect my privacy, my autonomy, and my sense of self rested solely in my hands.

The more I allowed these violations to go unchecked, the more I allowed my own sense of dignity to be eroded. I had to step into my strength and create clear, uncompromising boundaries to preserve what was left of my autonomy. This meant standing firm when others made me feel like I was asking too much, or when they tried to dismiss my needs as unimportant. Every moment I spent fighting for my privacy, for the space to be myself, was a step towards reasserting the right to exist on my own terms.

As I began setting those boundaries, I realized how much I had been conditioned to tolerate disrespect, to feel like I didn't deserve to be heard, seen, or protected. This experience at the workplace was a harsh lesson in the importance of not just protecting physical spaces but recognizing that emotional spaces require the same vigilance. I was learning that my privacy wasn't just about walls, cameras, or locked doors—it was about creating a system where I could demand respect, not just from others, but also from myself.

Boundaries vs. Autonomy: The Subtle Undermining of Independence

One of the most subtle and insidious forms of narcissistic manipulation is the deliberate undermining of a person's independence. When you're working hard to reclaim control over your life—whether it's through financial freedom, career growth, or simply becoming self-sufficient—it's not uncommon for people with narcissistic tendencies to view your progress as a threat to their control. The more you take charge of your own life, the more they push back, using tactics that may not always be immediately obvious but are incredibly damaging over time.

In the context of a workplace environment—a place where I was supposed to receive support and guidance—the manipulation and undermining of my independence were palpable. Early on, I was told that I would receive help with getting my car repaired—a fundamental need, as my car was critical for me to access therapy, work, and other important resources. But over time, the promises of help started to shrink and fade. Initially, I was told that the organization would take care of the diagnostic costs, but soon that turned into a "we can only help you with the diagnosis" offer. And then, when the emissions deadline passed, I was left with a car in violation of the law, unable to park it at the workplace environment because they deemed it a "hazard."

At first, it seemed like a small misstep, but as things progressed, I began to feel as though my autonomy was being stripped away with every step I took toward getting my car fixed. Instead of offering meaningful help, the workplace environment seemed more interested in making sure I remained reliant on them. They didn't want to help me become independent; they wanted to ensure I was always in a position of need—always having to turn to them for solutions, no matter how small. And when I finally did get the car fixed on my own, a victory that was a direct result of my own efforts, I was met with hostility rather than support.

The behavior from the workplace environment shifted noticeably after that. When I told them that I had resolved the issue on my own, I wasn't met with congratulations or acknowledgment of my hard work. Instead, I felt like I had violated some unwritten rule—they were resentful of my success. It was as if my ability to solve my own problems was a challenge to their authority and their control. They no longer saw me as someone in need of support; they saw me as a threat to the dynamic they had carefully curated. My independence made them uncomfortable.

This pattern of behavior mirrors the tactics often used by narcissists: first, they offer assistance, but that assistance is contingent on compliance with their control. The more you try to take independent action, the more they sabotage or minimize your efforts. And once you do succeed on your own, they undermine it, using passive-aggressive hostility to remind you that you still need them, even when it seems like you don't.

It wasn't just about the car; it was about the underlying power dynamic. The message they sent was clear: you're not supposed to succeed without our help, and if you do, it's going to cost you in emotional retribution. The more I tried to reclaim control of my life, the more they worked to remind me of my place—which was dependent on them.

In hindsight, this was about control. The more independent I became, the more they felt threatened by it. This behavior is not unique to abusive relationships or toxic work environments—it's the hallmark of a narcissistic personality. The goal isn't just to keep you dependent on them physically; it's to maintain psychological control, to ensure that no matter how far you move toward self-sufficiency, you will always need them.

When you're trying to regain your life, whether it's fixing your car, finding a new job, or simply asserting your autonomy, there will be people like this—who promise help, then subtly sabotage your progress, often with passive-aggressive tactics or by simply withholding the support you need. These are not just disagreements or misunderstandings; these are deliberate tactics used to maintain control, making it impossible for you to stand on your own two feet without facing obstacles.

And the more you try to stand up for yourself, the more these individuals will act out in ways that make you feel small, incompetent, and unworthy. This is the true impact of narcissistic manipulation—it forces you to constantly second-guess your own reality, making it incredibly difficult to trust your instincts or feel empowered to act on your own.

The Narcissistic Playbook: Undermining Independence and Autonomy

In the world of narcissism, control is the ultimate goal—whether it's control over your emotions, your decisions, or your independence. The narcissist, in their obsession with dominance, often uses subtle but powerful tactics to ensure that the victim remains in a state of dependency. Even when a victim takes steps toward becoming self-sufficient, the narcissist will go to great lengths to sabotage those efforts, not out of direct malice, but out of a deep-seated need to maintain control.

This dynamic was a daily reality for me, especially when it came to anything that reflected my independence or autonomy. My ex's manipulative tendencies would often manifest in ways that, at first, seemed like offers of help but were, in reality, calculated moves to keep me in a state of need. For example, take the situation with my car repair.

Initially, the offer to help with the repair costs seemed like a lifeline, a gesture that made sense within the context of me needing to have a car to meet basic needs. But when that offer started to shift into a vague promise of "just getting the car diagnosed," I began to notice the subtle undermining at play. The issue wasn't just that they were failing to deliver on the promises, it was the shifting of responsibility—making it clear that I was still reliant on their systems, their structure, their support, which wasn't really support at all.

Dr. Rachel Levitch

When the emissions deadline passed, and I was unable to legally park my car, the real impact of this tactic became evident. Instead of offering meaningful support or actively helping me resolve the issue, the response was passive-aggressive: disqualifying my car as a "hazard" and trying to prevent me from parking it where I needed to. This wasn't a concern for my well-being or safety; this was about asserting control over my situation, creating a barrier to my independence by making something simple into an obstacle.

As I went ahead and fixed the car on my own, the resentment from them became palpable. The fact that I didn't need them—that I took initiative, that I was able to solve the problem independently—struck at the very core of what the narcissist wants: they needed to feel as though my independence was an illusion, that their control was still essential. The moment I achieved something for myself, it was met not with congratulations, but with an emotional backlash, where their support became hostile and condescending.

This is the hallmark of narcissistic manipulation. The narcissist doesn't truly want to help; they want to maintain an ongoing sense of superiority, and that superiority depends on your need for them. When you begin to break free from this dynamic—whether it's by solving a practical problem, asserting your needs, or moving toward independence—the narcissist responds with covert sabotage, dismissive behavior, or withdrawal of support. They need you to remain in a state of dependence—even if it means you struggle longer than you should, even if it means they have to create new obstacles to keep you from succeeding.

In the context of someone who has been victimized by narcissistic abuse, these subtle forms of manipulation are extremely damaging because they force you to constantly question your self-worth and abilities. The more you try to stand on your own two feet, the more they push you back down. You're not just fighting to fix a car or solve a logistical issue; you're fighting against the invisible force that constantly undermines your confidence and autonomy.

This behavior is an example of how narcissism manifests not just in overt control but in covert sabotage—the type that is difficult to pin down and that can often leave you feeling gaslighted and confused. The narcissist uses your successes as threats to their control, and instead of helping you to become empowered, they drag you back into a state of dependence by diminishing your achievements and setting up new hurdles.

The result is a psychological trap: you're pushed to prove yourself, but the moment you do, it becomes clear that independence is the very thing the narcissist fears most. Their insecurity relies on your weakness—their need to see you as needy, dependent, and helpless. This dynamic is toxic, draining, and, over time, disorienting, especially when you are working so hard to regain control over your own life.

In the end, it's not about the car or the repair—it's about the psychological battle. It's about how the narcissist's need for control infiltrates even the most mundane aspects of your life. And every time you succeed on your own, every time you push past their manipulations, you reclaim a bit more of your autonomy and self-worth, no matter how many setbacks they create.

Connecting the Dots: Legal Implications and Moving Forward

Dr. Rachel Levitch

Understanding the psychological profile of someone like my ex has been a turning point in my healing journey. For so long, his behavior was an enigma—something I couldn't fully understand, even as I tried to make sense of the emotional, psychological, and even financial chaos he left in his wake. It wasn't just about the specific acts of abuse; it was the way his actions seemed to spiral out of control, with no clear pattern or reason. I questioned myself, wondering if I was overreacting, if I was imagining the extent of his manipulation. But as I started to piece together the puzzle—learning more about narcissistic behavior, emotional manipulation, and the psychological tools that people like him use—I began to understand that the problem wasn't me.

His behavior wasn't random; it was deliberate. It wasn't something I could control by simply being "stronger" or "better" at navigating our interactions. This was a pattern of behavior that went far beyond just a toxic relationship. It was a calculated campaign of emotional and psychological warfare, built on control, dominance, and a profound disregard for my autonomy. Narcissistic abuse operates on the principle of destabilizing the victim's sense of self, isolating them, and keeping them in a perpetual state of dependency.

Recognizing this has given me a crucial piece of the puzzle. It's not just about understanding the pain he caused, but about empowering myself. Now that I can identify the tactics he used to control me—whether it was the constant undermining of my confidence, the way he made me feel small, or how he blurred boundaries—I can start to fight back. This awareness isn't just a form of emotional clarity; it's an essential part of my legal strategy. By understanding how narcissistic individuals operate, I am better equipped to protect myself legally and ensure that his behavior is recognized for what it truly is.

This knowledge has helped me stop questioning myself, stop doubting my instincts, and stop allowing his manipulation to redefine my reality. It's a constant process of validation—both internal and external—where I acknowledge that I am not crazy for feeling the way I do. I am not imagining this abuse. I am responding to a very real, very harmful pattern of control and emotional manipulation. And understanding that has made all the difference in my ability to defend myself, both in the courtroom and in my own life.

For example, when my ex and I separated, I thought I had escaped. I assumed that distance, both physical and emotional, would be enough to break free from his influence. But the narcissistic need for control doesn't simply disappear with distance—it morphs into something even more insidious. His control didn't stop when the relationship ended; it just adapted. The narcissist's drive for dominance doesn't hinge on a conventional emotional connection; it thrives on dependency and control, and for him, the target was no longer just my emotions—it was my autonomy.

At first, the emotional manipulation continued in subtle ways— insisting that I needed him, that I couldn't make it without him, undermining my ability to stand on my own. But as time passed and I started to build a life outside of our relationship, his tactics shifted. Cyberstalking became his new weapon. He didn't need to be physically present to maintain control. Instead, he leveraged technology to infiltrate my life in ways that were almost impossible to escape. Digital harassment replaced late-night phone calls, and constant surveillance replaced the direct, face-to-face interactions we once had.

This transition from personal, emotional manipulation to a virtual invasion of my life was seamless for him. For me, it was overwhelming. I'd blocked him on every communication platform, changed my contact information, and tried to sever all ties—but the digital world gave him unlimited access. It was like the walls of my life had become transparent. I would see things that made no sense at first—things I'd posted privately, messages I hadn't sent, or comments made by others that he should never have known about. The more I tried to regain control over my personal space, the more I found myself being drawn back into his web.

The psychological toll was just as devastating, if not worse. I no longer had to endure just emotional manipulation—now I was being tracked, monitored, and controlled in ways I couldn't fully comprehend. The boundaries I had worked so hard to create were constantly violated, and there was no place I could retreat to that felt truly safe. I was no longer just in a relationship with a narcissist; I was entangled in a digital prison where every move I made felt monitored and scrutinized.

Narcissists like my ex thrive on dependency—their power comes from the victim's inability to function independently. It's not enough for them to control your emotions or dictate the terms of your relationship. They want to control every aspect of your life, even the spaces where you thought you had privacy and security. By engineering situations where I was forced to rely on him for information, for validation, or even for basic logistical support, he ensured that I remained dependent on him, even after we parted ways.

This is the nature of narcissistic control—it's designed to be all-encompassing. When I was too independent, too self-sufficient, he found new ways to break down those walls and undermine that independence. Cyberstalking became his tool of choice, and in doing so, he found a way to manipulate my perception of reality. It wasn't just about controlling my behavior; it was about creating a world where I couldn't trust my own experiences, my own thoughts, or my ability to act without his influence.

Every time I thought I had a handle on my life, I'd feel his presence again, not in a tangible way, but in an insidious, digital form. Whether it was an email notification, a strange message on social media, or even an odd comment on a post, he was always there—lurking in the background, making sure I couldn't truly live without him in some capacity.

Living in Uncertainty: The Weight of Constant Fear

There is a unique kind of terror that comes from living in a world where the ground beneath your feet feels like it's always shifting, where every decision, every action, every word you speak could potentially be manipulated or used against you. This is the reality of living with someone who thrives on control and chaos. Uncertainty becomes a constant companion, and with it, a pervasive sense of fear that you can never quite shake.

In the aftermath of narcissistic abuse, particularly when compounded by cyberstalking, it's not just about surviving the day-to-day—it's about navigating a life where the boundaries of safety and privacy are perpetually blurred. What once felt secure is no longer. There are no longer safe spaces—even in the quietest moments, the feeling of being observed, followed, or monitored lingers in the back of your mind. It's not just about what's happening now; it's about what might be happening in the background, beyond your line of sight.

When I started to feel the weight of this uncertainty, it wasn't just a passing anxiety—it was a crippling fear that would find its way into every corner of my life. I couldn't trust my own thoughts or perceptions because everything felt contaminated by the possibility that someone—specifically, my ex—was behind it, orchestrating it, manipulating it. It's the feeling that comes when you're constantly waiting for the next shoe to drop, but you have no idea when or where it will land. It's the knowledge that you're being watched, but you don't know exactly how or by whom.

One of the most insidious aspects of living with this kind of fear is that it often doesn't feel like fear at all. It's a low-grade sense of vigilance that wears you down over time. You might find yourself constantly checking your phone, looking over your shoulder, wondering if the comment on your social media is a coincidence or something more. You wonder if that odd email you received is just spam—or if it's part of a larger plan to destabilize you. And because the source of this anxiety is often hidden behind the screen, you can't confront it directly. It's this ever-present uncertainty that keeps you trapped in a cycle of hypervigilance, where every moment is tinged with suspicion.

For example, there were moments when I would go to sleep feeling exhausted from the day, only to wake up with a sense of dread that something had happened while I was unaware. I'd check my phone for messages, my accounts for signs of tampering, only to find nothing—but the unease remained. There's no clear enemy in these moments—no face, no voice. It's just the feeling that someone is out there pulling the strings, watching, waiting for you to slip up. That's the fear. The paranoia of never knowing if the next message, the next phone call, the next unexpected encounter is connected to something far darker, far more calculated than you could have anticipated.

It's a fear that makes even simple actions feel heavy with consequence. Leaving the house—once a simple and natural act—becomes a tense negotiation with your own sense of safety. Will they be watching me? Will something happen while I'm gone? And what happens if I make a mistake? The anxiety compounds when you realize that every decision is affected by the weight of this fear, and often, you make choices based on a distorted sense of urgency. This becomes your reality: constant fear of the unknown, of being perpetually on edge, of feeling like you can't relax because someone else has the power to invade your life at any moment.

Living with this level of uncertainty also deeply affects your sense of control. When you can't predict what will happen next, you start to lose confidence in your ability to navigate life without interference. There's a mental exhaustion that sets in, as you question every action, second-guess every decision, and wonder whether it's safe to trust your own judgment anymore. For me, it wasn't just about handling the logistics of daily life—it was about trying to maintain a sense of normalcy in the face of a constant threat, while also trying to keep my own identity intact.

The emotional toll is often underestimated. The fear isn't just a passing thought or a momentary jolt; it's a gnawing presence, one that you carry with you, whether you're working, spending time with loved ones, or even resting. It has a profound impact on your mental well-being, because you're forced to live in a state of perpetual readiness. This creates a cycle of anxiety, depression, and isolation, because the reality of living with uncertainty often means that you can't reach out for support in the ways you need to. You can't explain why you're fearful in a way that others would fully understand. The fear is subtle and insidious, and it's driven by something that's invisible, but always felt.

And yet, despite this constant cloud of fear, I learned that the only way to break free from this cycle was to acknowledge the uncertainty for what it was and to reclaim control in small, intentional ways. I couldn't let the unknown continue to dictate my reality. I had to set boundaries in ways that felt authentic to me, take actions that reinforced my sense of safety, and trust in the power of my own decisions. The process of facing and confronting the fear—acknowledging that it was real, but refusing to let it define my life—was one of the most important steps I took toward healing.

While the fear still sometimes lingers, I now recognize it for what it is: a consequence of living in a world where control was taken from me. But it doesn't have to govern my life. Each step I take in building a safer, more independent existence is a defiance against the fear that once ruled me. And with every decision to move forward, to act from a place of empowerment rather than fear, I reclaim the ground beneath me.

Fear But Not Fear: Hypervigilance and the Weight of Uncertainty

Living in the aftermath of narcissistic abuse, particularly in the context of cyberstalking and constant emotional manipulation, often feels less like living with fear and more like existing in a constant state of alertness. It's not that you're terrified at every moment, but your brain is rewired to anticipate the worst, to always be looking for threats, even when there's no clear enemy in sight.

It's like living with a shadow that never really leaves, one that follows you, not as an acute fear of an immediate danger, but as an awareness that something—someone—could be lurking just out of sight. The uncertainty of it all is what wears you down. You're aware that something is wrong, but you can't always pinpoint what it is or who is behind it. The feeling of being watched isn't a brief moment of panic; it's a subtle undercurrent that flows through your daily life, making you second-guess decisions, constantly check your surroundings, or monitor your digital presence.

This is where the distinction between fear and hypervigilance becomes clear. Fear is a short-lived, intense emotion triggered by a specific event. But hypervigilance is a constant, low-grade tension that affects your ability to relax. It's not the same as fear, but the cumulative effect is much more draining. There's a constant mental exhaustion from always being "on," always being aware of the possibility that something could happen, but you don't know exactly when or how.

For instance, when I would wake up in the morning, I didn't immediately feel fear—but I would feel an underlying tension, a low-level anxiety that something was about to intrude into my day. It wasn't a sudden terror, but it was the persistent thought: *What's going to happen today?* Will I find something in my email? Will there be a strange phone call? Will my car be tampered with again? These thoughts weren't about direct fear, but rather about expecting the unexpected, the feeling that no part of my life was fully within my control.

What made this worse was that, unlike a clear threat, these feelings were not grounded in a specific event or person. They were formed by the nature of my environment—by the constant manipulation, the deliberate chaos, the mind games. It was the uncertainty that was exhausting, because your mind constantly has to stay one step ahead, trying to predict what might happen next, even though the "next" is always intangible and unknown.

The most insidious part of living in this state was that it didn't always feel like fear. It felt like a shadow. A weight. A background hum that never fully goes away. It's not like a flash of terror when someone jumps out at you from behind a corner—it's the nagging feeling that someone is always there, just out of sight, watching, waiting for you to slip. The fear is subtle—it's not necessarily *terror*—but it's pervasive. It affects the way you think about everything, even the smallest decisions.

Take, for example, the feeling that would settle in when I realized I was about to leave the house. I wouldn't necessarily be afraid to go, but I'd feel a pull in my gut, as though something was off. I'd double-check everything. Did I lock the door? Was my phone charged? Were there any odd messages or emails that could have slipped by unnoticed? I wasn't afraid of what might happen, but I was hyper-aware of every potential threat, even if it was just the feeling of being watched or the idea that something I did could be scrutinized or manipulated later on.

And yet, despite all this, it wasn't true fear—at least not in the traditional sense. It wasn't the fear of an imminent threat. It was more like a knowledge—a knowledge that I was living in a world where my personal safety, my peace of mind, and my ability to control my life were never entirely my own. I didn't fear one specific event. I feared the unpredictability of it all—the idea that my life could be infiltrated at any moment by someone who had no regard for my boundaries, my needs, or my peace of mind. It's the feeling that you can't trust the space around you because it's always potentially contaminated by someone else's influence.

This constant state of hypervigilance is emotionally exhausting. Your brain never gets the chance to relax, and even when you think you're alone, the awareness of the potential for intrusion is always there. This perpetual state of alertness might not always feel like fear, but over time, it has the same toll. It's a quiet, draining mental fatigue that becomes difficult to escape.

But eventually, I realized that this wasn't the way I had to live forever. The key was acknowledging that this constant state of alertness was symptomatic of my experience—it was a result of the chaos that had been introduced into my life by someone who thrived on control. The first step toward reclaiming my life was not to deny these feelings, but to acknowledge them for what they were—the lingering effects of someone else's manipulation. Once I began to understand that, I could start taking active steps to rebuild my sense of safety and autonomy. The shadow didn't have to control me—and eventually, with time and effort, I could turn that vigilance into a force for my own healing, rather than continuing to live in its grip.

The Unraveling of Promises: How Independence is Threatened

Take, for example, the consistent failure to follow through on promises of support. It started with something as seemingly simple as getting my car repaired—an action that should have been straightforward and helpful, especially from someone who claimed they wanted to "help me get back on my feet." Initially, the promises were small but manageable—getting a diagnosis or just "checking out the issue." It seemed like a basic, supportive gesture. But as time passed, those small promises turned into vague assurances, followed by missed deadlines and a complete lack of tangible action. What was meant to be help became just another opportunity to create uncertainty and frustration.

This pattern of manipulation—giving a little, but never enough—became a familiar tactic. It wasn't about the actual help; it was about keeping me in a constant state of dependence, whether I realized it or not. In reality, what seemed like helpful gestures were part of a broader strategy to ensure I didn't feel capable of managing anything on my own. Narcissists don't want you to succeed independently—they need you to rely on them, whether it's for emotional validation, financial assistance, or even something as mundane as car repairs.

By withholding follow-through, the narcissist manufactures a crisis that only they can resolve. They don't want to see you succeed because your success threatens their control. So, when I took matters into my own hands and decided to fix the car on my own, it wasn't met with praise or support. Instead, there was hostility, resentment, and a passive-aggressive undermining of my efforts. My independence became a direct threat to his carefully constructed narrative of control, where I was supposed to remain dependent on him. My ability to manage on my own exposed his manipulation and put a crack in the power dynamic he had so diligently maintained.

This reaction—hostility in the face of my independence—was not about the car or even the logistics of fixing it. It was about maintaining dominance. Narcissists thrive in environments where they can control others' autonomy. When I stepped outside of that, when I found the strength to take action without needing his permission or assistance, it disrupted the delicate balance he had created. It wasn't just about the car; it was about sending a message: I wasn't supposed to succeed without him.

This type of behavior is a hallmark of narcissistic abuse. The narcissist will engage in control tactics at every turn, whether it's through neglecting promises or undermining your progress. The idea isn't to help; the goal is to maintain the illusion that they are the one who holds the power to solve your problems. When you no longer rely on them, they feel diminished, and the insecurity that comes with that leads to passive-aggressive behaviors designed to undermine your success. This was a direct reflection of the pattern of manipulation that would unfold over time: Promises were never meant to be kept; they were meant to keep you in a cycle of dependency.

By realizing this dynamic, I began to understand that his reluctance to follow through was not a failure of support—it was a deliberate strategy to break down my autonomy. And when I chose to act independently, it wasn't just my car that was repaired; it was my self-esteem, my agency, and my freedom. But this also marked a turning point—a realization that no matter how many promises he made or how many "attempts to help" he offered, his true goal was to make sure I never felt capable of standing on my own.

Each time I chose independence, it chipped away at the carefully constructed narrative he had built. The power was in my hands, not his, and that terrified him. He wanted me to rely on him. He wanted me to stay in the cycle of needing him. But with every small victory, every decision I made for myself, I reclaimed a little bit more of the life he'd tried to control. The car was just the beginning—a symbol of the larger struggle for autonomy and empowerment.

This pattern reflects a deeply embedded need for superiority and dominance. In his eyes, any action I took to stand on my own was seen as a threat to his sense of power. What I had once believed were attempts to "help" were really covert tactics to undermine my success and keep me dependent. His need to feel powerful in the relationship meant that my progress—whether it was physical, emotional, or even logistical—was consistently sabotaged. It wasn't about what I *needed*; it was about what he needed: control.

And then there's the gaslighting. Narcissists are experts at making their victims question their reality. They use indirect manipulation to create self-doubt and make you feel like you're incompetent or incapable of handling your own affairs. He gaslit me by creating confusion around boundaries—offering help, then withdrawing it, and then making it seem like I was ungrateful or too demanding when I pushed back. I was left questioning whether I was asking for too much, when in fact, all I was doing was attempting to meet basic needs—needs he had created and then ignored.

The psychological implications of these behaviors cannot be overstated. It wasn't just about maintaining a relationship; it was about obliterating my sense of self. Narcissistic abuse is designed to disorient and destabilize, and it's done through a combination of microaggressions, emotional neglect, and subtle sabotage. This manipulation doesn't just damage you in the moment—it wears away at your confidence, independence, and sense of control over time. When you're dealing with someone who thrives on power, your boundaries don't just get tested—they get violated, over and over again.

For me, recognizing these patterns wasn't just a matter of understanding the abuse—it was the key to empowering myself to fight back. When you're dealing with someone who is manipulative, narcissistic, and emotionally destructive, their behavior must be recognized for what it is. It's not just about getting justice for the harm that's been done—it's about preventing them from using the same tactics against someone else in the future. The legal system must take these psychological behaviors into account when assessing cases of cyberstalking and harassment.

In my case, the legal implications are clear: narcissistic manipulation and cyber harassment are crimes that need to be addressed, but only if they're recognized for what they are. This is where self-empowerment comes into play. I now have the knowledge and clarity to stand firm in my legal actions. I'm not just trying to right the wrongs that have been done to me; I'm fighting to make sure that these patterns are recognized and properly addressed so they don't continue.

The process isn't easy, but it is necessary. This fight is about more than just safety—it's about reclaiming my autonomy, my privacy, and my dignity. The narcissistic manipulation I've endured has impacted every part of my life—from personal relationships to my ability to function day-to-day. And while I'm continuing to heal, I know that the path forward lies in holding my ex accountable for his actions, while ensuring that no one else has to face the same kind of manipulation.

Significance of Cyberstalking and Harassment in Narcissistic and Antisocial Individuals

1. Cyberstalking as a Means of Control:

- Narcissists and individuals with Antisocial Personality Disorder (ASPD) often use cyberstalking as a tool to regain control over their victims after a relationship has ended. They may seek to punish the victim for perceived slights, real or imagined, and to maintain a sense of power and dominance.

- Key Feature: The primary aim is to instill fear, keep the victim in a state of emotional or psychological turmoil, and to reinforce the victim's sense of powerlessness. This is consistent with narcissistic tendencies to retaliate against anyone who does not submit to their needs or desires.

2. Repeating Patterns of Behavior:

Stalking and harassment are often part of a broader pattern of boundary violations. The perpetrator may have a long history of exploiting others, manipulating them emotionally, or engaging in criminal activity for personal gain, whether that's financial, emotional, or simply the need for control. The victim is frequently seen as an object for their narcissistic needs, which can lead to repeated harassment through multiple channels (email, text messages, social media, etc.).

- Continuous Cyber Harassment: This can range from digital threats, sharing sensitive information, sending unwanted communications, or even hacking into the victim's personal accounts. The lack of boundaries is a key characteristic of this behavior.

3. Psychological Impact on Victims:

- Emotional Trauma: Cyberstalking causes chronic emotional distress, which can lead to anxiety, depression, and PTSD. Victims often experience a sense of violation, which can be more intrusive than other forms of harassment because of the pervasive nature of technology. The harassment can occur any time, at any place, through digital devices.

- Power Imbalance: The perpetrator can take advantage of the victim's vulnerabilities, using the internet and digital tools to invade privacy and control aspects of their life, especially if they have access to personal or financial information.

Relevance of DSM-5 (Diagnostic and Statistical Manual of Mental Disorders, Fifth Edition)

The DSM-5 is the primary classification and diagnostic tool for mental health disorders. It categorizes and defines the behaviors and thought patterns that might contribute to pathological behaviors like stalking and harassment. Let's break down how cyberstalking and harassment connect to DSM-5 personality disorders, especially Narcissistic Personality Disorder (NPD) and Antisocial Personality Disorder (ASPD).

Narcissistic Personality Disorder (NPD):

- DSM-5 Criteria for NPD:
    - Grandiosity: A sense of entitlement, self-importance, and a need for admiration.
    - Lack of Empathy: Inability to recognize or identify with the feelings of others.
    - Exploitation: Taking advantage of others to achieve personal goals.
    - Entitlement: Belief that they deserve special treatment or attention.
    - Envy and Arrogance: A tendency to be envious of others or believe that others are envious of them.
- Cyberstalking/Harassment Connection:
  Narcissistic individuals may engage in cyberstalking as a method of punishing someone who has defied their sense of superiority. When a victim (like an ex-partner) breaks away or challenges the narcissist's control, the narcissist will often stalk, harass, or intimidate them to regain control or feel validated. This behavior fits the sense of entitlement and lack of empathy inherent in narcissism.

- Example of NPD in Cyberstalking:

  o A narcissistic ex-husband may hack into the victim's accounts, monitor their digital communications, or spread damaging rumors online to punish the victim for "leaving them" or "rejecting them." This is rooted in grandiosity, as they feel justified in doing so because they see themselves as above reproach.

Antisocial Personality Disorder (ASPD):

- DSM-5 Criteria for ASPD:

  o Disregard for the rights of others: Repeated violation of others' rights through deceit, manipulation, or coercion.

  o Lack of remorse: No guilt or regret for harmful actions.

  o Impulsivity and Irresponsibility: Tendency to act without regard for consequences.

  o Aggression: Physical fights or assaults as part of a pattern of violent behavior.

- Cyberstalking/HarassmentConnection:
  Individuals with ASPD engage in criminal and manipulative behaviors, including cyberstalking. These individuals feel entitled to manipulate or harm others to achieve their goals. They lack empathy and feel no remorse for the pain they cause. Therefore, they often engage in

illegal activities such as hacking, identity theft, or spreading malicious information online, all part of a larger pattern of exploitation and disregard for others' privacy and safety.

- Example of ASPD in Cyberstalking:

    o An individual with ASPD may hack into their ex-partner's social media accounts, using the information to threaten or blackmail them. This person may have no remorse for the damage caused to the victim's mental and emotional well-being.

Borderline Personality Disorder (BPD):

- DSM-5 Criteria for BPD:

    o Instability in relationships, self-image, and emotions: A pattern of unstable relationships, emotional outbursts, and fear of abandonment.

    o Impulsive behaviors: Engaging in behaviors like substance abuse, unsafe sexual practices, or self-harm.

    o Paranoia and dissociation: A tendency to feel paranoid or disconnected from reality, especially when faced with emotional distress.

- Cyberstalking/Harassment Connection:
    People with BPD may use cyberstalking as a means of

seeking validation or attention when they feel abandoned or emotionally rejected. If the individual feels betrayed by someone (often a romantic partner), they may engage in stalking behaviors to reassert control over the relationship or create a sense of emotional drama. These behaviors may be driven by an overwhelming fear of abandonment.

- Example of BPD in Cyberstalking:
    - An individual with BPD may repeatedly message, call, or stalk their ex-partner online, driven by fear of being abandoned. They might send threatening messages to get the victim's attention or to "test" if the victim still cares.

Psychological and Legal Implications of Cyberstalking and Harassment

1. Psychological Damage to Victims:

    a. Increased Vulnerability: Victims of cyberstalking may feel vulnerable and unsafe in their own homes, knowing that the stalker can reach them anywhere.

    b. PTSD and Complex Trauma: Prolonged exposure to this kind of harassment can lead to Post-Traumatic Stress Disorder (PTSD) or Complex PTSD, especially if the harassment occurs over a prolonged period and the victim has already suffered emotional or physical abuse.

    c. Anxiety and Depression: Cyberstalking can lead to constant anxiety, hypervigilance, and depression as the victim becomes trapped in a cycle of fear and distress.

2. Legal Challenges:

    a. Cyberstalking as a Crime: Many states have specific laws against cyberstalking, but these laws may not always take into account the full psychological profile of the stalker.

    b. Proving Malice or Intent: Proving that the perpetrator's intent was malicious or that their actions caused harm is often difficult, especially if the stalker uses tactics that are difficult to trace (e.g., anonymous online accounts, encrypted messages, or IP masking).

    c. Difficulty in Enforcement: The digital nature of cyberstalking makes it harder for authorities to pursue criminal charges, especially when the stalker uses technology to hide their identity or location.

1. The Emotional & Psychological Toll of Cyberstalking and Harassment

The most insidious aspect of cyberstalking isn't just the invasion of privacy—it's the deep, lingering psychological toll it takes on you, a toll that extends far beyond the initial violation. Unlike physical abuse, digital harassment has the unique ability to keep you on high alert 24/7. When your emotions are manipulated through invisible channels, when your personal boundaries are violated in spaces you once considered private, you feel like you're always being watched—even when you're alone. This was the reality I lived with every day after my ex's shift from physical control to digital harassment.

When he first started using technology to track me—monitoring my emails, infiltrating my personal life through social media, and even tampering with my devices—I was shocked by how seamlessly he blended into the background of my life. It wasn't like the in-your-face manipulation I had endured during our time together. Now, it was quieter, more subtle. He wasn't physically there in the room with me, but his presence was felt at all times. My phone, my computer, and even the corners of my mind were constantly invaded. It was like living in a digital shadow, where every action felt like it was being watched, every conversation overheard.

This shift from physical to digital manipulation wasn't just a tactical change on his part; it was a psychological warfare that made me question every aspect of my reality. I found myself walking a tightrope of hypervigilance, constantly scanning for signs that I was being watched. The worst part wasn't knowing he had the power to monitor me—it was the uncertainty, the not knowing when or how he'd strike next. It was like being in an emotional fog, never truly able to relax or be myself. Every email I opened, every text message I sent, and every social media post felt like a potential landmine.

Dr. Rachel Levitch

One moment that stands out was when I tried to take back some autonomy in my life—by fixing my car, for example. After I was promised support, the follow-through turned into an endless cycle of delayed promises and half-hearted efforts. First, they said they'd get the car fixed, then just "diagnose" the issue. Days passed, and when I finally took it upon myself to get the car repaired, the reaction from the people who were supposed to help me wasn't a sign of relief or pride—it was a cold hostility. The moment I became self-sufficient, the same people who had claimed they were there to help me suddenly turned on me. It was as if my independence threatened their control, and that was a boundary they weren't willing to tolerate.

The more I tried to free myself from these manipulations, the more my boundaries were tested—whether it was with the staff or through my ex's digital harassment. His need to control became increasingly evident, and it wasn't just physical; it was emotional and digital. He didn't want me to thrive independently; he wanted to keep me tethered to him in some way, even if it was just through the digital spaces where he could continue to monitor and manipulate my thoughts, actions, and decisions.

This wasn't just about the emotional exhaustion of being controlled, either. It was the fear of losing privacy and autonomy—the basic elements of self-respect. When your life becomes a constant cycle of questioning whether your emails or phone calls are being intercepted, when your boundaries are consistently violated by someone who sees them as an obstacle to their power, it has a profound effect on your mental health.

I started feeling isolated, like I was fighting invisible battles on multiple fronts: the battle for privacy in the digital space and the emotional battle with a person who had no respect for my autonomy. It's easy for people to say "just cut him off" or "block him"—but when you've spent years with someone who knows your vulnerabilities and uses them against you, it's not that simple. It feels like you're fighting not just a person, but a force that has infiltrated every part of your existence.

The feeling of not knowing whether my life was being scrutinized, whether he was watching me through my own devices or sending anonymous messages from other accounts, kept me constantly on edge. This wasn't a fear I could explain to anyone—it was a shadow, a lingering sense of dread that he was always lurking somewhere just out of sight. The emotional and psychological toll of being stalked online goes beyond the fear of someone watching you; it taps into a deeper fear of losing your ability to feel safe in your own life, even in spaces that used to be private and sacred to you.

In the context of a workplace, these same feelings would be amplified. If I had tried to function as I once did, without this intrusion, I'd be constantly wondering whether my every action would be scrutinized or distorted. Would I be punished for my success? Would my every attempt to move forward be sabotaged by someone with the power to remain invisible but omnipresent?

These subtle, unseen tactics erode your confidence. You're left questioning your own judgment and your ability to trust anyone. This uncertainty doesn't just impact your ability to function in a work environment—it chips away at your sense of self-worth and personal safety. It makes you feel like you are always walking on eggshells, never sure if the next step will be one of freedom or one more form of control.

2. The Narcissistic Need for Control: Beyond the Emotional Manipulation

The narcissistic need for control is one of the most insidious aspects of my ex's behavior. It's not just about emotional satisfaction or maintaining a relationship—it's about dominating every aspect of my life. For someone like my ex, the goal isn't merely to stay in a relationship or to co-parent; it's about owning my decisions, actions, and even my thoughts. The dynamics he orchestrated were never about love—they were about power.

When we separated, his efforts to control me shifted from traditional methods of emotional manipulation—gaslighting, belittling, and guilt-tripping—to something even more invasive: digital stalking and cyber harassment. At first, it felt like a subtle shift. It was hard to even notice, because it wasn't the overt physical manipulation I was used to. There were no more explosive arguments or intense emotional outbursts. Instead, there were silent intrusions into my private life, subtle manipulations that preyed on my vulnerability without me even realizing what was happening at the time.

My ex had long used emotional manipulation to keep me tethered to him. The manipulation was so intricately woven into our daily interactions that I didn't even notice it at first. He would offer help, promise to do something, and then fail to follow through. But the failure wasn't an accident—it was a calculated maneuver. The message was clear: I was dependent on him, and my autonomy was something he needed to undermine.

Take, for example, the promises about fixing my car. It wasn't just about the car—it was about controlling how I dealt with my own independence. First, it was the offer to help, which made me feel grateful and reliant on him. But when the promises unraveled, when the support didn't materialize, it became clear: I wasn't allowed to take care of things myself without his involvement. As the situation escalated, I realized that the lack of follow-through wasn't an oversight—it was a method of control.

The more independent I became, the more aggressive his attempts to pull me back under his influence. When I finally fixed the car myself, the reaction was one of hostility and resentment. This was no longer about a car—it was about me asserting control over my life. I wasn't supposed to be able to make these decisions without his interference. My success, my ability to fix the problem on my own, threatened the power he had spent years building up. It wasn't just about keeping me dependent on him—it was about keeping me subservient to his will.

In a work environment, this dynamic would have been no different. Someone like my ex thrives in spaces where they can manipulate others, where they can create a sense of dependence. When you're in a position of weakness, they prey on that vulnerability, making sure that you feel like you can't succeed without their involvement. They'll offer you help, but it's always on their terms—never on yours. The moment you step out of line, assert your own autonomy, or make a decision without their input, the reaction is hostility or punishment. You're made to feel like your success is something they control, not something you earn through your own effort and merit.

This is the essence of narcissistic manipulation. It's not about love or partnership; it's about ownership. The narcissist doesn't want to share space with you—they want to own it. They want to dictate the terms of your existence, from the big decisions down to the smallest details. For my ex, this meant using technology as a way to keep me within his control, not just emotionally but also digitally. By invading my privacy, monitoring my actions, and tracking my every move, he reinforced the idea that I was his property—not just physically but in every sense of the word.

The need for control goes beyond manipulation. It's not just about making me feel guilty or responsible for everything that went wrong—it's about erasing my autonomy, my ability to think independently, and my capacity to take control of my own life. Narcissists like him thrive when their victims are isolated, dependent, and afraid to act on their own. And that's exactly what he did: he created a system where my only way to function was through him.

3. The Weaponization of Information: Identity Theft and Digital Manipulation

One of the most powerful tactics my ex used was the weaponization of information—and this took on a particularly dangerous form through cyberstalking and digital manipulation. When you're dealing with someone like him, information isn't just a tool; it's a weapon to be wielded. The more he knew about me, the more he could twist, distort, and control my reality. Every detail, every piece of personal information, became a point of leverage, a way for him to maintain power even from a distance.

In a healthy relationship, personal information is shared with trust—it's something that connects you, binds you together in an environment of respect and mutual support. But for someone like my ex, information was always a currency he could use to control the narrative. And he didn't need to be physically close to do this. With the rise of technology, digital platforms gave him an open doorway into my life, and he seized it without hesitation.

What began as small, seemingly inconsequential acts of information gathering escalated quickly. Once he realized how deeply he could invade my privacy, it wasn't enough to just know things about me. He wanted control over my identity, over how I was seen in the world. It wasn't just about knowing my passwords or having access to my bank account—it was about shaping who I was, altering my reality, and making me feel unseen or unknown to the very people who were supposed to support me.

Take, for example, the identity theft issues I've faced. After our separation, my ex began accessing my private information in a way that felt more like an act of sabotage than a simple breach of trust. At first, it was subtle—incorrect changes to personal details, my address being updated without my consent, and my financial accounts being tampered with. But as the situation escalated, I realized this wasn't a random act of mischief. This was deliberate. This was his way of reasserting control.

The impact on my personal life was profound. It wasn't just about losing access to my accounts—it was about losing control over who I was and how I was perceived by others. Suddenly, I was left questioning whether I could trust anything about my own life. My financial stability, my ability to move forward with my own plans, were all called into question because of his digital interference. What had once been a tool for growth and independence became a source of fear and vulnerability, all because he knew how to manipulate the system to his advantage.

This is where the psychological toll becomes real. It's not just the loss of control—it's the disorientation that follows. You no longer know where your life ends and his influence begins. You become hyper-aware of every interaction, every piece of personal information, every communication. And even worse, you start to doubt your own ability to trust yourself, let alone the systems in place that should be protecting you. You become vulnerable in a way that feels almost unnatural.

At one point, I had to change my identity—a process that should have been empowering, a reclaiming of my life, but it became another battleground for control. I was still fighting an unseen war, one where my ex's grasp on my personal details was still long-reaching. Even as I worked to rebuild and reestablish boundaries, I was still trapped in a system that kept me tied to his manipulation. Every step forward felt like it was matched by a step backward.

What was most infuriating was how seamless the transition from physical to digital control was. At one point, I was dealing with gaslighting and emotional manipulation in real-time, and now I had to face it through digital means—where he had unlimited access to my private world. It was as if the boundary between our two worlds had been completely erased. What I thought were safe spaces were no longer mine. Even in places where I thought I could rebuild and start fresh, I was still being monitored, still being manipulated, and still being tracked.

And in a "workplace" environment—whether that be friends or a place that promised support—the weaponization of personal information was no different. In that space, my very identity was threatened. When I was given false promises, or when my needs were disregarded, the dissonance between what I was told and what was actually happening was deliberate. This wasn't just neglect—it was a calculated attempt to make me feel disempowered, just like the digital harassment. Every action taken—or not taken—was another means to make me feel small and unimportant. Even something as basic as a car repair became a power play, another instance where I wasn't allowed to maintain control without it coming with strings attached.

4. The Legal Landscape: Using the Narcissist's Tactics Against Him

The intersection of cyberstalking and narcissistic behavior creates a particularly dangerous and complex situation, not only from an emotional standpoint but from a legal perspective as well. Understanding the narcissistic and antisocial traits at play is not just helpful—it is critical. These traits shape how the abuser operates, how they manipulate, and how they interact with the world around them. For someone like my ex, the ultimate goal is not just to hurt or control but to reassert dominion over every aspect of my life. The digital abuse that followed our separation was not an accident; it was a strategic, calculated move to maintain power and control over me—by making my life difficult, uncertain, and constantly under surveillance.

The problem, however, is that the legal system is often ill-equipped to deal with the subtleties of this type of abuse. When dealing with narcissistic behaviors in a cyberstalking case, many courts may view it as an isolated act of harassment. But for those of us who've lived through it, we know this is a long-term campaign—one that spans years, evolves with new tactics, and has profound, lasting effects. The behavior is not an isolated incident; it is part of a broader pattern of control and manipulation.

For example, I've had to navigate the legal intricacies of trying to block my ex's access to my personal information. When it became clear that he was using my identity to maintain control over me, I tried to address it with the Social Security Administration (SSA) and other institutions. But the process was far more complex than I anticipated. Instead of a clear-cut solution, I was met with bureaucratic red tape, delays, and an ongoing struggle to maintain my privacy.

Even after I took steps to change my name, I was still left vulnerable because the legal changes weren't reflected quickly enough in the system. Meanwhile, my ex continued to use my personal data, accessing everything from bank accounts to children's benefits—trying to manipulate the very systems that should be protecting me. The legal system, as much as it attempts to offer recourse, is often a few steps behind the tactics of someone with narcissistic and antisocial tendencies, and it was clear that without proper evidence and legal strategy, I was left exposed.

The issue goes deeper, though, than just this battle over personal information. The psychological toll this takes on a victim is immense. When you're dealing with an abuser who uses the law—and legal loopholes—to further his control, the line between victim and perpetrator becomes blurred. I have had to fight not only to protect my identity but to preserve my mental and emotional health in the process. The stakes are so high that every interaction with the legal system feels like another test, another battle where I must prove that my life is being manipulated and that the damage he's causing to me is real.

And here is where the narcissistic behavior proves itself to be so insidious: it makes everything harder. It doesn't just remain confined to one area of your life. It infects everything. My relationships, my ability to trust the systems that are meant to help me, even my own mental health—all of these things are undermined by his constant need to control, manipulate, and break down my sense of security. So much of what he does is not just an attack on me, but a long-term erosion of my trust in others.

In the legal landscape, this behavior needs to be recognized and addressed for what it is. Narcissists use the law, technology, and even our public systems to inflict harm—whether it's through manipulation, cyber harassment, or identity theft. As a society, we must begin to understand the strategic use of these tools by narcissistic individuals in order to better protect victims.

For me, this is not just about achieving justice in the traditional sense. It's about fighting for a future where I can exist outside of his influence, where I am no longer bound to his tactics, and where my life is truly my own again. Every step forward in this legal battle feels like another reclaiming of my autonomy.

And while the battle with the legal system has been difficult, it is part of a larger fight for self-empowerment. The more I learn to navigate the intricacies of digital harassment and the more I become educated on the law, the more I am able to turn these narcissistic tactics of control against him. This isn't just about winning in court—it's about winning my life back, reclaiming my peace of mind, and removing him from the equation once and for all. It's the ultimate victory.

Cyberstalking and Narcissistic Manipulation: A Complex Web of Abuse

Dr. Rachel Levitch

The damage caused by cyberstalking and harassment can't be understated, especially when these behaviors are driven by narcissistic, antisocial, or borderline tendencies. For someone like my ex, the persistent, manipulative tactics aren't just about power—they're about control. Narcissistic individuals like him often need to ensure that the victim feels powerless, insecure, and dependent. Their behaviors are far from isolated incidents; they are part of a calculated effort to break down the victim's sense of autonomy and self-worth.

Understanding the psychological makeup of these individuals is essential. The DSM-5 offers critical insight into the mental health disorders that often drive these behaviors, including narcissistic personality disorder and antisocial personality disorder. These disorders are typically marked by manipulation, deceit, lack of empathy, and a tendency to engage in destructive, compulsive actions—like cyberstalking—that allow the abuser to maintain control over the victim.

For victims, this abuse is often compounded by severe emotional trauma. The constant surveillance, harassment, and digital manipulation create an environment of hypervigilance and fear. It's not just a violation of privacy—it's an attack on one's sense of self, one's mental stability, and ultimately, one's ability to heal. These behaviors are damaging, but they also need to be addressed both legally and psychologically to protect the victim and hold the perpetrator accountable.

Strategic Understanding of How I Protect Myself

One of the most important lessons I've learned throughout this journey is that protection—whether emotional, physical, or legal—

is a strategic endeavor. It's not just about taking action in the moment; it's about creating a framework that allows me to regain control over my life and protect myself from ongoing abuse.

The first and most crucial step in this process was to recognize that this wasn't just a phase or a series of misunderstandings—it was an ongoing campaign of control. This realization led me to begin documenting every incident, every breach of privacy, and every subtle act of manipulation. Whether it was through email, social media, or physical presence, I made sure I was able to track and record these behaviors. Documenting was my first line of defense—not just to hold him accountable, but to ensure that I wasn't silenced or gaslit into questioning my reality.

Technologically, my approach has been multifaceted. I took steps to secure my digital life by implementing stronger cybersecurity protocols, such as encrypted communications, more robust passwords, and multi-factor authentication across all accounts. But it wasn't just about changing passwords or using firewalls. It was about isolating my digital footprint to limit the data he could access. That meant closing down certain accounts, moving sensitive information to secure cloud storage, and ensuring that my personal details were only shared with trusted parties.

Additionally, I knew that boundaries needed to be set—not just emotionally but physically as well. In my case, I had to make sure that my physical space was equally protected. This involved installing security cameras, increasing lighting around my home, and taking measures to limit access to my living space. I also knew I had to take the extra step to report the ongoing cyberstalking to the authorities—working with legal professionals to understand how best to navigate these complex issues. Reporting wasn't just about drawing attention to the abuse; it was about making sure that there was a record—a trail of evidence that could later support my case if it came to legal action.

Psychologically, protecting myself meant maintaining self-awareness and learning how to respond to triggers in a way that kept me grounded. This was a long process, one that took a lot of emotional work and self-compassion. I had to break the cycle of fear and begin rewriting the script that my abuser had tried to impose on me. Instead of reacting out of instinct or panic, I began to consciously choose how to respond—sometimes with calm, other times with firm boundaries, and always with the understanding that I was not alone in this journey.

The key to this protection wasn't just in the immediate actions I took; it was about building a long-term strategy to ensure that this cycle of abuse would never happen again. It meant being vigilant without being overwhelmed by fear, creating a strong foundation of emotional resilience, and ensuring that I was never again dependent on someone who would use my vulnerabilities against me. This is how I've been able to shift from simply surviving to truly living again, and it's a journey I continue to take, one step at a time.

Chapter 3

Psychological Profile: The Illness Behind the Obsession

What drives an individual to engage in such persistent and damaging behavior, stalking their victims both in person and online? Understanding this starts with delving into the personality disorders that fuel these actions. At the core of such behaviors lies Narcissistic Personality Disorder (NPD) and Borderline Personality Disorder (BPD)—two conditions that can shape an individual's actions and interactions with others in deeply harmful ways. While these disorders are distinct, they often share traits that can lead to obsessive, controlling behaviors, especially when the person feels threatened, rejected, or loses control.

A narcissist is someone who sees themselves as superior to everyone else, deserving of special treatment and admiration. People become mere tools in their eyes, objects to be manipulated and controlled for the narcissist's own gain. Their sense of entitlement runs deep: they believe that they are above reproach and that their desires and needs should always take precedence. When this perceived superiority is challenged or when the narcissist is faced with rejection—whether from a relationship or a situation—they react with anger, rage, and desperation. Their emotional reactions, often disproportionate to the situation, trigger a dangerous need to reassert their control, even if it means resorting to harassment, digital surveillance, and emotional manipulation. These tactics are not only destructive but deeply ingrained in their personality, forming a cycle of abuse and control.

A person with NPD lacks the fundamental ability to feel empathy for others. This is perhaps the most dangerous element of narcissism—because without empathy, the narcissist views others solely as extensions of themselves, as tools to fulfill their needs. The narcissist's primary focus is on maintaining their inflated sense of importance, which often means monitoring and manipulating those around them to ensure they remain in control. This includes using digital tools—like social media, spyware, and fake identities—to constantly watch, track, and intrude on the victim's life. The invasion of privacy becomes a way for the narcissist to maintain power, to keep their victim within reach and under their thumb, even after the relationship has ended.

However, narcissism isn't a monolithic condition. Not all narcissists behave the same way or follow the same pattern. According to Dr. Ramani Durvasula's work on narcissism, there are eight distinct types of narcissists, each exhibiting different manifestations of narcissistic traits. While these types differ in their specific behaviors, their underlying need for control and emotional exploitation remains the same. By understanding these types, we can better identify how narcissism manifests in the world of cyberstalking, harassment, and online manipulation.

Below are the eight types of narcissists that Dr. Durvasula identifies, each with their own unique traits and behaviors. It's important to note that while some narcissists may exhibit multiple traits across these categories, each type interacts with technology in a distinct way—manipulating, controlling, and stalking their victims in ways that reflect their personality disorder.

1. Grandiose Narcissist

The classic narcissist who is confident, arrogant, and often visibly successful, the grandiose narcissist thrives on admiration and attention. They want to be seen as the best in everything they do. In the realm of cyberstalking, the grandiose narcissist may use their online presence to broadcast their achievements, drawing attention to themselves and seeking validation. They may spy on their victim's social media and feel entitled to interfere in their lives, ensuring that their victim is aware of their superior status. The grandiose narcissist can also be vindictive, posting negative information about their victim online to damage their reputation.

2. Covert/Vulnerable Narcissist

While the covert narcissist might seem humble or even self-deprecating, underneath they harbor a deep sense of entitlement and victimization. They are hypersensitive to perceived slights and often display passive-aggressive behaviors. This type of narcissist may engage in cyberstalking through subtle manipulation, such as sending cryptic messages, pretending to be concerned, or anonymously posting to social media to make their victim feel guilty. They often portray themselves as misunderstood, using their online presence to create a narrative where they are the victim and the target of unjust attacks. Gaslighting and projection are common tactics they use to confuse and control their victims.

3. Malignant Narcissist

This type is the most dangerous and toxic. Malignant narcissists blend narcissism with elements of psychopathy, sadism, and Machiavellianism—traits that make them manipulative, exploitative, and willing to go to great lengths to destroy their victims. They often use technology in harmful ways, such as hacking into emails, creating fake profiles to deceive the victim, and spreading lies to harm their reputation. The goal is not just control, but domination. Malignant narcissists will stop at nothing to cause fear, anxiety, and emotional devastation in their targets, using cyber tools to fuel the damage.

4. Communal Narcissist

A communal narcissist prides themselves on being the "giver", the one who is always helping others, but their actions are motivated by a need for praise and recognition. They often have an online presence that highlights their altruism—posting photos of charity work, community service, or social justice causes. However, when someone challenges their public image or calls them out, they can become retaliatory and use their online influence to manipulate the perception of others. They may use social media platforms to subtly undermine their victim's image, casting themselves as the saint and their victim as the villain.

5. Neglectful Narcissist

Neglectful narcissists are the least interested in direct control over others but still have an inflated sense of self-importance. They may neglect relationships until they need something from others. In the realm of online stalking, they might not engage as actively in monitoring their victim's every move, but when they do reach out or contact their victim, it's typically because they want something—validation, attention, or access to personal information. Flaky and self-absorbed, they may briefly pop into their victim's online world, only to disappear when they no longer need attention.

6. Benign Narcissist

The benign narcissist is typically viewed as someone who is simply immature or self-absorbed, but their lack of empathy and excessive self-centeredness still leads to emotional harm. They may post about their life on social media to gain likes and followers but may not engage in overt manipulation like other types. However, when they feel ignored or rejected, they may subtly escalate to passive-aggressive behavior, such as leaving cryptic messages or subtweeting their victim in an attempt to draw them back in. They often fail to understand the impact of their actions, dismissing any criticism as "overreacting" or "taking things too personally".

7. Entitled Narcissist

Entitled narcissists believe the world owes them everything. They demand special treatment and cannot fathom being told "no." They often use technology to demand attention or public recognition—posting rants online, sending multiple unnecessary messages to others, or attempting to control situations through social media. The entitlement becomes a tool for manipulation, using bullying tactics or intimidation to ensure that they are always in the spotlight, even if it means invading others' private lives to do so.

8. Generational/Cultural Narcissist

This type of narcissist operates within a specific cultural or familial context, where narcissistic traits are ingrained through generations. Their behavior might be less visible to outsiders because it's normalized within the family or community. They might use social media to boast about family traditions, ancestral pride, or social status, positioning themselves as the "heir" or "leader" in the family's legacy. The narcissistic control may manifest in exerting influence over the victim's decisions, isolating them from their broader social network, and subtly guilt-tripping them into compliance.

Psychological Profiles: Understanding the Narcissistic Stalking Behavior

Even though I have been away from my ex-husband for ten years, his obsessive behavior persists, especially whenever I attempt to reconnect with my children or use social media to reach out. This pattern of harassment is not coincidental—it is deeply rooted in narcissistic traits that drive him to stalk, control, and manipulate. The behaviors I endure are not just isolated incidents; they are a predictable cycle of abuse powered by his narcissistic personality disorder. Understanding the psychological profiles behind this behavior helps to contextualize the manipulation, control, and emotional harm I face.

In this section, we will delve into two of the most prominent narcissistic traits that fuel his ongoing harassment—the Grandiose Narcissist and the Entitled Narcissist—and connect them with the real-life experiences I've had over the years. These profiles will shed light on why, even after so much time apart, he still exerts such control over my life, using cyberstalking and digital manipulation to invade my privacy and punish me for daring to exist outside of his control.

The Grandiose Narcissist: The Need for Control and Admiration

A grandiose narcissist is the classic, overt form of narcissism—the one that demands attention, admiration, and control over others. My ex-husband embodies many traits of this narcissistic type, especially in the way he seeks to maintain an inflated sense of importance and superiority over me. It's not enough for him that we've been divorced for a decade. He cannot tolerate the idea of me moving on or having independent relationships outside of his influence.

Whenever I make the smallest gesture of reconnecting with my children—whether it's wishing them a happy birthday or posting a picture on social media to celebrate their lives—he becomes enraged. His narcissistic rage flares up, not because he loves them more than I do, but because I dared to interfere with his narrative of control. To a grandiose narcissist, control is paramount. The mere act of reaching out to my children is an affront to his perceived superiority, and he responds by escalating his harassment.

This is where his grandiosity plays out. He feels entitled to my attention, and even more so, he believes he has the right to monitor my every move. He will do anything to disrupt my life, from attempting to hack my social media accounts, to using various cyber tools to track my location and erase my digital presence. His belief in his superiority is reinforced by his actions—each violation of my privacy serves as a reaffirmation of his control over me. For him, my personal boundaries mean nothing; the act of wishing my children well is simply a reminder that I am still in his orbit, whether I like it or not.

The Entitled Narcissist: The Need for Special Treatment

Another striking characteristic of my ex-husband's behavior stems from his entitlement—the belief that he is owed special treatment, regardless of the situation. This is typical of an entitled narcissist, who expects the world to revolve around them, and anyone who challenges that perception must be punished. I've seen this time and time again. Every attempt I make to maintain some semblance of independence—whether it's interacting with my children, posting on social media, or simply enjoying life without his interference—triggers a sense of entitlement in him that knows no boundaries.

The entitled narcissist feels that their desires take priority over anyone else's, and in this case, my ex-husband believes that he deserves to know where I am, what I'm doing, and who I am interacting with—especially when I interact with my children. When he cannot physically control me, he uses technology to close that gap. His sense of entitlement leads him to monitor my every move, constantly searching for ways to infiltrate my digital life. If I post on social media, he immediately scrutinizes my every word, tracks my location, and removes any connection I've made with my children that doesn't meet his approval.

This relentless surveillance isn't just about curiosity or concern; it is about control. He believes that because I once belonged to him, he has an inherent right to dictate the terms of my existence. His harassment is not just an expression of possessiveness; it is rooted in a deep sense of entitlement. He sees any attempt I make to reclaim autonomy as a threat that must be eliminated. His behavior, while disguised as concern, is nothing more than a mechanism to reassert dominance in a relationship that has long been over.

The Devastation of Narcissistic Cyberstalking: A Continuous Battle

When narcissism intersects with cyberstalking, the result is a devastating, unseen assault on the victim's emotional and psychological well-being. For someone like me, who has been trying to rebuild her life and move forward, the constant intrusion into my private world is exhausting. Even after ten years, his obsession with controlling me has not waned; it has simply found new avenues to express itself.

These two types—Grandiose Narcissist and Entitled Narcissist—showcase how the narcissistic desire for control, admiration, and special treatment manifests in cyberstalking. The anonymity of the digital world allows narcissists like my ex-husband to hide behind screens, utilizing technology to manipulate, harass, and monitor their victims. Social media becomes a weapon for the grandiose narcissist, while the entitled narcissist uses technology to maintain their hold over others, feeling justified in violating personal boundaries at every turn.

For victims like me, this silent assault can feel endless. The psychological toll of being watched, monitored, and stalked digitally is immense. It is not just the invasions of privacy that hurt, but the constant reminder that someone is out there, seeking to control your every move. The experience leaves you feeling isolated, powerless, and shattered.

Conclusion: The Ongoing Battle for Control

The journey of living with narcissistic abuse—whether in person or online—is never easy. In my case, even a decade after separating from my ex-husband, his cyberstalking and obsessive behavior continue to shape my life. Narcissists like him will stop at nothing to maintain their sense of control, using every tool at their disposal to invade their victim's life, even when physical proximity is no longer possible.

Understanding the psychological profiles behind such behavior—particularly the Grandiose Narcissist and the Entitled Narcissist—provides much-needed clarity into why this cycle of abuse persists. Cyberstalking is a powerful tool in their arsenal, one that allows them to continue their harassment with little risk of repercussion, leaving their victims vulnerable and emotionally scarred.

It is crucial for those who face these challenges to understand the tactics narcissists use, document every instance of harassment, and seek legal and psychological support. Narcissistic abuse, whether offline or online, can be a lifelong battle, but with the right support, victims can regain control over their own lives and break free from the cycle of manipulation and abuse.

Children as Weapons: The Emotional and Psychological Toll

When my ex-husband engages in his endless cycle of cyberstalking, it is not just an attack on me—it is a calculated attempt to destroy the relationship I have with my children. Over the years, he has weaponized our children to maintain control over me, using them as pawns in a larger game of manipulation. The emotional toll this takes on me is almost indescribable, but the damage it does to my children—who are innocent in all of this—is a heavy, heartbreaking reality.

It's one thing to face the constant harassment and invasion of privacy as a victim of narcissistic abuse, but when children are involved, the stakes become even higher. Narcissistic individuals, especially those with traits of Narcissistic Personality Disorder (NPD) or Borderline Personality Disorder (BPD), often exploit family dynamics as a means of control, using their children as weapons to further emotionally and psychologically destabilize their ex-partners.

In my case, every time I try to reach out to my children—whether it's through a simple text message, a birthday wish, or a post on social media—he retaliates by using them to create discord and confusion. The children are often caught in the middle, torn between their own sense of loyalty and the pressure he places on them to stay aligned with his narrative. The emotional manipulation he uses, often masked as "concern" for their well-being, is just another tool to ensure I remain under his thumb, even when he is physically absent from our lives.

The Psychological Toll on Victims: Emotional and Physical Devastation

The impact of using children as weapons is deeply personal, and its consequences reverberate across all aspects of my life. When narcissists like my ex-husband attack their victim's relationships with their children, it creates a profound sense of helplessness. Studies consistently show that victims of domestic violence and cyberstalking often experience Post-Traumatic Stress Disorder (PTSD), Complex PTSD, and severe anxiety. These psychological conditions are triggered by prolonged exposure to trauma—especially in situations where the victim feels trapped and unable to escape the cycle of abuse.

Complex PTSD develops when trauma occurs repeatedly and over a long period. In my case, the emotional and psychological abuse that has extended over years—including the use of children as pawns—has created a pervasive sense of fear and helplessness. Symptoms of Complex PTSD include intrusive memories, hypervigilance, and dissociation, which often manifest in a constant state of emotional numbness or an overwhelming feeling of being powerless. For me, every day feels like walking through a fog of fear and anxiety, not knowing when the next emotional ambush will occur. My sense of safety has been eroded by the ongoing abuse, and even the most secure environments—like being with close friends or family—can feel unsafe.

The Narcissistic Use of Children: A Devastating Cycle

What makes the use of children in narcissistic abuse so insidious is the way it distorts reality. Narcissists like my ex-husband can manipulate children into believing that their well-being is at stake if they show affection or support toward their mother. The narcissist twists the narrative to present themselves as the victim, and in turn, the children are forced into a false allegiance, driven by guilt and confusion.

This constant manipulation affects their emotional development and creates confusion in their relationships with me. My children are not just dealing with the loss of a healthy relationship with their mother—they are dealing with the emotional burden of being pulled in two directions: one towards their father, who manipulates them to serve his narcissistic needs, and the other towards their mother, who is fighting to maintain a loving relationship despite the obstacles.

It is not just an emotional battle—it is a psychological war. The impact of these battles on children is often long-lasting. Children caught in the crossfire of narcissistic parental control suffer from an array of emotional issues, including low self-esteem, anxiety, and difficulty trusting others. The seeds of fear, confusion, and self-doubt are planted early, which can carry into their adulthood, affecting their ability to form healthy relationships.

An Ongoing Cycle of Abuse

The most difficult part of this entire ordeal is that my ex-husband will never stop. Narcissists do not respect boundaries, especially when it comes to their children. If anything, they use the children to maintain a psychological foothold, as if to say, "As long as I have the children, I have you." It is a power play disguised as concern for their well-being, but in reality, it is about control, not care.

For me, this battle feels endless. It is a constant cycle of emotional distress—trying to maintain a healthy relationship with my children while my ex-husband continues to manipulate, threaten, and disrupt that connection at every turn. The psychological toll is immense. I live in a constant state of hypervigilance, fearful that any communication with my children will trigger another wave of retaliation or harassment from him.

This is the reality for many victims of narcissistic abuse, and it is heartbreaking to witness the damage it does, not only to the victim but to the children who are used as tools for emotional manipulation. As they grow older, the scars left by these experiences can deeply impact their ability to trust, to form meaningful relationships, and to feel safe in their own skin.

The Ongoing Struggle: Living with Narcissistic Abuse and Cyberstalking

The emotional and psychological toll of narcissistic abuse is not a chapter that simply ends—it is a constant cycle of trauma that stretches on, even years after separation. For me, the ongoing battle against cyberstalking and emotional manipulation continues to this day. The tactics my ex-husband uses are as insidious as they are persistent, and the impact on my life is immeasurable. He doesn't stop. The use of children as weapons—manipulating them to isolate, control, and torment me—is one of the most devastating parts of this unrelenting abuse. It's a battle I fight daily, with no end in sight.

This is the reality for so many victims who are caught in the grip of narcissistic abuse: the fight for freedom, emotional well-being, and safety is not something that can be neatly packaged in a neat chapter with a definitive conclusion. In many ways, it's a journey without a clear endpoint, but rather a road of ongoing healing and survival.

Even though I have been away from my ex-husband for ten years, each time I reach out to my children or use my social media to celebrate a special moment, he retaliates—hacking my accounts, stalking my movements, and finding ways to disrupt my life. His narcissistic traits, his need for control, his rage at being "abandoned," all continue to fuel his actions. The cyberstalking tactics have only grown more advanced with time, and the emotional toll has become even more pronounced as he uses technology to monitor, invade, and violate my privacy. The anonymity of the internet, coupled with his deep-rooted narcissism, allows him to hide behind a screen, free from any immediate consequences.

But this isn't just about the digital world—it's about how these behaviors manifest in real life, creating an ongoing cycle of psychological warfare that takes a toll not only on the victim but on the children who are used as pawns. This use of children to manipulate and control the situation has caused irreparable damage to my relationships with them. The emotional manipulation he exerts leaves them torn, uncertain, and unsure of their own emotions.

Even as I work to maintain a healthy relationship with my children, he pulls them back into his narrative, sowing confusion and guilt, and pushing them further from me. This is what narcissistic abuse does—it wears you down over time. The psychological warfare it creates can shatter your sense of self, isolate you from those who would otherwise offer support, and continue to diminish your ability to heal. In many ways, these tactics are designed not just to control—but to destroy.

Psychological Warfare: The Subtle and Destructive Tactics of Narcissistic Control

Psychological warfare, in the context of narcissistic abuse, is not always loud or overt. It's insidious, deeply rooted in manipulation, and often invisible to those who aren't the target. At its core, it's about wearing down the victim's sense of self-worth, making them feel uncertain, powerless, and, ultimately, isolated. The narcissist's goal isn't just to control—it's to destabilize the victim's very understanding of reality.

For me, the effects of this psychological warfare became evident when I would try to re-establish a connection with my children or assert my own identity, only to have my efforts twisted and undermined. The narcissist doesn't just act to control—they act to discredit, making the victim question their own worth, their relationships, and even their sanity. This gaslighting—the manipulation of reality to make someone doubt their perception—often manifests as small, almost imperceptible actions. For instance, a comment in passing, a social media post manipulated, or a phone call where the narcissist pretends to be "concerned," but underneath it all is the underlying message: *You don't belong. You don't deserve peace. You're under my control.*

When the narcissist uses their children as weapons, it becomes a psychological battlefield that no one is prepared for. It's a constant tug-of-war, where the victim is left trying to protect their emotional well-being while keeping the children from being pulled into a narrative of guilt, confusion, and division. This creates a toxic environment where the children are made to feel like they are caught in the middle, unable to trust their own emotions or even their own thoughts about their mother. Narcissists are masters at division, creating wedges between the victim and their support system by spreading lies, manipulating conversations, and fostering suspicion.

This emotional manipulation isn't always seen as abuse by the outside world. To others, it might look like a benign interaction—an email, a text message, or even an "innocent" social media post—but to the victim, it's another layer of control. The narcissist doesn't have to physically be there to dominate your life. They do it through mind games, through the twisting of facts and actions, through the ongoing psychological pressure that wears you down over time. The consistent uncertainty, the never-ending doubt about your safety, your identity, and your place in the world, can take a toll that is as damaging as any physical assault.

Cyberstalking is just another extension of this abuse—a weaponized version of psychological warfare. The anonymity of the internet gives the narcissist an ability to harass, monitor, and control without the immediate consequences that would come from real-world stalking. As I have experienced firsthand, the fear of being found, watched, or even hacked is constant. He doesn't need to physically show up in my life anymore—he can simply tap into my devices, invade my privacy, and wreak havoc without me ever seeing his face. This digital dominance becomes a silent, constant presence that controls my movements and my state of mind.

For a victim, the constant surveillance—whether through social media or hacked accounts—becomes overwhelming. Every post, every comment, every like becomes a potential entry point for the narcissist to strike. And the toll it takes is cumulative. It's a psychic exhaustion that creeps in slowly, wearing down your mental health and creating a state of hypervigilance. You're always looking over your shoulder, wondering if the narcissist is lurking in the shadows of your life, waiting for the opportunity to strike again. The boundaries between personal space and online presence become blurred, leaving the victim feeling trapped, both physically and emotionally.

This tactic of constant monitoring also destroys the victim's sense of agency. When the narcissist is always watching, always commenting, always interfering, it becomes harder to live a life free from their influence. The simple joy of sending a message, posting a photo, or even engaging with others online feels like it could provoke a wave of retaliation. And that retaliation isn't always immediate—it's calculated. It's a slow burn, one that festers in the background until the victim begins to feel that they are constantly at the mercy of the narcissist's whims.

The use of technology in this way—by a narcissist—creates a prison of anxiety and fear. It's not just about keeping tabs on you; it's about maintaining control, making you feel as if there is no escape, no place that is truly safe. It's an unseen assault, but one that is no less damaging than physical violence. The worst part? You never know when it will stop. You never know when they will disengage or stop trying to invade your life. The narcissist is never done until they've worn down the victim completely, forcing them to live in a constant state of distress and survival.

The ripple effects of this type of warfare can be felt in every aspect of the victim's life. Mental health deteriorates, relationships become strained, and trust in oneself erodes. The victim is left wondering if they can ever truly reclaim their peace. Can they ever find their way back to a life where they feel secure—without the looming presence of the narcissist? These questions are often left unanswered as the cycle of abuse continues, with no clear end in sight.

The Hidden Battle: Narcissistic Abuse in the Digital Age

In the chapters to come, we will delve into the far-reaching implications of narcissistic abuse, especially in the context of the online world. We will look closely at how cyber tools—social media platforms, digital surveillance, and other online means—have not only amplified the narcissist's capacity to control but also created a new playing field for their manipulation and harassment. The internet has become an extension of the narcissist's reach, allowing them to stalk, monitor, and break their victim's sense of self without ever needing to leave the safety of their own screen.

The emotional devastation caused by these online tactics is more than just a ripple effect. It can feel like an overwhelming wave, eroding the victim's sense of safety, autonomy, and well-being. When the narcissist gains access to our personal lives through digital means, it doesn't just feel like a breach of privacy—it feels like an ongoing assault. This is no longer about physical boundaries; it's about psychological warfare, where the lines between the real and the virtual blur, and the victim feels trapped in an endless cycle of emotional turmoil.

At the core of this abuse is the narcissist's need for control. Whether online or offline, the narcissist's primary objective is to maintain dominance and power over their victim. This control is achieved through manipulation, surveillance, and the deliberate undermining of the victim's emotional stability. Narcissistic behaviors are insidious; they often manifest as seemingly benign interactions, like an email, a social media comment, or a phone call, but underneath, these actions are carefully crafted to induce fear, doubt, and a sense of vulnerability.

Victims of narcissistic abuse, like myself, often suffer from long-term emotional scars—scars that are not immediately visible but remain deeply embedded in the psyche. These scars are a result of prolonged manipulation, gaslighting, and emotional torment, and they cannot be healed through time alone. Healing from narcissistic abuse is not a passive experience; it requires action—and in many cases, this means turning to legal resources, seeking psychological support, and engaging in self-care strategies that help rebuild one's sense of identity and autonomy. The healing process can feel like a second battle—one that demands constant vigilance and the courage to take back what was stolen by the narcissist.

I've learned that no one should fight this battle alone. Support systems—whether through legal channels, therapy, or trusted loved ones—are essential in navigating the aftermath of narcissistic abuse. It is vital to recognize the narcissist's tactics for what they are: tools of control, designed to destabilize, disempower, and break the victim's spirit. Self-awareness is the first step toward regaining control, but action is equally necessary.

This chapter is only the beginning of my journey—and it could very well be the first step in yours as well. Understanding the narcissist's playbook, particularly when it comes to their use of cyberstalking, is not just about recognizing the abuse; it's about empowering yourself to fight back. The more we learn to identify these tactics, the more we can protect ourselves from them, and the stronger our sense of self and security can become. This is crucial, because reclaiming your life—whether it's your personal space, relationships, or emotional well-being—requires that we no longer allow narcissistic behavior to define us.

In the upcoming chapters, we will break down the psychological profile behind the narcissist's obsession with control and explore how these behaviors evolve online. We'll also examine the legal tools and psychological strategies that can help victims regain their lives, their confidence, and their emotional health. Narcissistic abuse is devastating, but with awareness, support, and action, it is possible to break free and reclaim the life that was almost taken.

Chapter 4

The Link Between Narcissistic Injury and Cyberstalking

When someone like my ex feels threatened—whether it's by a loss of control, a rejection of their power, or a shift in the dynamic of the relationship—they experience what is often referred to as narcissistic injury. I never understood this fully until I started reflecting on his behavior after I left. In my case, his reaction wasn't just emotional manipulation—it was a relentless, calculated effort to regain control. His injury didn't just hurt his ego temporarily; it drove him to dig deeper, and it became clear that everything I did was going to be scrutinized, twisted, and used against me.

At first, I couldn't make sense of it. Why was he so obsessed with my every move after we separated? Why was he constantly trying to force his way into my life, even after I had made it clear I wanted space? The more I looked at his actions—both online and offline—the clearer it became: narcissistic injury was at the root of it all. For him, it wasn't just about me leaving; it was about him losing the source of validation he had relied on to feel important. My departure was a blow to his sense of self. His reaction to that wasn't to process it as a loss, but to seek revenge—an effort to punish me for making him feel powerless, insignificant, and unimportant.

This is when the cyberstalking began. It wasn't just about him wanting to stay in touch or check on me—it was a tool he used to regain control, to punish me for rejecting him. I hadn't known it at the time, but my rejection was an existential threat to his fragile ego. And that kind of threat demands retribution. The more I stood firm, the more intense his harassment became. His attempts to reassert control escalated online, through constant surveillance, messages, and manipulations—each step a desperate attempt to maintain dominance.

What's even more devastating—and something I didn't fully understand at the time—is how this narcissistic injury continued to shape his behavior even in places where I wasn't physically present. At the "workplace" (a place that was supposed to offer me support), his emotional manipulation and control played out in ways I hadn't expected. I had started to take steps toward independence. Simple things, like handling my own home repairs. It wasn't just that I'd done it alone—it was that I didn't need him or anyone else to do it for me. And that seemed to threaten the very foundation of his existence.

He had promised to help me with these repairs, but after offering vague gestures like getting a diagnosis or "checking it out," the promises evaporated. The deadlines he set came and went without any follow-through. So, I took it upon myself to get the work done, hoping for a sense of relief, or at least for some recognition. But when I informed him that the repairs were complete, the reaction was not supportive—it was hostility. There was a palpable resentment, as though my ability to do something without his intervention was a direct challenge to his control over me.

This was the narcissistic injury playing out in real time. My independence wasn't just something he didn't understand—it was something he couldn't tolerate. The idea of me managing on my own threatened the narrative he had created for himself: the one where he was in charge, the one who was supposed to be the center of my life. By taking control of my own life, I wounded his sense of dominance. It was a deep insult to him. So, instead of celebrating my success, he responded with anger. It wasn't just frustration—it was humiliation, and that's what drove him to engage in further manipulative tactics.

Understanding the Narcissistic Cycle of Injury and Retaliation

For a narcissist, injury and retaliation form a destructive cycle. The narcissist's sense of self-worth is fragile, dependent on external validation, and when that validation is threatened—whether by rejection, independence, or refusal to comply with their demands—they experience what's known as *narcissistic injury*. This injury does not fade away easily. Instead, it spurs an immediate need for retaliation.

The key to understanding this cycle is to recognize that it's not just about hurting the person who caused the injury; it's about the destruction of the victim's sense of self-worth and autonomy. For my ex, when I left or asserted my independence, this rejection was seen as a direct affront to his superiority and control. The more I resisted, the deeper his rage grew. But the focus of his retaliation wasn't merely to punish me—it was to strip me of any remaining sense of power or independence. The act of leaving him wasn't just a break-up; it was an existential threat to his need for validation and control.

In my case, the retaliation took on the form of digital harassment and manipulation. The cyberstalking was the perfect tool for a narcissist like him to reassert dominance—after all, it's easier to manipulate from a distance. But beyond the digital space, the cycle of retaliation continued, whether it was through undermining my efforts to succeed on my own, withholding promised support, or creating situations where my independence felt like a direct challenge to his authority.

The narcissistic cycle of injury and retaliation isn't just about getting even. It's about re-establishing control over the victim, making sure they never forget who is "in charge." Every step I took toward independence—whether it was fixing my car or simply seeking housing options—was met with hostility. The narcissist sees these steps as a rebellion against their control, and so, they punish it.

In the context of my experience, this ongoing cycle—pushing me down, then pulling me back in—wasn't just emotionally draining, it was a strategic form of control. As the attacks escalated, I realized that the goal wasn't just to harass or punish me—it was to ensure I would never break free completely, no matter how hard I tried.

The Dangers of Leaving: How Narcissistic Injury Fuels the Cyberstalking Cycle

When you leave a narcissist, it's not just a relationship that's over—it's a psychological battle. For someone like my ex, rejection wasn't something to simply accept or move past. It was an existential threat to his carefully constructed sense of self-worth. For him, rejection wasn't just about me choosing to leave—it was about a direct challenge to his dominance, his control, and his idea of himself as the one who dictates the terms of every relationship. Humiliation is the ultimate trigger for a narcissist, and when they are humiliated, they will stop at nothing to regain their sense of power.

Leaving a narcissistic individual is often the most dangerous thing you can do. This is the point at which they are most likely to lash out, and the behaviors that were subtle before now escalate into something much more destructive. The cyberstalking I experienced wasn't just an act of surveillance—it was a weapon used to punish me, to remind me that he could still control my life, even from a distance.

The moment I walked away, I unknowingly delivered a narcissistic injury. It's like hitting a nerve. The narcissist's ego is fragile, and when their image of superiority is shattered, they feel a deep, painful sense of loss. It's not just about losing a partner or relationship; it's about losing control. My leaving wasn't just a breakup. It was a rejection of everything he had come to rely on—my dependence on him, my vulnerability, and my emotional neediness. Once that was gone, his sense of self was compromised, and humiliation flooded in.

And that humiliation had to be dealt with.

For my ex, that meant retaliation. The cyberstalking began almost immediately, not just as an effort to stay connected, but as a revenge tactic. His need to punish me for rejecting him was overpowering, and his online harassment wasn't merely about keeping tabs on me—it was about asserting control. He needed me to know that no matter how far I went, no matter how much I tried to rebuild, he could always reach me. This wasn't about closure or communication—it was about him reminding me that I couldn't escape him.

Each time I showed independence, like when I started handling my own business—whether it was repairs to my home or other aspects of my life—his response was hostility, not support. This is classic narcissistic behavior. The need to assert power over every aspect of the victim's life, even in small ways, reflects their belief that the victim exists only in relation to them. My ability to act independently, to take control of my situation, was a direct affront to the narrative he had built around himself: the idea that he was the one who controlled me, that I was weak without him.

In one instance, when I sought help with home repairs, it was clear that this wasn't about genuine concern—it was about him controlling the process, dictating how things should be done, and inserting himself into the equation at every turn. When I eventually took matters into my own hands and got the repairs done myself, it wasn't met with congratulations or relief. No. It was met with resentment, hostility, and even sabotage in some cases. The very idea that I could function without him—or anyone else for that matter—was a threat to his authority. His pride, which had been injured by my growing independence, was now being reinforced through his digital harassment. His narcissistic injury couldn't be allowed to exist unchecked.

I started to realize that every time I took control of my life, even in small ways, it was like poking a hole in his carefully constructed world. The more I stood my ground, the more he came at me, using cyberstalking as a tool to undermine my confidence, punish me, and force me back into submission.

The reaction to my independence wasn't just disappointment or concern—it was rage. A rage that manifested through more invasive actions, more manipulative tactics, and greater levels of digital harassment. He couldn't accept that I didn't need him. He needed me to need him. That's the core of narcissistic injury: the inability to tolerate the loss of control, the fear of being irrelevant or powerless.

What's dangerous about this dynamic is that, at first, the cyberstalking didn't feel like harassment. It felt like something benign—a few texts here and there, a couple of social media posts, maybe even attempts to stay "friendly." But as I started to take control of my life, the cyberstalking grew more malicious, more relentless. And I realized that what I was facing wasn't just a man trying to get back in my life—it was a person who saw my very existence as an extension of his own ego. When I left, he didn't just lose a partner. He lost an object of validation. And in the narcissistic world, that's unacceptable.

The most dangerous aspect of leaving a narcissist isn't the initial pain of the breakup or the arguments that follow. It's the psychological warfare that begins after. The narcissist's rage is triggered, and the way they use tools like cyberstalking to regain control becomes a strategy to punish the victim. For someone like my ex, this wasn't just about harassment—it was about restoring his shattered sense of power. It was about showing me that I couldn't escape. No matter how far I ran, no matter how much I tried to heal, he would always be there, watching, manipulating, undermining my efforts at every turn.

In the end, the cyberstalking wasn't just about monitoring my life. It was about ensuring that I would always feel his presence, always be reminded that he was still in charge. And even if I thought I was free, the reality was that I was never truly able to escape the clutches of someone who couldn't tolerate being rejected.

The Narcissist's Need for Control and How it Manifests in Cyberstalking

When you think about the behavior of someone with narcissistic traits, it all boils down to one thing: control. Narcissists are not merely interested in having their way—they require dominance over the lives and actions of those around them. The need for control is not just about managing situations, it's about establishing a psychological hierarchy, where the narcissist sits at the top, and everyone else is beneath them. If you start to step outside the parameters they've set for you, it's not just a breach—it's a threat.

In my case, this manifested in several ways that were often disguised as "help" or "concern," but in reality, it was a means of asserting ownership over me. One of the most subtle ways narcissists maintain control is through manipulation of communication. Take, for instance, how I'd ask about updates regarding housing or my basic needs, like food or repairs. Each time I asked, I was met with delayed responses or vague answers. This wasn't because they were busy or disorganized; it was because they were controlling the narrative. They were putting me in a position where I had to depend on them for something so basic, and that dependence gave them leverage.

The psychological play here is key: by keeping the control over the narrative and limiting information, narcissists create a sense of helplessness in their victims. Every time I reached out for help or clarification, it was like I was begging for permission to act. And if I pushed too hard, I was accused of being too demanding or difficult—effectively silencing me and ensuring that their authority wasn't challenged.

This kind of control isn't just about withholding information—it's about establishing a power dynamic where I felt small and dependent. It wasn't simply about keeping me from acting independently. It was about making me feel inadequate when I tried to take charge of my own life. The lack of communication became a weapon, and when I asked too many questions or expressed too much initiative, it would often be met with passive-aggressive behavior or overt hostility.

For someone with narcissistic traits, their need to assert control doesn't stop at the physical world. Once they realize that their influence over their victim is slipping, they'll go digital. The cyberstalking then begins as a way to continue controlling the narrative online. For me, this manifested as unsolicited messages, tracking of my social media, and even indirect commentary that was clearly designed to keep me in line—reminding me that no matter how much I tried to move forward or control my own space, I could never fully escape.

This wasn't just harassment—it was the narcissist's final attempt to keep the power. The texts, the emails, the online comments—they weren't about engaging with me. They were about creating a digital presence that was impossible to ignore, a constant reminder that he was watching, that he could still influence me, even from a distance.

The irony in all of this is that my independence, whether in small actions or big decisions, was always perceived as a challenge to his control. The more I took charge of my life, whether by handling repairs or organizing housing on my own, the more aggressively he lashed out. Each act of self-reliance triggered a psychological injury. To him, it was no longer about simply breaking up. It was about maintaining control, making sure that I would never have a life outside of his reach. The cyberstalking was just a tool to keep me under surveillance, to ensure that I was still playing by his rules—even if he was doing it from a screen.

The Narcissist's Manipulation of the Narrative in Public and Private Spheres

One of the most insidious tactics a narcissist uses to regain control is their ability to manipulate the narrative—both in public and private spaces. Their need to control how they are perceived by others is paramount, and this extends far beyond the relationship at hand. For someone like my ex, the narrative was everything. How he was seen by the outside world—and more importantly, by me—was crucial to maintaining his sense of superiority and power.

This need to control the narrative goes hand-in-hand with narcissistic injury. When I left, the story he was telling about himself was suddenly interrupted. In his eyes, he was no longer the victim of a failed relationship, but the perpetrator of a broken bond. In his mind, I was now the one who had abandoned him, which didn't fit the perfect image he had created for himself. So, in an effort to reclaim control, he began to manipulate the story around our separation—spinning it to his benefit at every opportunity.

In public, this manipulation often came in the form of spreading false narratives. There were times when I would hear from mutual acquaintances or through social media that my ex was telling people stories about me that were completely fabricated—painting me as an unreasonable or difficult person. It didn't matter if what he was saying was true or not. What mattered was that he was painting me as the problem and himself as the victim—the classic narcissist tactic.

For example, when I showed signs of independence and made efforts to take care of myself without his involvement (such as dealing with housing or completing the necessary repairs in my home), it was immediately framed as something that wasn't good enough or something he should have been in charge of. It was as if I had somehow failed in his eyes, no matter how much progress I made. His goal was to control how others perceived me—especially in front of mutual friends, professionals, or acquaintances. If he couldn't control me directly, he would do it through others by painting me as someone incapable of taking care of myself.

On a more intimate, private level, this manipulation continued but in more covert ways. When I would push back or question his actions—whether it was about finances, custody, or anything else—he would quickly turn the tables, accusing me of overreacting or being too emotional. The conversation would shift away from the real issues and into the realm of my behavior, which was always framed as the problem. It wasn't about discussing facts or resolving issues—it was about making me doubt myself and my perception of reality.

One example stands out. When I asked for updates about housing or even the most basic information about my case, I would get either vague answers or no response at all. When I sought clarity, I would be accused of being difficult or demanding, but that wasn't the real issue. The real issue was that I was challenging his narrative. He didn't want me to see through the manipulation or to make decisions on my own. And by painting me as someone overly dependent, unable to handle things on my own, he was effectively trying to control the perception of me in the eyes of others.

This wasn't just about a failed relationship—it was about protecting his fragile ego and his need to be in control of how I saw myself and how others saw me. Narcissists cannot tolerate being viewed negatively, and so, when faced with the reality that someone was challenging their version of the truth, they fight back. This can come in the form of rumors, lies, or social manipulation, all designed to undermine the victim's credibility and autonomy. And of course, the cyberstalking was part of this strategy—ensuring that no matter how far I tried to run from the toxic dynamic, he could still control the narrative online. Whether it was through messages, subtle social media posts, or even just tracking my movements, he was reminding me that he had a finger on the pulse of my life—and that would never change, in his eyes.

The Narcissist's Need to Control and Punish for "Winning"

The most dangerous time in any narcissistic relationship is when you "win"—whether it's escaping, asserting your independence, or simply standing your ground in the face of their tactics. For someone like my ex, this was never a part of the script. Narcissists cannot handle being perceived as losers or as being weak. When I made decisions for myself, broke free from the mental hold he had on me, or handled things like my housing situation without his intervention, it became a personal affront.

One of the most glaring examples was when I made strides in taking control of my housing situation. For months, I had begged for updates—whether it was on housing, case management, or support. Every time I asked, I was met with vague responses, delays, or no response at all. The lack of communication was a tactic to create dependency and maintain power. Waiting for answers became a form of emotional manipulation, where I was forced to constantly check in, reaffirming my position as someone who needed help.

But once I took charge, once I finally figured it out on my own and made strides without their intervention, it was like I had flipped the script on him. The narcissist cannot tolerate this. The idea that I could succeed without him, that I didn't need his validation or assistance, went directly against the narrative he had tried to control. This was his narcissistic injury—the painful realization that I could exist independently, without his constant manipulation, or worse—succeed without him.

That's when the cyberstalking began intensifying. He had to punish me for this independence. He couldn't let it stand that I had "won" by managing things on my own, so he began to sabotage my peace by tracking me online, sending threatening messages, and monitoring my social media to ensure I knew he was still in control of my narrative. Every time I took a step forward, every time I did something that reaffirmed my independence, the response was more digital harassment. This wasn't just random stalking—it was a targeted, emotional punishment.

The need for control in a narcissistic environment is insatiable, and when that control is challenged, the response can often be vindictive. I saw this firsthand with the "support" I received at the workplace. Their resistance to my autonomy was palpable, but once I started meeting my own needs and stopped relying on their help, they didn't acknowledge my progress. Instead, they turned hostile. It was as if my ability to take care of myself somehow made me less worthy of the support they were supposed to offer. This wasn't about whether I needed help or not—it was about their sense of loss, their narcissistic injury, from seeing me succeed without their intervention.

The Narcissist's Need to Control the Narrative: Public vs. Private

A narcissist's need to control the narrative extends far beyond one-on-one interactions; it seeps into both private and professional spheres. The narcissist is constantly working behind the scenes to reshape reality—whether by manipulating how others see them or by distorting the image of their "target." In my case, this wasn't just about maintaining control within the confines of our past relationship—it was about ensuring that everyone, including external parties like the "workplace" staff, saw my failures and their successes. They would orchestrate situations where the narrative was carefully controlled, where I was positioned as the problem, and where they could retain control over how I was perceived by others.

One of the most striking examples of this occurred during my time at the workplace. When I first started, I was still in the process of rebuilding my professional life. I was vulnerable and unsure, but I was also determined. I had always maintained a high standard for honesty and transparency, but I quickly realized that this wasn't a place that valued those qualities. At first, my colleagues and I got along well, but as I began to assert my independence, take responsibility for my own progress, and show more autonomy, the dynamic began to shift. I started receiving subtle messages of control: "You're doing fine, but we're concerned about your next steps," or "Maybe you need to let us take the lead more." The more I stood on my own and took initiative, the more their need to control the narrative became clear.

This subtle shift in dynamics mirrored what my ex had done before. The need to control the narrative became paramount when I began to take action on my own terms. Whether it was making my own decisions about housing or finances, or even deciding to take care of my own car repairs without waiting for their "help," every move was scrutinized. They would ask questions designed to subtly undermine my confidence: "Are you sure you can do that yourself?" or "Don't you think it's too soon for you to handle this alone?" It was gaslighting, trying to create doubt and confusion about my ability to take care of myself. And when I succeeded without their intervention, there was no recognition. In fact, there was even hostility, as if my success without them was a direct challenge to their authority and their role in my recovery.

Public vs. Private Control:

The narcissist's need to control the narrative extends into both private and public spheres. The private sphere is the one-on-one relationship they seek to manipulate—where they shape your identity and influence your perception of reality. The public sphere, however, is where they attempt to shift the narrative with external people or groups, ensuring that the public perception aligns with their self-image. When I began to succeed on my own, despite the undermining from my colleagues and the control they tried to exert over me, it shifted the public narrative. It was no longer about me being the vulnerable, needy individual that everyone assumed I was. I was becoming someone stronger, someone independent—and that was the last thing my ex or some of my coworkers wanted to see.

At the core of this was a toxic dynamic—the narcissist's need to control the story they've constructed around the "target." My ex felt that if he could manipulate the narrative surrounding our breakup and my life afterward, he would maintain control. The more I took charge of my own life and independence, the more he felt the need to sabotage it—through digital harassment, gossip, and manipulation of the story, so that everyone, including the workplace staff, would see me as weak, in need of his "guidance," or incapable of moving forward on my own.

The Narcissist's Control Mechanism: Using Public Perception to Reinforce Power

For a narcissist, nothing is more important than the way they are perceived by others. In some ways, they live off the feedback they receive from people, using that feedback to reinforce their sense of power. When I managed to make progress without their involvement, it was as if they couldn't allow that new narrative to exist. It wasn't just about them feeling rejected—it was about losing control of how others viewed me and how others would interpret their role in my recovery. They wanted the narrative to stay the same: that I was a victim, that I needed help, and that they were the ones in charge of my journey.

This tactic of controlling the public narrative was a mirror of what my ex had done in our relationship. When he felt like I was breaking free, his next move was always to discredit me, whether through cyberstalking, harassing emails, or manipulating others to see me as unfit or ungrateful. The danger was not just in the manipulation itself, but in the way it distorted my public image—making it harder for me to ask for help or be taken seriously when I needed support.

5. The Narcissistic Cycle: Sabotaging Independence and Reinforcing Dependency

One of the most insidious behaviors a narcissist exhibits is their ability to sabotage their victim's independence while simultaneously reinforcing their dependency. This cycle of control creates an environment where the victim cannot function without the narcissist's involvement—at least, that's the illusion the narcissist tries to create. The subtlety of this tactic makes it especially dangerous, as it's not always obvious to those on the outside or even to the victim until it's too late.

In my case, the more I tried to stand on my own two feet, the more my ex—and later, those at the "workplace"—seemed to undermine my progress. Take, for example, the broken promises that became a constant source of frustration and helplessness. Initially, I was told that I would receive support in areas like housing and home repairs. These were promises I relied on, hoping they would help me rebuild my life and get back on my feet.

But as time went on, it became clear that those promises were either delayed or outright sabotaged. With the home repairs situation, I was told I would get assistance with fixing things. Yet, when I took initiative and arranged repairs myself, there was resentment rather than support. Instead of celebrating my independence and progress, the response was hostility—as if my ability to manage without their intervention threatened their control over me.

Similarly, when I asked for updates on housing, I was met with vague answers and empty promises. The lack of clear communication was another subtle way they reinforced my dependence. It was as if they didn't want me to find a place on my own, or if I did, they didn't want me to succeed without them. These delayed promises made me feel as though I wasn't capable of progressing without their approval or help. Every time I tried to move forward, they subtly pushed me back into the role of someone who couldn't function without their intervention.

This is the crux of the narcissistic cycle: the narcissist needs to keep you dependent on them, so they will sabotage any effort you make toward independence. This sabotage is never direct or obvious; it's subtle, cloaked in manipulation and misdirection. When I fixed my own home repairs, when I started actively searching for housing without their help, I wasn't met with pride or encouragement. Instead, they subtly reinforced the idea that I couldn't thrive without them.

Narcissistic Tactics: Sabotage in Plain Sight

The cycle of sabotage and reinforcement works in several ways:

1. The Delayed Promise: Narcissists are often experts at delaying promises in a way that creates uncertainty. When promises are made but not kept, it creates a sense of helplessness. I experienced this time and time again, particularly when I was told that certain goals—like securing housing or getting home repairs done—were "on the way," only for nothing to materialize. The more I tried to take action on my own, the more the narrative shifted to one of failure, despite the fact that the only thing standing in the way of my success was their lack of support.

2. Reinforcing Dependency: Every time I tried to make progress or gain independence, I was met with subtle tactics designed to reinforce my dependency. In the workplace, for instance, when I asked about housing updates, I was met with vague responses or no answers at all. The message was clear: I wasn't capable of navigating my own path forward without their guidance. When I began to handle tasks on my own, like home repairs or making housing decisions, the underlying threat was obvious—they didn't want me to become self-sufficient, because it took away the power they held over me.

3. The Backlash of Independence: The narcissist's rage at my independence became palpable when I didn't fall into the role they had scripted for me. This was particularly clear when I handled personal matters independently, like managing home repairs or making housing arrangements without their interference. Instead of congratulating me on my progress, I was met with passive-aggressive remarks and even hostility. The narrative shifted to one of criticism

and undermining, as if my actions were a personal attack on their authority.

The Role of Cyberstalking in Reinforcing Dependency

The cyberstalking became a critical component in this cycle. It wasn't just about keeping tabs on me; it was about reminding me that I couldn't escape. By using digital tactics to undermine my autonomy, the narcissist sought to make me feel as though I was constantly under their control. If I tried to step outside the lines they had drawn for me—whether it was with my career, my personal life, or even my online presence—they would strike back with passive-aggressive messages or manipulative online behaviors.

This constant surveillance served as a form of psychological control, reinforcing the idea that no matter how far I tried to go, they were always watching, always ready to sabotage my efforts. The cyberstalking was a tool for re-establishing power by using the fear of being watched to reinforce dependency. It kept me in a state of constant vigilance, where I felt like I could never truly break free from their influence.

Creating Confusion and Mistrust (Communication as a Tool of Control)

From the very start, I didn't realize that something as simple as a lack of clear communication was a tool being used to control me. In the environment I was in, the lack of follow-through on promises wasn't an innocent oversight—it was deliberate. I would ask for housing updates, or for clarity on a particular situation, only to be met with vague responses like, "We'll get back to you," or "We don't have an answer yet." The absence of real communication wasn't about lack of information; it was about the power of silence.

The aim wasn't just to keep me uninformed—it was to keep me in a constant state of uncertainty, to create a mental fog where I would feel powerless and dependent on them. I would leave conversations feeling frustrated and confused, but never with enough clarity to make any real decisions on my own. Instead, I was left with feelings of self-doubt and insecurity, questioning whether I was being too demanding, or if I simply wasn't understanding things correctly.

For example, when I asked about updates on housing, I would be given a timeline or a date, but when that time came, I would receive no updates. Days would pass, then weeks, and I was left in limbo, with no answers. The staff, who I had hoped would support me, seemed to either ignore my questions or provide vague responses that only deepened the confusion. When I tried to press for clarification, I was met with irritation, as though my need for certainty was a problem. This lack of clear communication was a way of asserting control—by keeping me in a state of perpetual waiting, they kept me off balance and emotionally vulnerable.

This pattern of inconsistent communication mirrored my ex's behavior when we were together. Whether it was about simple requests or serious discussions, he would often be vague or elusive with his answers. I remember asking for clear timelines on important decisions, but he would deflect or give promises that were never kept. I would wait, and when the time came, nothing had changed. In this way, the lack of communication itself became a form of emotional manipulation, a way to keep me dependent on him for answers and uncertain of my own ability to make decisions without his influence.

Key Takeaways: Understanding the Intersection of Narcissistic Injury, Cyberstalking, and Control

The complex relationship between narcissistic injury, cyberstalking, and control tactics is essential to understanding the ongoing manipulation and emotional abuse that victims experience. As we explore the key behaviors, it's clear that narcissistic individuals rely on a range of subtle yet devastating tactics to regain control, punish those who defy them, and maintain dominance over their victims. Below, the key takeaways offer a concise overview of the patterns of behavior that contribute to both emotional and psychological harm, highlighting the ongoing struggle for autonomy in the face of relentless manipulation.

1. Narcissistic Injury and the Need for Control

- Narcissistic Injury: A rejection or challenge to their perceived superiority triggers profound emotional pain in a narcissist, often manifesting as rage and the need to reassert control.

- Use of Cyberstalking: Cyberstalking becomes a tool for narcissists to regain control after a narcissistic injury, often in response to rejection or abandonment.

- Punishment for Independence: Any attempt to regain autonomy or self-sufficiency (like handling a car repair or making decisions independently) is perceived as an affront, triggering hostility and punishment from the narcissist.

- Need for Validation: Narcissists rely on external validation to feel powerful. When that source of validation is taken away (e.g., through rejection), they react with increasing hostility and revenge.

- Dominance and Control: The narcissist's goal is to dominate and control the victim, making sure they feel powerless and unable to escape.

2. Creating Confusion and Mistrust (Communication as a Tool of Control)

- Power of Silence: The narcissist controls the flow of information and communication to maintain control over the victim's understanding of their situation.

- Lack of Communication: Vague responses, lack of updates, and deliberate omissions are used to keep the victim uncertain, dependent, and emotionally destabilized.

- Perpetual Waiting: By withholding key information or failing to follow through on promises, the narcissist keeps the victim in a state of limbo, increasing their sense of powerlessness and self-doubt.

- Manipulating Uncertainty: The goal is to create confusion and make the victim feel unsure of their decisions, reinforcing the narcissist's control over their life.

- Reflection of Ex's Behavior: Similar patterns of vague communication, missed deadlines, and unfulfilled promises from the ex mirror these tactics, reinforcing emotional manipulation and dependence.

3. Controlling the Narrative in Public and Private

- Narrative Manipulation: The narcissist actively works to control how others perceive the situation, painting themselves as the victim and the other person (e.g., the victim) as the aggressor.

- Self-Perception: The narcissist constructs a story where they are seen as the hero or martyr, ensuring their public image is protected and the victim is discredited.

- Twisting Reality: Public and private narratives are manipulated to present the narcissist as the wronged party, and the victim's autonomy is framed as problematic or unacceptable.

- Gaslighting the Victim: By controlling the story and shifting blame, the narcissist gaslights the victim into doubting their own reality, further diminishing their sense of self-worth.

- Example of Rewriting History: The victim's actions (e.g., asserting independence) are framed as rebellion or difficulty, making them appear unstable or high-maintenance to others.

4. Controlling the Environment (Physical and Emotional Space)

- Subtle Sabotage: Narcissists often sabotage or obstruct the victim's ability to maintain a sense of security or normalcy. For example, denying assistance for necessary repairs or labeling independent actions (e.g., fixing things on one's own) as "hazardous."

- Dismissing Boundaries: The narcissist will disregard the victim's personal boundaries or needs in order to force compliance with their agenda or expectations.

- Shifting Responsibility: When the victim asserts their independence or takes initiative, the narcissist responds with hostility, as it threatens their control and dominance over the situation.

- Opposition to Self-Sufficiency: Narcissists feel deeply threatened by any sign that the victim can survive without their intervention, making any independent act (e.g., fixing a car or managing personal affairs) a perceived affront to their authority.

- Creating Dependency: Through passive-aggressive behavior and manipulation, the narcissist creates an environment where the victim feels dependent on them, even when they are capable of managing their own life.

Chapter 5

The Unyielding Presence of Cyber Stalking and Harassment: "Check-ins" and "Pop-ups"

Throughout our marriage and long after our separation, there was no escape from his need to control my every move. It wasn't enough for him to be a part of my life in the ways that a partner should. No, his obsession went far beyond that. Even in the most mundane moments, whether I was at the grocery store, out with friends, or trying to enjoy a quiet night at the movies, he would somehow find a way to insert himself into my life.

This wasn't a one-time thing; it was constant. The "check-ins" became routine—his need to know where I was, what I was doing, and who I was with was insatiable. The pop-ups, which began subtly, turned into a terrifying normalcy. It didn't matter if I was miles away, or if I had just told him I needed space—he'd show up unannounced. Sometimes, it was at my work, other times at places I had no idea he knew I'd be. I'd be mid-conversation with a friend or family member, and there he was, lurking, silently observing from a distance. His presence was never just an inconvenience—it was a reminder that he was watching, always lurking, always in control.

The most eerie part? He would appear at places I frequented, almost as if he had some hidden knowledge of my schedule. It wasn't just limited to the home, where the daily surveillance already felt like a prison. He had a way of showing up unexpectedly, at places I would consider safe, places where I thought I could have peace. The grocery store, the gym, even a quiet café—his shadow loomed wherever I went. These encounters were not just creepy—they were calculated. Every "pop-up" was a subtle way of reminding me that no matter how far I tried to distance myself, he was always one step behind. It was as if his need to control my movements, my choices, was a force he couldn't relinquish, even when I physically removed myself from the situation.

And let's not forget his double standards. He would check in on me constantly, but I wasn't allowed the same freedom. He was a serial cheater, and everyone knew it. The trust that should have been foundational in a marriage had been long destroyed by his constant lies and infidelity. But still, he couldn't stand the thought of me being free, of me living a life without him. While I was forced to account for every second of my day, he was out, without a care in the world, creating chaos and devastation.

I found out about one of his many affairs the hard way—when the nurses at the hospital where I gave birth to our child informed me that he had been there the week before, not with me, but with another woman. As if it wasn't enough that I was in labor, vulnerable and exhausted, I was confronted with the sickening truth about the man I was married to. While I was preparing for one of the most important moments of my life, he was planning another betrayal behind my back. The nurses told me that he had been there with a woman who wasn't me, and he'd even been boasting about his escapades. That was the moment when the reality of his emotional and psychological abuse hit me square in the chest—his betrayal wasn't just physical; it was psychological too. While I was giving him my trust, my energy, and my life, he was living in a world of lies and deceit. And yet, the most twisted part of it all was that he still wanted to monitor *me*. He was still holding on, even when he had no right to.

The lies he told about his actions—the cheating, the gaslighting—were constant. He would tell me I was imagining things or that I was overreacting when I confronted him. But the reality was, his actions were a constant barrage of emotional manipulation, meant to confuse and destabilize me, to keep me questioning my own worth and reality. While I was left to pick up the pieces of his deceit, he played the role of the "victim" in front of others, painting himself as the misunderstood partner while I was made to feel crazy for calling out the obvious.

His obsessive checking, the "pop-ups," and the relentless surveillance didn't just continue after we separated—it worsened. After we broke up, he still found ways to keep me tethered to him, through technology, through manipulation of our shared responsibilities, and by continuing to invade my personal life. He would track my online activity, create fake profiles to get close to my friends, and monitor any relationship I had with others, whether personal or professional. The deeper I tried to go into my own life, the more deeply he tried to follow. No matter how many times I blocked him or moved to a new city, he found a way to breach the walls I put up.

This toxic cycle of control and manipulation, the constant reminders that I could never fully escape him, left me emotionally scarred. I couldn't make a decision without second-guessing myself, unsure if he was watching, listening, or waiting for the moment I would falter. The anxiety of being constantly monitored—the worry that every move I made could be scrutinized by him—destroyed any sense of personal freedom.

His inability to let go, combined with his need to constantly check on me, became a weapon of emotional warfare. It wasn't just harassment. It was psychological entrapment. And it wasn't just about his obsessive need for control—it was about his desire to always be *there*, even in the most intimate, private moments of my life, like the birth of our child. That was a moment I should have shared with my support system, with people who cared about me. But instead, I was forced to face his presence even then, a constant reminder that he would never let me go, never let me heal, never let me move on.

The Pop-Ups: The Insidious Web of Surveillance and Emotional Control

The "pop-ups" were not random incidents; they were meticulously planned and executed, a reflection of his obsessive need to control and monitor my every move. At first, they seemed like mere inconveniences—annoying, but not enough to raise alarms. However, over time, it became evident that they were an extension of his surveillance, a chilling tool he used to remind me that no matter how far I moved, no matter how much I tried to live my own life, he was always *watching*.

Every time he showed up unannounced, it sent my mind spinning. I could be at the store, trying to pick up a few things, and there he was. Or I might be at a friend's house, simply trying to enjoy a rare moment of peace, and suddenly, he'd appear at the door. The places I once considered "safe" were no longer free from his control. His need to know where I was, who I was with, and what I was doing had turned these seemingly ordinary moments into emotional battlegrounds. The fear of seeing him, the dread of what he might say or do, created an atmosphere of constant tension. It was as if my life wasn't my own—it belonged to him, and he'd invade it on his terms, regardless of how I felt.

This constant, unpredictable intrusion was part of a bigger pattern of behavior rooted in his Narcissistic Personality Disorder (NPD) and Obsessive-Compulsive Personality Disorder (OCPD). People with NPD often see others as extensions of themselves, not as separate individuals with autonomy. My movements, my relationships, my very sense of self were something he believed he had a right to monitor, control, and dictate. When I would make even the smallest decision—whether to go out with a friend, take a different route home, or spend a quiet evening alone—it would trigger an obsessive need in him to assert his dominance, to *check* on me, to make sure I wasn't doing anything that might jeopardize the fragile narrative he had built around his control.

His OCPD tendencies were evident in the rigidity of these "pop-ups." He couldn't allow any deviation from his perception of how things *should* be, so I could never relax or live freely. His obsessive need to know where I was at all times, what I was doing, and even who I was interacting with, stemmed from a deep-rooted fear of losing control. This wasn't just about checking in; it was about maintaining a level of dominance in our relationship that he couldn't relinquish. The "pop-ups" were part of a larger pattern of surveillance, where every action I took, every place I went, was a potential violation of his need to control.

The emotional destabilization from these intrusive behaviors was profound. It shattered my sense of privacy and security, eroding the very foundation of my personal well-being. Every time he showed up unexpectedly, it felt like a violation—not just of my space, but of my autonomy. I never knew if I was being "watched," if he was lurking in the background, tracking my every move. This lack of control over my own life was overwhelming, and it began to take a toll on my mental state.

The unpredictability of his actions made it difficult for me to trust anything or anyone. I would start to question if he was watching me when I wasn't aware, and even the people around me started to feel like potential spies in his game. The constant anxiety of waiting for the next pop-up left me on edge. There was no way to relax, no way to reclaim my own boundaries. I became hyper-vigilant, always on alert, and it took everything in me just to feel like I could live a normal, peaceful life.

This wasn't just harassment—it was a mind game. It was psychological warfare designed to destabilize me, to make me feel like I couldn't trust my own decisions, my own space, or my ability to be free. He wanted me to feel small, to feel controlled, to feel like I could never fully break free from his grasp. His emotional manipulation was a tool of control, a tactic meant to keep me in a perpetual state of anxiety, doubt, and fear.

Each pop-up, each unexpected appearance, was a reminder that I was never truly free. Even if he wasn't physically in the room, his influence was everywhere. His presence, even when absent, was a shadow that loomed over my every move. I was living under constant surveillance, and the worst part was that I couldn't escape it. It wasn't just about him showing up—it was about how that act made me question my own agency, my own right to exist without his interference.

His behavior was calculated, designed to maintain control, and it wasn't just disruptive—it was destructive. Every uninvited pop-up and every breach of my privacy chipped away at my ability to trust, to heal, and to regain any sense of normalcy. Over time, this erosion of trust and autonomy led me to question everything around me, making it harder to rebuild my life, my identity, and my peace.

Disruptive Intrusions: A Narcissist's Need to Dominate and Control Social Spaces

One of the most unsettling behaviors he exhibited was his ability to infiltrate situations, not through constant presence, but through disruptive intrusions. These weren't mere appearances; they were calculated moves designed to assert control and manipulate the environment around us. It wasn't about him simply showing up to be involved—no, it was about him asserting his dominance and forcing the world to recognize his presence and power, even in spaces where he didn't belong.

A perfect example of this was during my child's baseball games. I'd try to enjoy watching my child play, surrounded by other parents and families. But once he appeared, everything changed. It wasn't that he came to watch the game—it was that he needed everyone to see him. His sudden intrusion into this seemingly neutral space had an immediate, unsettling effect. People who once treated me with warmth and respect now began to look at me differently. I could feel the shift in the air, the subtle, but undeniable change in how I was perceived. The other parents, once friendly and engaging, suddenly became distant, almost as if a veil of suspicion had been cast over me.

He didn't need to speak to anyone directly—his mere presence, the way he would stand apart, just enough to be seen but not engaged with, was enough. It was as if he was sending a message: *"I'm here, and I'm in charge."* And somehow, the others received that message, because shortly after his appearance, the dynamics between me and the parents would subtly shift. Their treatment of me and my children would become colder, more distant, even though I had done nothing to provoke this change.

This was no accident. His narcissistic need to dominate was at play. He wanted to show everyone that he controlled the narrative of our family, that he had the power in our lives, not me. In these moments, his tactic wasn't to stay around and build relationships—it was to make sure people knew *he* was the one pulling the strings, that *he* was the figure to be reckoned with, even if it was in subtle, insidious ways.

Once his presence had been felt, he would vanish. He didn't care about attending games or practices regularly. His goal wasn't to be a father in the traditional sense, but to create a spectacle—an event where the other parents would be forced to acknowledge his power and control. He was never there long enough to engage meaningfully with anyone; he didn't stick around to be a supportive presence. No, he was there to disrupt, to alter the social order, and then to disappear, leaving behind confusion and a changed dynamic.

The emotional toll of this behavior was profound. After his disruptions, I felt isolated, alienated from the very community I had once felt a part of. The other parents, many of whom I had connected with, suddenly seemed colder, as if they were following an unspoken script of mistrust and suspicion, as though they had been influenced by something—his manipulation, his ability to make others question the truth. It didn't matter that the reality was that he was the one who had always been the problem. The fact that he had created this disruptive intrusion—this moment where he forced others to take notice of him, even without saying a word—was enough to shake my sense of belonging, and it cast a shadow over my children's experiences too.

He wanted to strip me of my identity as a mother in those spaces, to make me feel disempowered in front of others, to rewrite the narrative about our family in his own image. By showing up, making a brief, dominating appearance, and then leaving, he ensured that I was left holding the emotional fallout. The parents, who had once been supportive, now treated me as though I was somehow "tainted," "difficult," or "unstable." And in doing so, he accomplished his goal: he turned normal social spaces into sites of emotional control—places where I was no longer just a parent, but a person who had to defend her integrity and sanity against a backdrop of confusion and alienation.

But this wasn't the first time, and it wouldn't be the last. He used these disruptive intrusions to keep me on my toes, to ensure that I would never have a chance to breathe, to heal, to feel like I had a community. Each new social situation became another battleground where I could never predict when or where the next disruption would occur. And the fallout? It was always the same: my isolation, my feeling of being "othered," and the emotional toll that came with having to constantly fight for my truth in the face of his manipulations.

His tactic of disruptive intrusion wasn't just about creating chaos in my life—it was about making sure that no matter where I went, no matter how far I tried to move on, I would always be reminded of his control. He wanted the world to know that I couldn't escape him, that he was still in charge. And the worst part was that it worked. For a time, it made me question whether I could ever truly rebuild my life without his shadow constantly looming over it.

His tactic of disruptive intrusion wasn't just about creating chaos in my life—it was about making sure that no matter where I went, no matter how far I tried to move on, I would always be reminded of his control. He wanted the world to know that I couldn't escape him, that he was still in charge. And the worst part was that it worked. For a time, it made me question whether I could ever truly rebuild my life without his shadow constantly looming over it.

The emotional destabilization from these disruptions was profound and insidious. The anxiety wasn't just about him showing up, it was about what came after—the ripple effect. Every time he created a scene or shifted the way others viewed me, it left me feeling like I was constantly walking on a tightrope. He didn't need to stay long to make his impact; in fact, the shorter his intrusion, the more damaging it was. Because after he left, I was left to pick up the pieces, left to explain myself to others, and left to bear the weight of the emotional fallout.

It wasn't just that he showed up and made things uncomfortable. It was that his presence redefined every space he entered. My once peaceful moments—whether it was watching my child at a game or talking with other parents—became contaminated by the fear of what would come next. The relationships I was trying to build, the trust I was trying to cultivate with those around me, were subtly undermined every time he showed up. I could feel it in the way people started to look at me differently, as if I had something to explain. As if I was the one who was at fault. I didn't just feel isolated—I felt discredited, misunderstood, and alienated. It was like he was planting seeds of doubt wherever he went, even in people who barely knew me. His interruptions made my world feel smaller and smaller, turning it into a place where I couldn't even trust the spaces that once felt safe.

And the most twisted part was that he was never there to build anything—no, he wasn't there to create meaningful connections with our children or the people around us. He didn't stay for the game, the practice, or the community. His goal was to disrupt—to shift the power dynamics and keep me always on the defensive. Once he'd made his appearance, he would disappear, leaving me to face the fallout, to explain, to navigate the uncomfortable silences, and to rebuild my sense of self amidst the confusion he'd sown.

This wasn't just emotional manipulation—it was a deliberate attempt to destabilize my entire world. No matter how much I tried to *move forward*, no matter how hard I worked to regain a sense of peace, he would find a way to push me back into a corner. Each intrusion felt like a reminder that I couldn't escape his reach, and the longer I lived under his shadow, the more I began to doubt myself. I began to question my own worth, my own decisions, even my right to be free of his influence. His disruptions were designed to break me down, to make me feel like I was living in a world of uncertainty, where nothing was stable, and I was always a step away from another emotional setback.

It also affected my children. Though they were young, they too could feel the tension that followed each one of his disruptive intrusions. After he'd come and gone, they would see the shift in the way people treated us, and their world would be altered, however subtly. They were forced to witness this ongoing manipulation, this emotional warfare, which in turn shaped how they understood relationships, trust, and family. His ability to make me feel isolated also impacted them, forcing them to navigate a confusing landscape of divided loyalties and shifting dynamics, where he created tension and left me to clean up the emotional mess.

The longer this continued, the more the emotional toll became unbearable. The anxiety from living under this constant manipulation, the mental exhaustion of having to constantly recalibrate my own reality, took a toll I wasn't prepared for. There were days when I would feel paralyzed, unsure of what to do or who I could trust. The lack of a stable emotional foundation left me grasping for any sense of control, but he had already ensured that every space I inhabited felt like it was under siege. And what made it worse was that this wasn't just happening in my private life—it was happening in front of people, in shared spaces where I should have been able to have peace. These weren't just fleeting moments of discomfort—they were psychological intrusions that rewired how I viewed myself, my relationships, and my world.

I couldn't even trust the most basic aspects of life—the social fabric I once had, the friendships, the sense of belonging. Every time he inserted himself into my life, he fractured the pieces of me that were still trying to heal. This emotional destabilization didn't just affect me in the short term; it was a long-term strategy to wear me down, to exhaust me, to ensure I never fully recovered. The constant shifts in my emotional state became my new normal, and as much as I tried to fight against it, I realized that the longer I lived in his shadow, the more I was giving up my sense of self.

In this way, his need for dominance wasn't just about control—it was about ensuring that I would never feel fully at peace. That I would always be in a state of flux, of instability, emotionally shackled to the chaos he created. Even when he wasn't physically present, the scars of his disruptions remained, embedded in the way people saw me, in the way I saw myself, and in the fractured sense of peace I could never seem to rebuild.

The disruptive intrusions I faced in the years after our separation didn't come from nowhere. They were the culmination of a much deeper, more insidious pattern that had started years earlier. At the time, I didn't have the words or the understanding to describe what was happening. His behavior was manipulative, erratic, and confusing, but it didn't always seem sinister in the beginning. Our relationship started like any other—marked by the typical highs and lows, moments of joy, and the intimacy of shared connection. But beneath the surface, there was something more—a seed of control that, over time, would grow into a suffocating obsession.

It's easy to dismiss early signs of unhealthy behavior as "just a phase" or "something small," especially when you're in the midst of a relationship. But looking back, the subtle signs of possessiveness and jealousy were the first cracks in the foundation of what would become a destructive and controlling dynamic. These behaviors were initially trivial—comments about who I was spending time with, questions about where I'd been, accusations when I didn't respond to a message immediately. But with time, these small cracks widened. The things that felt like *normal relationship concerns* in the beginning gradually revealed themselves to be tactics of control—the kind of manipulation that starts slowly, but escalates relentlessly.

As his obsession grew, so did the intrusiveness. It wasn't just jealousy anymore; it was a complete disregard for my autonomy. His need to know where I was, who I was with, and even what I was doing when I wasn't with him became all-consuming. It was a slow, relentless erosion of my personal freedom, and I didn't even realize I was losing it until it was too late. These behaviors were precursors to the emotional abuse that would define our relationship, and they were a reflection of the deeper psychological issues that I could not have understood at the time.

In retrospect, his actions were a clear expression of the symptoms of obsessive behavior and emotional manipulation, patterns that often characterize Narcissistic Personality Disorder (NPD) and Borderline Personality Disorder (BPD). People with these disorders lack empathy and see others as extensions of themselves, rather than as separate, autonomous beings. For someone like him, I wasn't a partner—I was an object to be controlled. The more I tried to assert my independence, the more aggressive his tactics became. The need to break me down emotionally and make me dependent on him was his only focus, and it became more intense as time went on.

The behavior escalated. He began calling me incessantly, demanding answers to questions about my whereabouts, my activities, and the people I spoke to. Every action I took—every step I made to reclaim my independence—was met with hostility. This narcissistic rage, a term used to describe the intense anger and frustration someone with NPD feels when they perceive abandonment or criticism, was always just beneath the surface, ready to explode.

When I finally made the decision to leave, his reaction wasn't one of understanding or acceptance. Instead, it triggered a torrent of rage, setting off a chain of manipulative actions that would grow more damaging by the day. What began as subtle signs of control and jealousy had now turned into full-blown obsession—a dangerous, suffocating need to keep me tethered to him no matter the cost.

The Symptoms of Obsession: How It All Began

Looking back, the patterns of behavior that would eventually evolve into full-blown obsession didn't start with dramatic, life-altering events. In fact, the early signs were deceptively subtle—small actions, often dismissed as typical relational quirks, that laid the foundation for a much darker reality. At first, it was the small things: a comment about what I was wearing, a question about where I had been, a suggestion about who I should or shouldn't spend time with. These behaviors didn't raise alarms initially, because they appeared to be simple displays of affection or concern.

But as time went on, what seemed like normal jealousy in a relationship began to morph into something much more controlling. He would start making subtle demands for my time and attention, not out of care, but out of a need to maintain control. His behavior was not driven by love, but by a deep-rooted need to dominate and possess. The early signs, though easy to overlook, were an insidious creeping into my life—a slow, careful erosion of my personal freedom.

What began as trivial accusations would soon escalate into full-blown surveillance. I didn't realize it at the time, but his need to control and monitor my every move was a red flag. His obsessive tendencies weren't obvious from the start, but they took shape in ways that were hard to ignore. I would get random phone calls, asking where I was, who I was with, and what I was doing. If I didn't answer fast enough or if I was with a friend or colleague, his tone would shift, and I could sense the undercurrent of anger and disappointment. The more I tried to be independent, the more his behavior escalated.

But it wasn't just in the form of questioning or monitoring. His attempts to control my actions and interactions extended to the point of manipulating my relationships with others. I found myself questioning my own decisions about friendships, because he would plant doubts in my mind, subtly accusing people of being untrustworthy or a bad influence. The more I tried to distance myself, the more aggressive he became. It was like being trapped in a never-ending cycle where the more I tried to escape, the tighter his grip became.

These behaviors took root early in our relationship, masked as "concerns" or "protectiveness," but in reality, they were signs of something far more sinister. Looking back, I see them now as the precursors to the emotional abuse that would define our relationship. At the time, I couldn't fully grasp what was happening—I just thought it was a part of the ups and downs of being in a relationship. But in reality, this was the slow buildup of a psychological prison.

As the relationship continued, his behaviors became more erratic and intrusive. It was no longer about asking a few questions here and there—it became about complete control. I started noticing he was spying on me. Checking my phone when I wasn't around, making sure to monitor my social media accounts, even going through my emails to track my every interaction. His obsession with knowing everything I did, even down to the people I spoke with, became all-consuming.

Every time I tried to assert some form of autonomy, it was met with resistance. He wasn't willing to let me have any independence, and the more I pushed back, the more his behavior became erratic. He would call me incessantly, sending texts demanding answers to questions that seemed innocent but carried a deep weight of control. Where are you? Why haven't you answered? Who are you with? These weren't questions born of love; they were born out of a narcissistic need to dominate. The more I pulled away, the more I was punished. Every ounce of personal space I tried to create was met with guilt trips, arguments, and accusations.

This is when I began to see a shift—a change in the dynamics of the relationship. It wasn't about being loved, it was about being controlled. And when I finally left, when I had the courage to break free, it became clear that his obsession wasn't just about keeping me in the relationship. His need to maintain power and control was far stronger than his desire for love. It was narcissistic rage in its purest form—a reaction to the rejection he couldn't accept.

His obsession didn't stop when we separated. In fact, it intensified. The moment I attempted to live a life separate from him, he reacted like someone who had lost his sense of self. His behavior became even more intrusive, more controlling. He wasn't just mad that I left; he was incensed by my autonomy. The idea that I could live without him, could exist as my own person, was a direct assault on his fragile ego. His need for control became a consuming obsession, and it would only continue to grow over time

As his obsession with controlling my every move grew more pronounced, it didn't take long for the digital world to become the next battlefield. The early signs were there—the incessant checking of my phone, the unwanted intrusion into my private space, the constant surveillance—but nothing prepared me for what would come next: the weaponization of technology. The rise of social media and online communication opened up a new frontier for him, one he quickly adapted to, turning it into a tool for stalking, manipulation, and harassment.

Initially, his attempts to engage with me online seemed trivial. Liking posts, commenting on pictures, sending me casual messages. It wasn't enough for him to just watch from afar—he needed to make his presence felt, to insert himself into my digital world as he had done in my physical one. But what I initially chalked up to simple annoyance quickly spiraled into something far more calculated and insidious.

It wasn't just that he was following me online; he was creating an entire digital persona, one that allowed him to interact with me under false pretenses. He began creating fake accounts—accounts designed to make it look like I was being contacted by a friend or acquaintance when, in reality, it was him. Under these false identities, he would comment on my posts, send me private messages, and try to lure me into conversations that felt benign at first but were actually just ways for him to assert his control. These fake interactions were designed to throw me off guard, to make me question who was truly in my corner and who wasn't.

What seemed like harmless overreach at first soon became a deliberate strategy to dominate my digital life. His presence was omnipresent. Every time I posted something, I would feel the anxiety of wondering if he was watching, if he would leave a comment or like something in a way that would make me uncomfortable. I felt like I was constantly being monitored, even in spaces that should have felt private. He made sure that I knew he was always there, lurking in the background, waiting for the opportunity to make himself known.

This wasn't just an emotional annoyance—it was a psychological tactic meant to destabilize me. The constant reminders of his presence online—the likes, the comments, the messages—became a digital chain around my neck. It wasn't just a nuisance; it was a form of emotional control. Every notification on my phone, every ping on social media, felt like a little stab of anxiety. The fear of what might come next—the fear of another message, another attempt to manipulate or gaslight me—became all-consuming.

As his behavior escalated, it became clear that he wasn't just stalking my online presence for validation; he was using it as a means of control and punishment. Cyberstalking is often seen in individuals with Narcissistic Personality Disorder (NPD) and Antisocial Personality Disorder (ASPD), and it's exactly this kind of emotional manipulation that marks the tactics of someone like him. The desire to assert control, to make me feel isolated, to constantly remind me of his presence, is at the core of cyber harassment. For someone with NPD, any form of autonomy is a threat. The moment I sought to break free from his influence, he retaliated by invading my digital world.

And this was only the beginning. As he grew more emboldened by the power he had over my online life, his actions became more destructive. He started posting derogatory messages about me—lies and half-truths that sought to ruin my reputation and turn others against me. He used his online presence to spread rumors, to create a false narrative about who I was and what had happened in our relationship. His need to control went beyond me; it was about ensuring that everyone around me saw me as he wanted them to see me—a liar, unstable, untrustworthy.

He didn't stop there. His obsession expanded into the realm of identity theft. He hacked into my devices, monitored my communications, and made decisions about my financial life without my consent. I would discover strange transactions on my accounts or realize that my passwords had been changed, and I would be left scrambling to regain control. His digital stalking didn't just invade my social life; it destroyed my financial security, and worse, it left me constantly questioning whether I could ever truly have a safe space again.

The emotional toll of living under constant digital surveillance was overwhelming. It wasn't just about the harassment; it was about the psychological burden of knowing that my every move could be monitored at any time. He was no longer just in my house; he was in my inbox, in my phone, in my online identity. And every time I tried to reclaim control, he would find a way to reassert himself.

For years, this digital assault became a constant companion—an invisible presence that followed me wherever I went. The emotional distress it caused was profound. It wasn't just about being harassed online—it was about living in a constant state of vigilance, always on edge, always wondering when the next attack would come. And when he made these attacks, they were calculated, meant to keep me off balance, to ensure I never fully regained control of my life.

This cycle of online harassment became so much more than just an extension of his obsession—it was his primary tool for maintaining power over me. The digital prison he built was just as suffocating as the one he'd created in our physical relationship. It was a constant reminder that no matter where I went, no matter what I did, I could never truly escape him.

The Symptoms of Obsession: How It All Began

Our relationship started, like so many others, with excitement, passion, and the typical ups and downs of early romance. But there were cracks beneath the surface, small signs that would eventually reveal themselves as clear indicators of deeper, more insidious issues. In the beginning, they seemed like typical relationship issues—jealousy, possessiveness, and minor arguments. It wasn't until much later that I began to see these behaviors for what they truly were: signs of emotional manipulation and obsessive control.

In the early days, his actions were subtle enough to pass as "normal" relational concerns. He'd ask questions about my whereabouts, question who I was spending time with, or express concern over certain friendships. These things seemed benign at the time, especially when wrapped in the guise of "care" or "concern" for my well-being. But soon, these small gestures started to accumulate—becoming demands, and then, rules. The more I tried to create space for myself, the more I was met with resistance.

The subtle shift into possessiveness and control grew clearer. He wanted to know where I was at all times, what I was doing, who I was talking to. Everything became a reason for scrutiny. If I spent time with friends without him, I was met with guilt. If I spoke to someone at work, I was questioned about the details. Even my hobbies and interests, which had once been part of my own identity, were scrutinized. His constant checking-in, like an invisible hand on my shoulder, left me feeling uneasy and like I could never fully be myself.

This controlling behavior escalated as time went on, and soon, it wasn't just about questioning me—it became about following my every move, checking my phone, reading my emails, and even monitoring my social media accounts. This wasn't random curiosity—it was an all-encompassing need for control, a compulsion to track and dominate every aspect of my life. His obsession began to consume him, and that's when I realized how deeply rooted his desire for control had become.

I wasn't allowed to simply exist in peace—I had to live in constant awareness of his presence. It didn't matter where I was or what I was doing. If he could show up unannounced, he would. I could be at the grocery store, out with a friend, or at my child's baseball game, and there he was, appearing out of nowhere, as if to remind me that I was never fully free. These weren't coincidental "chance" encounters—they were deliberate disruptions meant to reinforce his power over me.

I would later recognize these moments for what they were: the classic moves of someone deeply entrenched in narcissistic behavior. His need to assert dominance wasn't limited to physical spaces; it extended into social settings, too. I remember taking my children to a baseball game, a space where I had hoped to enjoy a sense of normalcy and community. But once he showed up—uninvited and unannounced—it changed everything. It wasn't just that he appeared; it was how he commanded attention. The other parents, who had once treated me like a fellow mother, now seemed to look at me differently. Their behavior subtly shifted after meeting him. He'd inserted himself into my circle, letting everyone know that he was the "authority", and from that point forward, I became the outsider.

What started as his occasional interference soon morphed into a broader, ongoing campaign to disrupt my life and emotional stability. After these encounters, I would often be left with a feeling of alienation and anxiety. The people around me, influenced by his presence, would start treating me differently, which only deepened my sense of isolation. His sporadic appearances were never about being a part of my life—they were strategic disruptions to undermine my confidence and manipulate the social dynamics around me.

This tactic of control wasn't just about physical proximity—it was about turning every situation into a psychological battleground, one where I never felt truly free. Even in spaces where I should have been able to find peace, such as at a child's sporting event or in the company of friends, his need to take over and undermine my autonomy was always lurking just beneath the surface.

He wasn't content with just controlling our relationship. He needed to assert that he was in charge of everything—from the way people saw me to the way I viewed myself. It wasn't just about possession of my time; it was about possession of my identity, my sense of self. The goal was clear: to make sure I knew that no matter where I went, no matter who I was with, he was always there, pulling the strings, subtly or overtly, to remind me that I was never truly free.

But it didn't stop there. When I finally found the courage to leave him, to carve out space for myself, the obsession didn't diminish—it grew more toxic and aggressive. The emotional manipulation and psychological games didn't end with the relationship. In fact, they only escalated, taking new and more dangerous forms. And that's when his need to control began to merge with a new force—technology—which would become his primary tool for harassment and coercion.

The Cyber Harassment Begins: The Digital Prison

With the advent of social media, my ex-husband found new tools to extend his reach into my life. Where before, his control was physically present, it now had an invisible, omnipresent quality that no one could see but I could feel at all times. It wasn't enough for him to just show up unannounced anymore—he needed to be everywhere. He found ways to infiltrate the digital spaces where I thought I could finally breathe. Social media became his new playground, a place for him to stalk, manipulate, and emotionally assault me from afar.

He began stalking my online presence—liking and commenting on my posts, sending me harassing messages, and even creating fake accounts to interact with me under the guise of someone else. What seemed like a harmless form of overreach at first soon became clear: it was a calculated, deliberate strategy to reassert control over me. Cyberstalking—the use of technology to stalk or harass someone—became a frequent and dangerous tactic in his arsenal. He wasn't just watching my life; he was inserting himself into it in ways that felt as suffocating as his physical presence ever had.

Cyberstalking is often seen in individuals with Narcissistic Personality Disorder (NPD), and my ex's behavior mirrored the exact traits of someone with this disorder. People with NPD view others not as autonomous beings, but as extensions of themselves. Any attempt to break free, any action that threatened their control, is met with rage and retaliation. For him, my attempts at independence were perceived as an attack on his ego, and he reacted by launching a covert campaign to track my every move, not just in the physical world but in the digital one as well.

For years, he used the internet as a weapon—posting derogatory messages about me, spreading lies to friends and family, attempting to smear my reputation in every way possible. Each lie he spread became a part of the digital prison he had constructed around me. He hacked into my devices, accessed my personal accounts, and even manipulated my financial life without my consent. His digital harassment didn't just affect me emotionally—it had real-world consequences, making it nearly impossible for me to regain control of my finances or even my personal identity.

The combination of online harassment and identity theft left me with long-lasting scars. I would find myself constantly on edge, checking my accounts, changing passwords, and trying to keep up with the never-ending disruptions he orchestrated from behind a screen. The worst part wasn't just that he could reach me anywhere, at any time; it was that I couldn't escape him. Even when I tried to rebuild my life, he was always there, hiding behind a fake profile, lurking in the background, pulling the strings of the narrative he'd created about me.

The Stalking Escalates: From Online to Real Life

Despite the control he had already established over my life through digital means, it was not enough. Eventually, his obsession seeped into my physical world. It wasn't just an occasional "check-in" or an email; this was the kind of relentless stalking that no victim should have to endure. I started noticing signs that he was following me—appearing unexpectedly at my workplace, outside my home, and at places I frequented with friends or family. He would act like it was just a coincidence, that he just "happened" to be there. But the encounters were too frequent, too perfectly timed, to be anything but deliberate interference in my life.

The signs started off subtle. Notes would appear on my car, messages on my phone, or strange interactions with people I knew. At first, I convinced myself that I was just being paranoid, but with each incident, the reality set in: this was real. His stalking wasn't just an emotional tactic—it was now physical. He was present, not just in my digital life, but in the spaces I thought were safe. Under the law, stalking is defined as a course of conduct directed at a specific person that causes them to fear for their safety or the safety of others. This behavior wasn't just inconvenient—it was a violation of my peace, and, according to the law, it was illegal.

The challenge, however, was that while certain legal protections—such as restraining orders or no-contact orders—could limit his physical presence, they did little to address the digital intrusion that had become just as pervasive, if not more so. These legal measures were primarily designed to keep him from physically approaching me, but they didn't account for the complexities of modern technology. For example, while a protective order could keep him from showing up at my doorstep or my workplace, it didn't prevent him from harassing me online, from creating fake profiles to follow my every post or from using social media to continue his assault on my reputation.

The challenge was that his narcissistic behavior wasn't bound by the physical world. His need for control didn't stop just because a court said he couldn't show up at my house. He needed to dominate every space I entered, even the spaces that were supposed to be private and personal. His behavior intensified when he realized that, despite these protections, he could still use technology to infiltrate my life and continue to make me feel like I had no control. It was an extension of his narcissism—the need to be present, to have an audience, to manipulate the narrative, regardless of the boundaries that were legally imposed on him.

His stalking tactics became more covert. He would constantly monitor my social media accounts, sending me unsolicited messages under different accounts, and creating fake profiles to engage with my posts as if he were someone else. He did this to keep me in a perpetual state of fear—fear of being watched, of being judged, of knowing that no matter where I went or what I did, he was always lurking, always trying to manipulate my image and interactions from a distance. He wasn't just physically stalking me—he was actively involved in destroying my sense of peace and self in every corner of my life, including the digital one.

The problem was that the legal systems in place weren't equipped to handle this kind of harassment. The digital world, where so much of our daily lives now unfold, wasn't protected by the same laws that governed physical boundaries. While there were tools to protect me in real life, such as restraining orders or programs that shielded my address from public records, there was no equivalent to protect my social media presence or the constant surveillance I endured through the very devices that were supposed to help me stay connected to the world.

This discrepancy allowed him to continue to stalk and control me, just in a different form. His narcissism thrived in the digital age, where he could remain anonymous yet still have power over my life. The law could shield me from his physical presence, but it couldn't stop him from invading my personal space online, from manipulating my reputation, or from forcing me to live in a constant state of anxiety.

The Digital Cage

This was the new reality I found myself trapped in—a digital cage, invisible yet inescapable, where my every move, every word, every interaction was being monitored and manipulated by a man who believed he had a right to control me, even from a distance. Despite the legal barriers placed between us, the protections meant to guard my body and mind from his reach, he found a way in, time and time again. His narcissistic obsession with me—his need to assert power over me even when he was miles away—was a constant reminder that no matter how many court orders or restrictions were placed on him, he could still make my life a nightmare.

The law could stop him from physically showing up at my door, but it couldn't stop him from hacking my life, from destroying my reputation in the very spaces where I was supposed to feel safe. The digital realm became his playground, a place where he could continue to assert control without consequence. His presence was a shadow that stretched into every corner of my existence, haunting my thoughts, my relationships, my ability to rebuild a life.

It's difficult to describe the emotional toll of such constant intrusion—how it feels to never be truly free. I didn't just lose my privacy; I lost my peace. Every ping from my phone, every unexpected notification, was a trigger, a reminder that I was never truly alone. I couldn't even escape into the virtual world without his fingers creeping into the spaces that should have been mine. And that was his ultimate victory: he didn't need to be physically present to control my life. His narcissism—the belief that I was an extension of him, that my life could never truly be mine—turned every part of my existence into a battleground, a place where he could strike without warning, without limit.

But as devastating as this has been, as painful as the erosion of my sense of self and safety has been, it is also a cautionary tale. A reminder of how easy it is to be trapped by someone who believes they have a right to your existence, your thoughts, your identity. And it's not just a personal story—it's a shared reality for so many people who suffer in silence, who feel like they have no escape because their abuser has found a way to reach them wherever they are, whenever they want.

The truth is, this kind of abuse—this kind of digital manipulation—is not just a victim's issue; it is a society's problem. We live in a world where the lines between physical and virtual are blurred, where someone with the right tools and the wrong mentality can invade another person's life and leave them feeling powerless. And the law, though helpful, is still playing catch-up.

As I continue to fight for my right to live a life free from his influence, I am reminded that freedom—true freedom—is more than just escaping his physical reach. It's about breaking free from the chains that are invisible but just as real. It's about reclaiming my identity, my peace, and my autonomy from a man who never saw me as a person but as an object to be controlled. I will not let him continue to define my reality. The digital chains he's tried to lock around me will not be the end of my story.

I refuse to be silenced, to be erased, or to live in fear of a man whose world was built on control. My life is mine again. And no matter how much he tries to invade it, I will keep fighting to keep it free.

Moreover, the lack of treatment for these conditions only exacerbates the damage. A person with Untreated Narcissistic Personality Disorder or Borderline Personality Disorder can continue their obsessive behavior indefinitely, often without recognizing the harm they are causing.

Dr. Rachel Levitch

Chapter 6

Narcissistic Abuse in the Digital Age: How Technology Empowers Stalkers and Manipulators

In today's increasingly digitally connected world, the patterns of narcissistic behavior—whether overt or covert—are not only amplified but also empowered by the tools technology provides. Narcissists have long used manipulation and control tactics to dominate their victims, but now, with the anonymity and reach of the internet, these individuals can hide behind screens, making it easier to stalk, harass, and torment without consequence. The emotional devastation that comes with such abuse is not just a personal experience—it's a widespread crisis that leaves victims feeling isolated, shamed, and disempowered, often unsure of where to seek help or how to fight back.

Drawing from Dr. Ramani Durvasula's insights on narcissistic personality traits, this section will explore the eight primary types of narcissists—each with their own manipulative tactics—and how these traits manifest both in real-life interactions and, more dangerously, in cyberstalking. We'll delve into how narcissists utilize online tools to maintain control over their victims, highlighting the specific online behaviors that correspond with each narcissistic type. These behaviors aren't just theoretical or psychological concepts—they are the lived realities of those who have experienced narcissistic abuse, like myself. The chart in Appendix 1 will provide a detailed breakdown of these narcissistic types and the cyberstalking behaviors that often accompany them, offering a clear look at how these individuals manipulate, deceive, and hurt others in ways that are difficult to detect and even harder to escape.

The emotional toll of such narcissistic cyberstalking is difficult to put into words. Victims often face constant anxiety, self-doubt, and a gradual erosion of their mental well-being as they navigate a world where their privacy is constantly invaded and their reputation is under attack. The anonymity of the digital world gives these narcissists an advantage—they can manipulate, control, and cause harm without facing any immediate repercussions. For those caught in this cycle of harassment, it can feel like an endless assault on their emotional resilience.

For victims like me, the experience is a silent battle—one where it feels as though the systems of support are often insufficient to provide the necessary legal, psychological, and emotional safeguards. While there have been some improvements in the protection of victims, there's still much more to be done. The psychological toll—from loss of self-trust to overwhelming anxiety—is something that continues to affect those who are trapped in the toxic orbit of narcissistic abuse.

This is not just an individual struggle. It is a societal issue that demands attention. As the prevalence of cyberstalking and online abuse grows, we must collectively demand accountability and stronger protections for victims. No one should have to endure this kind of torment in silence. For those in this battle, it is crucial to recognize the tactics, document every incident, and seek both legal and psychological support. Most importantly, it is vital to know that no one should fight this fight alone.

While the digital age offers many conveniences, it also gives narcissists a platform to manipulate, control, and exploit their victims in ways that were once unimaginable. Social media, texting, email, and tracking software have all become tools in the arsenal of those who thrive on power and control. Narcissists, empowered by this anonymity, can reach into their victim's life without being seen—making it harder to catch them, harder to stop them, and even harder to prove what's happening. It's a modern-day battle that feels like you're constantly fighting ghosts—there but never quite within reach.

These narcissists are often skilled at deceptive manipulation, taking advantage of the vulnerability that the digital world provides. They may create fake profiles or impersonate others to get closer to their target. They'll use social media platforms to monitor every aspect of a victim's life—liking posts, commenting on photos, sending unsolicited messages, or even fabricating interactions with friends and family. It becomes a web of control, and every click, every post, every interaction feels like a calculated move designed to pull the victim back into the narcissist's orbit. These seemingly innocent actions are not benign; they are part of an intricate psychological strategy designed to break down the victim's emotional defenses.

This behavior is particularly insidious because narcissistic stalking is often hidden behind a veil of plausibility. A narcissist may seem like they're just "checking in" or trying to "help," but the intention is to assert dominance. For example, they might show up at a workplace event, or message you under the guise of concern. At first, these actions might seem harmless—perhaps even flattering—but over time, they create a sense of constant surveillance, leaving the victim feeling watched and trapped. Even in a virtual world, they make their presence known, effectively erasing the concept of boundaries and personal space.

One of the most damaging tactics of narcissistic cyberstalking is the repeated violation of privacy. Victims find that their phones, emails, and even home networks are no longer safe. Personal information is no longer just theirs—it is something to be exploited, twisted, and used against them. In some cases, narcissists might go so far as to hack devices, monitor communications, or access sensitive information—all to maintain control. It is a calculated process that wears down the victim's sense of security and self-worth. The constant intrusion erodes trust in every relationship and every facet of daily life.

What's even more troubling is the feeling of powerlessness that accompanies narcissistic cyberstalking. The digital tools that these individuals use to control and manipulate are often so subtle that the victim might not even realize how far they've been infiltrated until it's too late. Fake accounts, tracking tools, and manipulated communication can make it seem like the narcissist is everywhere, even when they're not physically present. The victim's life is turned into a game of cat and mouse, where every action is monitored, every decision is questioned, and every move is followed.

As we explore the types of narcissists in Appendix 1, we can see how their online stalking tactics align with their personalities and traits. From the grandiose narcissist, who might boast about their achievements while secretly watching their victim's every move, to the covert narcissist, whose silent manipulation and gaslighting tactics can drive a victim to madness, each type utilizes technology in a unique way to assert control.

But it's important to note that the harm caused by this cyberstalking is not just about invasion of privacy or manipulation; it's about the emotional toll on the victim. The constant fear, the anxiety, the sense of being unable to escape—these are the lasting effects of being relentlessly stalked in both the physical and digital realms.

The need for better protections for victims of cyberstalking is paramount. While some legal frameworks are in place to address online harassment, they often fail to capture the full extent of the damage inflicted by narcissistic manipulation. Laws may punish physical stalking or harassment, but what happens when the harassment is digital? What recourse does a victim have when the perpetrator can hide behind multiple aliases, fake profiles, or even deepfake technology?

More needs to be done to ensure that victims of narcissistic abuse in the digital age are not left vulnerable. Legal protections must catch up to the realities of our hyper-connected world, and we need to advocate for policies that can help address both cyberstalking and online harassment in ways that are comprehensive and protective.

The journey toward healing from such cyberstalking abuse is not easy. It requires constant vigilance, self-empowerment, and a network of support from both loved ones and professionals who understand the nuances of emotional abuse and narcissistic manipulation. But it also requires systemic change—from stronger legal actions to better resources for mental health support for victims who are often left to pick up the pieces of their shattered sense of self.

No one should have to fight this battle alone. The more we speak out, the more we expose these hidden forms of abuse, and the more we advocate for victims, the closer we come to creating a society that protects everyone from the devastating effects of narcissistic cyberstalking.

Grandiose Narcissist: The Arrogant Manipulator

Grandiose narcissists are the archetypes many people think of when they hear the term "narcissism." They are often charismatic, but their grandiose self-image hides a deep lack of empathy. These individuals love to be the center of attention and will do anything to maintain their superiority. In the digital age, their online presence is loud and assertive, constantly flaunting their successes and achievements.

Online, these narcissists engage in behaviors like:

- Excessive self-promotion through social media, regularly posting about their accomplishments while monitoring others' profiles for signs of competition or slights.

- Creating fake accounts to subtly undermine others by spreading rumors or gossip, often with the goal of damaging their reputation.

- Using online harassment tactics such as public shaming or slandering to maintain their perceived dominance.

Victims of grandiose narcissists are often left feeling small, insecure, and inadequate. The constant barrage of manipulative content can chip away at their self-esteem, leaving them constantly questioning their own worth.

Covert (Vulnerable) Narcissist: The Passive-Aggressive Saboteur

The covert or vulnerable narcissist is much less obvious than the grandiose type. While they may appear humble or shy, their deep-seated sense of entitlement and lack of empathy remain unchanged. These individuals often feel victimized, interpreting the world around them as hostile or unjust. Their online behaviors reflect this internal conflict, and they may use passive-aggressive methods to control and manipulate others.

Online tactics include:

- Creating fake identities or personas to gaslight victims into believing they are overreacting or imagining things.

- Engaging in projection, where they accuse others of the very behaviors they are guilty of (e.g., accusing the victim of being manipulative while being manipulative themselves).

- Subtle trolling or emotional blackmail disguised as concern, such as sending cryptic messages to make the victim feel guilty for not complying with their demands.

These narcissists rely heavily on emotionally abusive tactics, such as guilt-tripping and confusion, leaving the victim feeling paranoid and unsure of their own reality.

Malignant Narcissist: The Dangerous Predator

Among the most destructive and toxic types, malignant narcissists combine traits of grandiosity, Machiavellianism, and at times, psychopathy. They are cold, calculating, and exploitative. Malignant narcissists have little to no remorse for the damage they cause, and they often take pleasure in tormenting and manipulating their victims.

In the online realm, these individuals may:

- Hack devices to monitor the victim's every move, from personal messages to social media activity.

- Use their anonymity online to create fake accounts for harassment, sending threatening messages, or spreading malicious lies.

- Engage in escalating behavior, from psychological abuse to threats of physical violence if the victim attempts to break free.

Victims of malignant narcissists often feel trapped and terrified, knowing that the stalker is relentlessly watching them and willing to use any means to regain control. The emotional damage can be profound, often leaving the victim with extreme anxiety and paranoia.

Communal Narcissist: The 'Do-Gooder' With an Agenda

Communal narcissists are individuals who use altruism as a mask to gain admiration and admiration online. They frequently share their charitable actions and acts of kindness on social media, but the underlying motivation is not to help others—it's to gain validation and status. They use the appearance of goodness to disguise their deep-seated sense of entitlement and superiority.

Their online behaviors may include:

- Excessive self-promotion in the form of philanthropic posts, often with the aim of gaining praise and attention.

- Publicly displaying charity work, while secretly undermining others or engaging in subtle manipulations behind the scenes.

- Using their social media presence to gather followers who validate their image of being a "good person," all the while denigrating the victim in private.

For victims, the emotional toll is confusing. They may see this narcissist as a "giver" but find that their genuine emotional needs are constantly dismissed. The superficiality of the relationship makes it hard for victims to trust others and build real connections.

Neglectful Narcissist: The Disconnected Manipulator

Neglectful narcissists are emotionally disengaged and distant. They use people as tools to serve their own interests but fail to nurture genuine relationships. When these narcissists engage online, it's often out of necessity, and their indifference to others is evident.

Their online tactics are typically:

- Minimal engagement, only reaching out when they need something or want to assert control.

- Manipulating or ghosting the victim through delayed responses, leaving them in a state of uncertainty.

- Breadcrumbing—offering just enough attention to keep the victim hopeful, but never truly investing in the relationship.

Victims of neglectful narcissists often feel ignored and unimportant, questioning if the narcissist even cares. The lack of consistency in communication can leave them feeling empty and unfulfilled.

Benign Narcissist: The Childlike Attention-Seeker

Benign narcissists display superficial immaturity in their behavior, often coming across as self-centered or shallow. They may engage in narcissistic behaviors but on a smaller, less harmful scale. In an online setting, they may act in ways that are annoying or irritating rather than dangerous.

Their digital presence often includes:

- Oversharing personal details, with a focus on attention-seeking behaviors.

- Superficial posts that focus only on themselves, without considering the feelings of others.

- Incessantly seeking likes, comments, and validation from their online followers.

While they may not be as malicious as other types, the emotional toll can still be significant, as victims may feel like their authenticity is being dismissed or overshadowed by the narcissist's constant self-centeredness.

Entitled Narcissist: The Self-Absorbed Taker

The entitled narcissist believes they deserve special treatment—often at the expense of others. They feel that they are entitled to recognition, praise, and privileges without having to work for them. Their entitlement can drive them to use any means necessary, including online tactics, to force others to validate them.

Their online behaviors often look like:

- Trolling or flaming: Posting inflammatory content to trigger responses from others.

- Manipulating online platforms to demand attention, often at the expense of others' emotions.

- Ignoring boundaries, such as sending unsolicited messages or invading privacy in ways that make the victim feel violated.

Victims of entitled narcissists often experience a sense of helplessness and anger, as they are forced to endure someone else's self-righteous demands for attention and admiration.

The types of narcissists outlined above are not just theoretical—they represent the real-world behaviors that many victims of cyberstalking and online harassment experience every day. By hiding behind digital anonymity, narcissists can easily manipulate, harass, and torment their victims without facing any immediate consequences. This digital veil allows them to escape detection while leaving their victims feeling isolated, powerless, and devastated.

The chart in Appendix 1 breaks down the types of narcissists and their online behaviors more comprehensively, but one thing remains clear: narcissistic abuse is damaging, both online and offline. Recognizing these traits and behaviors is the first step toward breaking free from the cycle of abuse and rebuilding one's emotional well-being.

Dr. Rachel Levitch

Chapter 7

The Fragmented Self: Emotional and Physical Effects

The aftermath of narcissistic abuse is not just about emotional trauma; it fragments your very sense of self. You begin to lose sight of who you are, what you want, and what you deserve. The narcissist's constant manipulation, gaslighting, and emotional abuse carve deep wounds into your psyche, leaving you in a constant state of self-doubt and confusion. Over time, the victim is no longer able to separate their true identity from the image the narcissist has forced upon them.

For years, I had lived in a world where my worth was constantly being undermined. My ex-husband's emotional abuse, coupled with his need to control every aspect of my life, made me feel small, powerless, and invisible. I was constantly questioning my thoughts, my decisions, and even my perceptions of reality. This confusion—the inability to trust myself—was the core of my fractured sense of identity. I had become a shadow of the person I once was, and the longer the abuse continued, the harder it became to reconnect with the person I had been before him.

The Psychological Toll: Cognitive Dissonance and Self-Doubt

Cognitive dissonance is one of the most insidious psychological effects of narcissistic abuse. In my case, I would constantly experience internal conflict between my lived reality and the narrative my ex-husband was pushing onto me. He would twist situations, rewrite history, and manipulate the facts to make me doubt everything I knew to be true. This created a constant state of mental dissonance, where I couldn't reconcile the two conflicting realities: the man I had once loved and the man who was now relentlessly abusing me.

This self-doubt was compounded by gaslighting—a form of psychological manipulation where the narcissist systematically denies the victim's reality. I began to question whether my thoughts were valid or if I was just overreacting. Every emotional response, every instinct, felt foreign. I had been trained to invalidate myself, to ignore my own needs in favor of his. It wasn't just about feeling lost—it was about feeling unmoored, as if my entire sense of self had been eroded.

Fragmented Sense of Identity: Loss of Autonomy and Self-Worth

For someone who has endured narcissistic abuse, the fragmentation of self is often irreversible without significant intervention. Narcissists view their victims not as autonomous individuals but as extensions of themselves. They believe that they have the right to define their victim's identity, to control their actions, thoughts, and emotions. In doing so, they essentially erase the victim's autonomy and sense of self-worth.

When my ex-husband's abuse escalated—through identity theft, cyberstalking, and manipulation—my sense of who I was became even more distorted. I wasn't just confused; I felt like I had no control over my own life. Each attempt to assert my independence was met with resistance. Self-worth wasn't something I could find within myself; it was something I was constantly seeking in the form of his approval, his validation, or his affection. But the more I sought these external validations, the less I recognized who I truly was. My desires and boundaries were lost in a sea of confusion, leaving me disconnected from my own needs.

The Emotional and Physical Toll: Chronic Stress and Health Consequences

The effects of narcissistic abuse aren't just mental; they also take a severe physical toll. Living in a constant state of emotional turmoil, anxiety, and hypervigilance leaves the victim's body in a chronic stress state, activating the body's fight-or-flight response over and over again. This prolonged state of heightened alertness—always anticipating the next blow, the next emotional attack—takes a toll on both physical and mental health.

I personally experienced physical symptoms I had never dealt with before: insomnia, digestive problems, panic attacks, and even chronic fatigue. My mind was constantly racing, processing the aftermath of his manipulations, while my body was exhausted from the tension. Over time, these stress-related symptoms became so pervasive that I couldn't ignore them any longer. I had become a prisoner of my own body, unable to find relief from the constant emotional and physical strain of his abuse.

Research has shown that prolonged exposure to stress hormones like cortisol can weaken the immune system, increase inflammation, and even affect brain function, leading to cognitive decline and difficulty concentrating (Sapolsky, 2004). In my case, the cognitive fog I experienced, the inability to make decisions, and the constant worry were all signs of how deeply the abuse had affected me—not just emotionally, but physically as well.

Dissociation: The Escape from Emotional Pain

As the abuse continued, I resorted to dissociation—a coping mechanism that allows the mind to detach from the overwhelming emotional pain. Dissociation is often described as the feeling of being "outside of one's body" or as though you are observing your life from a distance. It is a protective mechanism, but one that ultimately leaves you feeling more disconnected from your true self.

For years, I existed in this fragmented state. On the outside, I appeared fine, going through the motions of daily life, but on the inside, I was barely hanging on. I couldn't feel anything—I was numb. My emotions were dulled, my sense of self was distorted, and I felt like I was living in a fog. Dissociation was my mind's way of protecting itself from the abuse, but it also created a barrier between me and the reality I so desperately needed to confront.

Rebuilding the Self: The Long Road to Healing

Reclaiming my sense of self after years of narcissistic abuse has been the most challenging part of my recovery. The journey of healing is not linear, and it's not immediate. For someone who has suffered from narcissistic abuse, rebuilding your identity means dismantling the beliefs the narcissist instilled in you—beliefs that you are unworthy, unlovable, and incapable of trusting yourself.

It involves reconnecting with the person you were before the abuse, or perhaps creating a new version of yourself that isn't tied to the toxic narrative the narcissist forced upon you. It requires learning to trust yourself again, to set healthy boundaries, and to affirm your worth every single day. It also means recognizing that the fragmented self—the version of yourself that felt lost and disconnected—was never your true self to begin with. It was the result of someone else's manipulation.

Physical Healing: Caring for the Body

The body also needs healing. Once I recognized the toll the abuse had taken on my health, I made a conscious effort to prioritize my well-being. This included regular exercise to reduce stress, improving my sleep hygiene, and seeking out medical care for the physical symptoms of chronic stress. Therapy helped me understand how deeply the abuse had impacted my mind and body, and slowly, with time, I began to rebuild my health—both physical and emotional.

In the end, healing from narcissistic abuse is a multifaceted process. It's about reconnecting with the self, rebuilding trust, and giving yourself the compassion and care that you were never afforded during the abuse. It's about recognizing that you are worthy of love, respect, and care—not just from others, but from yourself. The fragmentation may have been deep, but it is not permanent. With the right support, patience, and commitment to self-care, the shattered pieces of your identity can be reassembled, and you can learn to live freely and fully again.

The Fragmented Self: The Psychological Devastation of Narcissistic Abuse

The psychological toll of narcissistic abuse doesn't just leave behind temporary scars; it fragments your very sense of self. Victims of narcissistic abuse often experience a profound disintegration of their identity, where who they were before the abuse feels distant, unreal, or even entirely lost. The trauma doesn't just happen in isolated moments; it happens in the constant, unrelenting psychological warfare that manipulates, distorts, and shatters every part of your being. Complex PTSD (C-PTSD) is the most accurate way to describe the mental and emotional devastation that comes from prolonged exposure to narcissistic manipulation.

While traditional PTSD may occur after a single traumatic event, C-PTSD is the result of sustained, repeated trauma that occurs in environments where the victim feels trapped and helpless—usually at the hands of an abuser (Herman, 1997). The narcissist's tactics—gaslighting, emotional invalidation, and constant psychological undermining—turn you into a shell of who you once were, with your identity constantly in flux. You become an invisible prisoner, forced to question everything you once trusted: your memories, your instincts, and your reality.

Gaslighting: The Weapon of Self-Doubt

Gaslighting is one of the most insidious tools in a narcissist's arsenal. The narcissist doesn't just manipulate events or conversations—they manipulate your sense of reality. Through repeated lies, contradictions, and twisted explanations, they cause you to doubt your own perceptions and question your memory.

Over time, this wears down your ability to trust yourself, making you feel crazy, confused, and disconnected from your own mind.

In my experience, gaslighting was a daily experience. I would remember things clearly—events, conversations, and feelings—but my ex-husband would systematically deny them. He would insist I was mistaken, that I was "too sensitive" or "imagining things." It wasn't just about making me feel bad in the moment—it was about breaking my grip on the truth. Narcissists seek to obliterate your sense of self by questioning the very foundation of your thoughts and emotions.

The danger of gaslighting lies in its cumulative effect. It is not just one lie or one moment of doubt—it's a relentless barrage of manipulation. The narcissist doesn't let up, they keep chipping away at your certainty until you're left feeling fragmented, unsure of where the abuse ends and where your truth begins.

Fragmented Identity: Living in a Constant State of Self-Doubt

The psychological effects of narcissistic abuse are not isolated incidents—they create a shattered identity. Narcissists treat their victims as mere objects to be controlled and manipulated, and in doing so, they reshape their victim's sense of who they are. What was once a strong, independent person becomes a shell of their former self, struggling to navigate the emotional chaos inflicted by their abuser.

Living in a constant state of emotional manipulation caused me to feel as though I was losing pieces of myself every day. My ex-husband didn't just want to control me physically or emotionally—he wanted to reshape my very being, forcing me into roles that didn't fit. I was either the "perfect" partner or the "worthless" failure, never allowed to exist in between. This constant cycle of idealization followed by devaluation made me feel as if I could never be "enough." The more I tried to meet his expectations, the more fragmented my identity became.

Each time I tried to assert myself, he would cut me down, undermining my confidence and planting seeds of doubt. My sense of self-worth eroded because I never knew which version of myself was real. Was I the person he wanted me to be? Or was I the person I remembered before the abuse? I lost sight of what I truly valued or wanted. It wasn't just the emotional damage that hurt—it was the realization that I no longer knew who I was, or who I was meant to be.

Emotional Paralysis: The Struggle with Self-Trust

As the emotional abuse escalated, my self-doubt grew to the point where I became emotionally paralyzed. I couldn't trust my own feelings or instincts. The narcissist had successfully eroded my confidence and gutted my ability to make decisions. Every situation felt clouded by his presence, even when he was physically absent. I began to question even the simplest things: Did I make the right choice? Was that decision truly my own? Or was it the result of his manipulation?

This emotional paralysis is one of the most profound impacts of narcissistic abuse. When you no longer trust your own mind, it's difficult to move forward. You become stuck, afraid to make choices, because you're no longer sure what's real. You lose the ability to act in your own best interest. I lived in a fog for so long—second-guessing myself, unsure of who I was or where I was headed.

The Emotional and Physical Toll: Chronic Stress and Health Effects

The emotional damage inflicted by narcissistic abuse doesn't just affect your mental well-being—it physically manifests in the form of chronic stress. The constant barrage of emotional abuse triggers the body's fight-or-flight response, sending stress hormones like cortisol surging through the system. When the body is in a constant state of heightened stress, it begins to show physical signs of strain.

In my case, the stress manifested in severe sleep disturbances, digestive problems, and constant anxiety. It wasn't just a matter of feeling overwhelmed—it became a physical burden I could no longer ignore. My body had become a battleground, constantly responding to emotional abuse with symptoms of illness and exhaustion.

Studies show that prolonged exposure to stress not only weakens the immune system but also damages cognitive functions, leaving victims unable to focus or think clearly (Sapolsky, 2004). The toll on my body was undeniable. I felt physically drained, unable to rest, and unable to escape the emotional chaos that had become my new normal.

Rebuilding a Fragmented Self: The Path to Recovery

The journey to healing from narcissistic abuse is long and complex. Recovery from C-PTSD and the fragmentation of self requires reclaiming the parts of yourself that were stolen and distorted by the narcissist's abuse. It is about restoring your sense of autonomy, trusting your instincts again, and recognizing that you are not the sum of the abuse that was inflicted upon you.

Rebuilding trust in yourself—learning to feel again, to believe that your emotions matter—is one of the hardest parts of recovery. It's about reminding yourself that your thoughts, feelings, and memories are valid, even if the narcissist tried to convince you otherwise. It's also about rediscovering your true identity, the person you were before the abuse took hold.

For me, therapy was the first step in reclaiming my fragmented self. Through the process of healing, I learned to recognize the patterns of narcissistic abuse and understand how they had shaped my sense of identity. It took time—years, in fact—but each day I began to feel more like the person I was before. A person who was worthy of love, respect, and peace.

Healing from narcissistic abuse is not a linear process, but fragmentation is not permanent. With patience, support, and self-compassion, it is possible to heal, reclaim your identity, and rebuild the life that was taken from you. It is not about forgetting the past—it's about recognizing it, accepting it, and choosing to move forward with a renewed sense of self.

Shattered from the Trauma Bond

Breaking free from a narcissistic relationship is never as simple as walking away. The emotional and psychological damage doesn't vanish with physical separation. Instead, it lingers in the form of deeply ingrained trauma bonds that can feel impossible to break. The concept of a trauma bond—coined by Dutton and Painter

(1993)—describes a psychological attachment that victims form when they are subjected to a cycle of love and cruelty. The narcissist's manipulation is so subtle yet profound that it fosters a sense of dependency on the occasional bursts of affection they offer, making it nearly impossible for victims to see their own worth outside of the narcissist's validation.

This cycle of intermittent reinforcement, where moments of love or attention are alternated with cruelty or neglect, traps the victim in a perpetual state of emotional confusion. It creates a distorted sense of reality, one where love feels like control, and rejection feels like an emotional death. The narcissist becomes the center of the victim's emotional universe—and in that twisted dynamic, the victim believes they need the narcissist to feel whole. The victim's self-worth becomes so closely tied to the narcissist's validation that leaving feels like an impossible feat.

Even after physically separating from my ex-husband, the grip of the trauma bond remained tightly woven around my heart and mind. I was emotionally tethered to someone who had broken me down for years, yet I couldn't shake the belief that maybe, just maybe, his behavior would change. It wasn't until I truly understood the nature of narcissistic abuse and the psychology behind intermittent reinforcement that I began to unravel the mess he had made of my sense of self.

The Power of No Contact: A Critical Tool in Breaking the Bond

One of the most powerful tools in my recovery has been No Contact. The process of cutting off all communication with the narcissist—whether it's through phone calls, text messages, social media, or even indirect communication through third parties—

creates a necessary emotional and psychological distance that allows healing to begin.

When I first initiated No Contact, it felt like a drastic and even painful step. The trauma bond made every instinct in me want to reach out, to seek validation, or to just know if he was okay. These urges stemmed from years of being conditioned to believe that my worth was tied to his perception of me. But, as I moved further into recovery, I began to understand that No Contact was not about revenge or punishment. It was about protecting myself from further harm and establishing the boundaries I had never been able to enforce before.

The absence of contact allowed me to disconnect from the constant manipulation and confusion. It was a process of reclaiming my autonomy, a way of telling myself that I no longer had to subject myself to his emotional chaos. At first, it was extremely difficult. The narcissist uses many tactics to break No Contact—whether it's hoovering, guilt-tripping, or creating drama to draw the victim back in. But I had to remind myself daily that breaking free meant staying free, even when the pull of the bond felt unbearable.

Embracing Boundaries: Reclaiming Control

No Contact isn't just about avoiding the narcissist—it's about setting new boundaries. Narcissistic abuse thrives in an environment where boundaries are non-existent or constantly violated. Victims are conditioned to feel guilty for asserting their needs, and even when boundaries are set, they are often ignored, belittled, or punished by the narcissist. In my recovery, I had to relearn how to set firm boundaries with myself and others. It was about learning to say "no" without fear of repercussions, without apologizing for existing as a person with needs, desires, and limits.

One of the biggest lessons I learned during this process was that boundaries are an act of self-love. Prior to my escape, I had believed that love meant sacrificing my own needs to make someone else happy. But in recovery, I realized that true love—real self-love—requires the preservation of self. I had to trust that I was worthy of love that didn't come at the expense of my dignity and peace of mind.

Therapy and Support: The Role of Healing Communities

Therapy and support groups became a vital part of my healing process. Narcissistic abuse is isolating, and without proper support, victims are at risk of falling back into the cycle of manipulation. One of the first steps in therapy was recognizing that the abuse wasn't my fault, that I wasn't crazy for feeling the way I did, and that the narcissist's behavior was not the result of "mistakes" or "misunderstandings," but a deliberate and destructive pattern of manipulation.

Support groups also played a crucial role in my recovery. For years, I had felt alone in my experience, like no one could truly understand the depth of emotional turmoil I had gone through. But in support groups, I found a community of people who had gone through similar experiences—people who knew what it was like to feel shattered by another person's cruelty. Being able to share my story, to listen to others, and to be heard, helped me begin to rebuild the fractured pieces of myself.

The Path Forward: Rebuilding Trust and Rediscovering Self-Worth

Healing from narcissistic abuse is not a linear process. It's a journey filled with ups and downs, victories and setbacks. I had to learn that recovery wasn't about "getting over it" quickly—it was about gradually rebuilding who I was before the abuse. For years, I had been emotionally stunted, unable to trust my instincts, my choices, or my feelings. But over time, I began to reconnect with the woman I once was—the person who was confident, strong, and whole.

Rebuilding trust—especially trust in myself—was a major hurdle. For so long, I had doubted my own perceptions, my thoughts, and my decisions. I had been gaslighted, manipulated, and controlled. But as I moved through therapy and continued to practice No Contact, I began to rediscover my ability to trust my intuition. I started to make decisions based on what was right for me, not what would please someone else. My self-worth, once destroyed, began to revive, piece by piece.

Recovery from narcissistic abuse is ultimately about reclaiming your identity and finding strength in your own voice. It's about looking back and recognizing the damage, but also looking forward and knowing that the pieces of yourself that were lost can be found again. The trauma bond might feel impossible to break at first, but with the right tools—No Contact, therapy, support, and time—the shattered self can slowly be pieced back together.

The Shattered Self: Breaking Free from the Trauma Bond (Continued)

Recovery from narcissistic abuse is ultimately about reclaiming your identity and finding strength in your own voice. It's about looking back and recognizing the damage, but also looking forward and knowing that the pieces of yourself that were lost can be found again. The trauma bond might feel impossible to break at first, but with the right tools—No Contact, therapy, support, and time—the shattered self can slowly be pieced back together.

While No Contact was an essential tool for distancing myself emotionally and psychologically from my ex-husband, there were other techniques that helped me solidify my sense of autonomy and reduce the emotional toll narcissistic manipulation continued to take. One such method was Gray Rocking—a powerful yet subtle way of protecting yourself from a narcissist's emotional entanglements.

Introducing Gray Rocking: Emotional Detachment as Self-Preservation

Gray rocking is a technique that involves becoming as emotionally unresponsive, dull, and boring as possible when interacting with the narcissist. The idea is to be like a "gray rock"—something so dull and lifeless that it no longer piques the narcissist's interest. For someone who thrives on emotional reactions, whether positive or negative, the narcissist will often discard a person who no longer offers them a "reactionary supply." By adopting a neutral, unemotional stance, you create an emotional barrier, making it harder for the narcissist to control or manipulate your feelings.

When I first used gray rocking, it was difficult—years of emotional dependency and the urge to "fix" things made it challenging to resist engaging. But as time passed, I realized that my silence and lack of emotional response were far more effective in breaking his hold on me than any argument or plea for understanding. Every time he tried to provoke a reaction—whether through manipulation, guilt-tripping, or baiting me into emotional chaos—I simply refused to play along.

This emotional detachment wasn't about being cold or callous, but about creating a protective shield. It helped me regain control over my emotional landscape by ensuring that I didn't feed into the narcissist's drama or cruelty. Over time, this strategy became a powerful way to minimize his ability to drain my energy and steal my peace. He stopped getting the emotional supply he craved, and without it, his attempts to manipulate became less effective. Gray rocking became my subtle yet resilient way of saying, "I will no longer let you control my emotions or thoughts."

Hiding Emotions: Mastering Emotional Boundaries

Another technique I employed, especially during times when direct contact couldn't be avoided, was hiding my emotions or keeping them private. Narcissists excel at reading others, often using this ability to exploit vulnerabilities or gain leverage. By guarding my emotions, I made it impossible for him to use my feelings against me. This wasn't about repressing my emotions or pretending they didn't exist, but about not allowing him access to them.

Keeping my emotions hidden from the narcissist allowed me to protect my vulnerability and maintain a sense of control over my internal world. Instead of allowing him to fish for emotional responses, I learned to filter my reactions and emotions, keeping them internal—a strategy that limited his power to manipulate me. As I practiced this more, I grew increasingly aware of how much control he had gained over my emotions in the past. By refusing to let him see my inner turmoil or happiness, I regained a sense of self-sufficiency and autonomy.

Empowerment Through Emotional Independence

Ultimately, the key to healing from narcissistic abuse is learning how to maintain emotional independence. Both No Contact and techniques like gray rocking and hiding emotions are powerful strategies for creating the emotional distance that is essential for recovery. But beyond these tools, the real shift comes when you realize that you are no longer dependent on the narcissist for your sense of self-worth, validation, or even emotional stability.

Every time I successfully implemented these techniques, I found myself reclaiming more of the self that had been fragmented. I had been conditioned to believe that I needed him to validate my existence, to prove my worth, and to make me feel seen. But through consistent use of these methods, I began to recognize that my worth was never tied to his approval—it was intrinsic, and I had the power to protect it.

As I continued to grow stronger emotionally and psychologically, I learned that healing is a process of rebuilding—not just cutting ties, but rediscovering the pieces of yourself that had been lost. The strategies I employed were not just tools for survival, but stepping stones toward wholeness—toward becoming the person I was always meant to be, without the weight of narcissistic abuse holding me down.

Emotional Depletion and Dysregulation: The Road to Transformation

The emotional toll of narcissistic abuse is not only measured by the immediate trauma but by the gradual depletion of one's emotional resources over time. The long-term effects of manipulation, gaslighting, and psychological warfare create an environment of emotional dysregulation—a state where you no longer know how to process or manage your emotions in a healthy, balanced way. You become caught in a cycle of fear, guilt, confusion, and self-doubt, all while the narcissist's control tightens around you. It's as though every emotional response becomes a reaction, driven by the narcissist's manipulation, instead of being an expression of your own authentic self.

In my own experience, this emotional depletion was often felt in waves. There were days when I felt completely drained, unable to summon any emotional energy to take care of myself or even function normally. The emotional dysregulation that resulted from prolonged abuse made it nearly impossible to trust my own instincts or decisions. I would feel anxious, overwhelmed, and, at times, completely numb, unsure of whether I was overreacting or if my feelings were valid at all. The narcissist's emotional manipulation had left me fragmented—unable to access my true emotions without the filter of fear or insecurity.

But through this process, I learned that emotional depletion is not a permanent state—it is a result of being constantly drained by external forces. Recovery came not just through detachment but through rediscovering the ability to regulate and trust my emotions again. This process took time, and it involved healing emotional wounds, developing boundaries, and redefining my relationship with myself. It was not about "fixing" what had been broken, but about learning to rebuild from a place of understanding and self-compassion.

Reclaiming Self-Regulation: Moving Toward Emotional Empowerment

The ultimate goal in recovering from narcissistic abuse is emotional empowerment—the ability to not only recognize and manage your emotions but to create an internal space where self-regulation becomes second nature. Over time, I came to understand that my emotions weren't a reflection of the narcissist's control, but an important part of who I am. Healing from the emotional depletion caused by narcissistic abuse is about reclaiming that internal compass—learning to trust your gut, validating your own experiences, and taking back the emotional authority that was once hijacked by the narcissist.

The process of self-regulation isn't just about managing emotions in the moment. It's about changing the way you interact with your emotions on a deeper level—allowing them to inform your choices without letting them dictate your actions. This transformation is incredibly liberating because it shifts the focus from the external chaos created by the narcissist to the internal sense of calm and empowerment that comes from taking charge of your emotional world.

For me, this was a gradual process that involved therapy, self-reflection, and the consistent practice of tools like No Contact, gray rocking, and emotionally distancing myself from triggers. But it wasn't just about surviving the abuse anymore—it was about creating a new way of being. I moved from a state of emotional chaos to one where I could make choices based on self-trust and personal alignment, rather than being controlled by external manipulation.

Emerging from the Ashes: A New Emotional Foundation

Emerging from emotional depletion is like rising from the ashes of your old self. In many ways, the narcissist's influence had left me feeling like a shell of who I once was. But with each day that I took small steps toward healing, I began to rediscover parts of myself that had been buried beneath the weight of abuse. I found my confidence again, my voice, and my sense of self—all of which had been systematically taken from me over the years.

The emotional depletion I once felt slowly turned into an emotional resilience that I hadn't known I was capable of. What had once been broken became a source of strength—each scar a reminder that I had endured, and each lesson a tool for creating the life I had always deserved.

This shift from emotional depletion to empowerment is the true transformation that comes with recovering from narcissistic abuse. It's about changing the way you relate to yourself and others. Where once you may have been drained by external forces, you now build an inner strength that comes from a healthy relationship with your emotions. The more I practiced self-care, self-compassion, and emotional regulation, the more I could see the future I wanted to create—one that wasn't dictated by fear, guilt, or narcissistic control.

Chapter 8

The Breaking Point: The Narcissistic Injury and Psychological Breakdown

The day I decided to leave him was the day I realized I could no longer sustain the emotional, psychological, and mental toll his manipulation had on me. For years, I had sacrificed my own sense of self to appease his volatile temperament, desperately seeking to meet his needs in the hope that he would love me back. But I reached a point where I could no longer keep up the facade. I had tried everything to make the relationship work, but his manipulation and control had suffocated me. Leaving him became a necessary act of self-preservation—a chance to reclaim my autonomy and rebuild my life.

What I didn't understand at the time, however, was how deeply my departure would affect him. To me, leaving was a courageous step toward healing. But to him, it was a direct challenge to his narcissistic sense of superiority. In his eyes, I was not just leaving a man, but attacking his ego and shattering the control he had meticulously built over me. For someone with narcissistic traits, control is everything. My exit wasn't just a breakup; it was a blow to the fragile self-image he had spent years constructing.

This emotional wound he experienced can be understood as what psychologists call a narcissistic injury (Kernberg, 1975). Narcissists derive their self-worth from external validation and their ability to dominate others. When they feel rejected or devalued, it triggers a profound emotional breakdown. For my ex-husband, my departure wasn't a personal choice—it was an affront to his ego, a reflection of his inability to maintain his perfect narrative of control. The illusion of superiority he had built around himself crumbled, and in its place, a raging need to reassert his power took over.

This is where narcissistic rage comes in. Once his ego was injured, it unleashed a torrent of intense emotional volatility (Dutton & Painter, 1993). This was not a temporary anger. His rage wasn't fleeting—it was all-consuming, leaving no room for reason or empathy. He couldn't accept that my leaving was simply a personal decision to regain control of my life. Instead, it became a direct attack on his self-image. To him, it wasn't just about the end of a marriage—it was about his loss of control and his fear of being rejected by someone he believed should never leave him. The emotional fallout from this injury became the fuel for his obsession with revenge.

The rage and obsession that followed my departure were not merely about getting me back—they were about his need to reclaim control over the narrative and punish me for daring to challenge his authority. The more I distanced myself, the more his obsession grew, and what followed was a relentless cycle of retaliation, fueled by his fragile ego.

The Descent Into Obsession: Narcissistic Rage and Coercive Control

Once I left, my ex-husband's behavior escalated into unrelenting obsession. The behaviors he exhibited were not simply random acts of emotional instability; they were part of a deliberate and calculated strategy to reassert dominance. Narcissistic individuals often feel entitled to the lives of those they abuse, believing that they are owed something—especially after a breakup (Miller, Lynam, and Hyatt, 2011). To him, my leaving was unacceptable, a betrayal that required punishment.

The stalking, the relentless harassment, and the emotional manipulation were all part of his attempt to reclaim control over me. Narcissists are deeply insecure and, therefore, cannot tolerate rejection—and when it happens, they react with obsessive vengeance. For him, this wasn't about closure—it was about reasserting the dominance that had been stripped away. In his mind, I wasn't just a person he once loved; I had become an object to be possessed, a prize that needed to be reclaimed.

This behavior is closely tied to the concept of coercive control (Stark, 2007). Coercive control is a pattern of manipulation and domination where the abuser seeks to limit the victim's freedom, isolate them, and break their will. While coercive control doesn't always manifest as physical violence, it creates an environment of psychological terror that keeps the victim in a constant state of fear, vulnerability, and emotional depletion. In my case, his obsession became a psychological warfare that aimed to diminish my sense of self and re-establish his control over my life.

It wasn't enough for him to just manipulate me into coming back. No. He used every weapon in his arsenal—stalking me, violating my privacy, and even attempting to undermine my sense of safety by getting others involved. The constant psychological attacks left me on edge. I lived in a perpetual state of hypervigilance, unable to trust myself, my surroundings, or the people around me. The fear of what he might do next haunted me at every moment.

The Psychological Fallout: Trauma and Complex PTSD

Living in that environment of constant emotional manipulation and control took a toll on my mind and spirit. The trauma was not only psychological—it was emotional and physical as well. After years of enduring his relentless tactics, I began to feel like a shell of myself. I had lost touch with who I was—my identity had become so entangled with his control that I no longer knew how to define myself.

I developed Complex PTSD (C-PTSD) (Briere & Scott, 2015), a condition resulting from prolonged exposure to relational trauma. C-PTSD is often caused by abusive relationships where the victim is subjected to constant emotional turmoil. Unlike traditional PTSD, which often arises from a single traumatic event, C-PTSD stems from the cumulative effect of long-term abuse. The emotional toll was overwhelming—my sense of self became fragmented.

Living under constant threat, I found myself questioning everything—my emotions, my perceptions, and even my memories. This confusion and instability were central to the emotional devastation I felt. I was emotionally paralyzed, always second-guessing myself, unsure of whether my feelings were real or just a result of his manipulation. I had no stable foundation on which to stand.

The emotional turmoil was profound: my hypervigilance kept me constantly on edge, living with an ever-present sense of fear. Even after I physically left, the mental scars from his abuse lingered. The psychological manipulation had rewired my brain, leaving me unable to trust my instincts, my perceptions, or my emotions. The damage went deeper than I had realized—it wasn't just about escaping the relationship; it was about reclaiming my own mind.

The Narcissistic Cycle: Gaslighting and Control

One of the most insidious tools my ex-husband used to destabilize me was gaslighting. Gaslighting is a form of psychological manipulation that causes the victim to question their perception of reality. It's a tactic that narcissists use to undermine the victim's sense of self and confuse their perception of events.

My ex would frequently twist the narrative, convincing me that my memories were flawed or that I was overreacting to his actions. He would attempt to distort reality in such a way that I began to feel like I was the problem. He would claim that I was being irrational or too sensitive, all the while undermining my self-esteem and making me doubt my own sanity. This emotional manipulation is exhausting—it makes you feel like you're walking on shifting sands, never knowing what's real and what's part of his distorted version of the truth.

His gaslighting was just one of many psychological warfare tactics. The goal was never to come to terms with the breakup; it was about reasserting his dominance over my life, mentally controlling me. It wasn't about love—it was about him keeping me in his orbit, continually doubting my reality to make me dependent on his version of things.

The Unraveling: Obsession and Entitlement

The deeper his obsession grew, the more his entitlement became apparent. Narcissists often feel that they are owed something—that they deserve admiration, respect, and even the right to control others. In my ex-husband's case, he believed that, even though I had left him, I was still somehow his to control. He was consumed by the belief that he had the right to dictate the terms of our relationship—whether I was physically with him or not.

The obsession wasn't just about getting back together. It was about possessing me—owning me, mind and soul. I was no longer a person to him; I was an object—an extension of his ego that he couldn't bear to lose. This constant sense of entitlement and ownership led to his escalating obsession. The more I tried to escape, the harder he fought to pull me back into his world. It was as if I was never meant to be free—my freedom threatened the very foundation of his narcissistic identity.

The Cycle of Abuse: A Never-Ending Battle

What makes narcissistic abuse so insidious is the cyclical nature of the violence. It never truly ends—it lingers, festers, and continues to unfold over time. Even though I had physically left, my ex-husband's need for control had not dissipated. In fact, it intensified. This was not simply a relationship; it was a battle over power.

His desire to control me remained unbroken, even after the relationship ended. My independence became a direct threat to him—a challenge he could not, and would not, tolerate. The ongoing gaslighting, stalking, and psychological warfare continued for years after our separation. It was as if my exit had unleashed a fury that could only be quelled by breaking me down, piece by piece, until I no longer existed outside his narcissistic narrative.

This journey through narcissistic abuse was an emotional and psychological war, fought on the battlefield of my mind. Every moment of freedom was met with retaliation, every step toward independence shadowed by his constant need to control. Leaving him was not the end of the story—it was merely the beginning of a long, painful fight to reclaim myself from his grasp.

The road to healing is ongoing, but the battle for freedom, autonomy, and mental peace continues. The psychological scars may never fully disappear, but they have forged a resilience within me that no longer allows me to be defined by his obsession, his control, or his narcissistic rage. The only way out of this endless cycle is through it—and reclaiming my life, one step at a time.

More Digital Tactics: The New Frontier of Control

When the relationship physically ended, my ex-husband's desire for control didn't dissipate; instead, it evolved into something more pervasive and elusive: digital manipulation. In the digital age, where communication happens at the speed of light and online presence is nearly inescapable, the narcissist's tactics take on a new form. What once required physical proximity could now be executed from a distance, through messages, social media, and a constant barrage of virtual engagement. This gave him the freedom to stalk, harass, and undermine me without ever having to leave his home.

The abuse was no longer visible to the outside world, making it even more insidious. While in the past, he had used in-person threats, angry outbursts, and public displays of control, he now resorted to subtle and calculated online tactics that blurred the line between normal communication and manipulation. It became clear: just because I physically left didn't mean I was free.

The Reach of Digital Control: Beyond Physical Boundaries

With the rise of digital communication, the tools of narcissistic control extended well beyond the confines of a face-to-face interaction. Text messages, phone calls, and social media became his new avenues for manipulation. Where I had once been able to physically distance myself from him, the constant digital presence allowed him to remain a part of my life in ways that were almost impossible to escape.

At first, these interactions were mostly text messages or emails, which I initially dismissed as "just communication." I thought he was trying to talk, to clear things up, or perhaps make amends. But soon, the messages began to grow more insistent, more subtle in their manipulation, always casting me as the problem, or implying that my decision to leave was a mistake. This was gaslighting in its most covert form. The digital tools made it easier for him to infiltrate my emotional boundaries, especially when I had already felt so emotionally fragile.

The harassment didn't stop there. He began creating fake social media profiles to monitor my life, pretending to be someone else so he could see what I was doing and who I was interacting with. The creation of these false identities was another method of control—he could keep tabs on me, access my private moments, and monitor my relationships under the guise of anonymity. The deception and surveillance became part of his repertoire, making me feel like I was constantly being watched, even when I was supposed to be alone.

The Watcher: Narcissistic Smear Campaign and the Digital Hunt

Leaving him physically didn't mean I was free. In fact, the digital hunt began the moment I thought I had escaped his grip. With the rise of digital communication, narcissists like him can now extend their reach far beyond the confines of physical presence. Text messages, phone calls, and social media became his new playground, a place where he could constantly infiltrate my life and erode my privacy. Where I once had the semblance of physical distance, I now found myself tethered to him in the most invasive way possible—through my phone, my social accounts, my emails, my every online interaction.

I felt like I was always being watched, like I could never completely escape him. It wasn't just about the texts or emails anymore. At first, I thought these were just his attempts to communicate—perhaps to clear things up or maybe to make me feel guilty for leaving. But over time, the tone shifted. The messages became more insistent, subtle in their manipulation, and ultimately gaslighting in nature. He would try to convince me that my decision was wrong, that I was the one causing chaos. He was trying to warp my reality again, to make me second-guess myself, and the most insidious part was that he now had the tools to do this 24/7.

The digital space became his new form of control, and it wasn't limited to mere messages. He began to create fake social media profiles—anonymous accounts designed to monitor my every move. He pretended to be others—friends, acquaintances, even family—and stalked my online presence. His surveillance was methodical. He knew what I posted, who I spoke to, what I was doing at any given moment. His ability to be a "shadow" in my digital life created a constant, suffocating pressure. The idea that he was still out there, watching, even when I thought I was safe in my own space, made it feel as though there was nowhere I could hide.

But his digital presence didn't stop at watching. He launched a full-scale smear campaign. The damage wasn't just psychological; it was social, relational, and emotional. My children, my friends, even my old social groups became his target. He spun a narrative that painted himself as the hero, the victim, and me as the deadbeat mom or the unstable one—whatever it took to maintain his inflated sense of self-worth. He sought to turn everyone against me, to create doubt, and to make sure that I had no support. No ally was safe from his manipulation.

His goal was clear: to obliterate my reputation and make sure that even when I wasn't there, people were still talking about me in a negative light, questioning my character, and ultimately isolating me. What made this particularly chilling was how believable his version of events was. Narcissists have a way of twisting the truth, painting themselves as flawless victims while casting their targets as unworthy, malicious, or erratic.

The Smear Campaign: Narcissistic Manipulation and Digital Warfare

When I finally left him, it wasn't just a breakup—it was a declaration of independence. Yet, what followed was an insidious, calculated attempt to destroy me, not physically, but emotionally, socially, and psychologically. He immediately launched a smear campaign aimed at undermining my reputation and isolating me from my support system. It was as though the moment I became free, he saw my autonomy as a threat to his fragile self-image, and he had to do everything in his power to tear me down in the eyes of others.

Narcissists, by nature, need to be in control. When they lose that control, especially in situations like a breakup, their sense of self is threatened. To compensate for this, they often resort to smear campaigns—a form of psychological warfare aimed at ruining their victim's reputation and casting themselves as the innocent party. He did exactly that. As I pulled away and tried to rebuild my life, he went on the offensive, trying to make sure that no one saw me as a victim, but instead as the problem, the unreasonable one, the unstable partner.

The smear campaign wasn't just about gossip or isolated comments; it was a calculated, manipulative strategy meant to isolate me, control how people viewed me, and ensure I had no support when I needed it most. The online platforms—social media, text messages, emails—became his new playground for manipulation. His goal wasn't simply to hurt me; it was to obliterate my reputation, to ensure that anyone who came into contact with me would view me with suspicion, question my character, and see me as the person who had failed, who had abandoned, who had destroyed what we had.

Twisting the Truth and Playing the Victim

One of the most chilling tactics he used was twisting the truth. He would take small, isolated incidents, distort them, and present them as evidence of my so-called unpredictability or unstable behavior. Anything that painted him in a bad light was buried, and anything that painted me as unreasonable or emotionally volatile was inflated to make me seem like the villain. He was the victim in every narrative, the one who was wronged—and I was the selfish one who had torn apart a family.

This was his favorite tactic—playing the victim. It allowed him to divert attention from the emotional abuse he had inflicted on me, the constant gaslighting, and the manipulative behavior that had driven me to leave in the first place. Instead of acknowledging his part in the breakdown of our relationship, he framed the entire situation as though I had been the one causing all the problems. He sought sympathy from friends, family, and even our children by spinning a narrative in which I was a deadbeat, someone who didn't care about the family or the future we had built together.

Projection: The Accusation of Behaviors He Himself Was Guilty Of

What he didn't realize—or perhaps he didn't care—was how his own tactics of projection began to expose his true character. Narcissists often project their flaws onto others, accusing their victims of the very behaviors they themselves are guilty of. So, as he painted me as the manipulative one, the abusive one, or the unstable one, I began to see how much of his own behavior was mirrored in these accusations. It was a classic narcissistic move—accuse the victim of the very tactics and actions that the narcissist themselves regularly engage in.

In his smear campaign, he cast himself as the martyr—someone who had been wronged by a partner who didn't appreciate him, didn't understand his needs, and had the audacity to leave. In reality, he had abused and manipulated me for years, gaslit me, and created a toxic environment. But the smear campaign was his way of deflecting attention from his own flaws, and of protecting his fragile ego by redirecting blame toward me.

The Role of "Flying Monkeys": Turning Others Against Me

A truly disturbing part of his smear campaign was his ability to turn others into his "flying monkeys"—people he had manipulated or coerced into doing his bidding. These were not just acquaintances; these were family members, old friends, and people in my circle whom he had twisted into supporting his narrative. He groomed them to view me as a threat, a crazy, emotional person who could not be trusted.

The flying monkeys would then repeat his lies, amplify his version of events, and further isolate me from anyone who might have supported me. These people were instrumental in ensuring that his manipulation continued to work and that his false narrative spread. They became his army, spreading rumors, confirming his victimhood, and essentially pushing me into emotional exile.

The Emotional Toll: Isolation and Betrayal

This smear campaign was emotionally devastating. It wasn't just the public humiliation; it was the emotional isolation. The people who once knew me, who once stood by my side, were now questioning my character, turning against me because of the lies he spread. He knew that by controlling the narrative, he could systematically remove any source of support I had. The worst part was that many of these people believed his lies—they didn't see the behind-the-scenes manipulation, the psychological abuse, or the years of emotional turmoil. They only saw his charming façade, his ability to manipulate their emotions, and the "evidence" of my "craziness."

What hurt the most was knowing that the people I loved were now part of this campaign, not because they were inherently bad, but because they had been manipulated by him. I was left to defend myself, often feeling like I was fighting an unwinnable battle—because no matter how much truth I spoke, his lies had already taken root.

How to Cope: Protecting Your Peace

Dealing with a narcissist's smear campaign is emotionally exhausting. The best way to handle it, as hard as it is, is not to engage. Responding, trying to justify myself, or fighting back only fed his need for control and gave him more ammunition to continue the manipulation. The more I defended myself, the more he would twist the facts, accuse me of being defensive, and reinforce his narrative that I was unhinged.

The most important tactic was disengagement—using the "Grey Rock" method: becoming emotionally unavailable and non-responsive to his attempts to provoke me. I had to learn to withdraw from the cycle and stop responding to his emotional bait. By refusing to engage in his drama, I slowly removed the fuel for his fire.

I also kept a detailed record of everything—every email, text message, and interaction. This documentation wasn't just for me; it was for proof if the situation escalated to legal matters or if I needed to defend myself. It helped me stay grounded in facts, which became my anchor in the midst of his emotional chaos.

Finally, leaning on my real support system was crucial. As the smear campaign progressed, I had to distance myself from the flying monkeys and turn to the people who knew me for who I truly was. Over time, the truth started to surface as inconsistencies in his story became clear to others. But the emotional scars of his smear campaign were deep, and they took time to heal.

The Endurance of a Narcissistic Smear Campaign

Narcissists like my ex don't stop at just spreading lies; they aim to break you, to isolate you, and to ensure they maintain power over your life—even after the relationship ends. His smear campaign wasn't just an attack on my reputation; it was an extension of his need for control. The emotional cost of this campaign was immense, but over time, I began to see that his narrative was crumbling.

Truth has a way of surfacing, and while it may take time, the narcissist's lies will inevitably be exposed. The damage he caused may not be easily undone, but by holding firm to my integrity, refusing to participate in his manipulation, and leaning on those who knew the truth, I eventually began to rebuild. Reclaiming my life, my reputation, and my peace of mind took time, but it was worth it.

Cyberstalking: The Shadow That Never Left

What I hadn't anticipated when I left him physically was that his influence wouldn't end with the separation. As much as I distanced myself from him in person, he found new ways to infiltrate my life. The cyberstalking began in earnest, and it was relentless. It wasn't just the occasional text message or phone call, or the passive-aggressive emails where he'd try to emotionally bait me—it was the constant, creeping presence in my online world. Every time I posted something personal on social media, he was there. Sometimes, he commented publicly, other times he messaged me privately under his own name or, more disturbingly, under fake accounts he had created to mask his identity.

At first, I tried to brush it off, telling myself it was just his way of trying to maintain a connection. But the more I saw, the more I realized how deep his obsession went. The messages varied in tone. Some were subtly passive-aggressive, dripping with condescension. Others were outright accusatory, suggesting that I was "wrong" or "ungrateful" for leaving him. And then there were the fake concern messages, where he would pretend to be caring or worried about my wellbeing, feigning empathy in the hopes of reopening lines of communication. But no matter the guise, the goal was always the same: to remind me of his presence, to make me feel like I could never truly escape him.

The digital harassment felt like a constant reminder that no matter how far I ran, I was still tethered to him. Each new message, each new comment, felt like another brick in a wall he was building around me. It made me question my safety, my autonomy, and worst of all, my own perception of reality. I started to feel paranoid, wondering if everyone around me was somehow unknowingly complicit in his games. Were my friends secretly feeding him information? Were my new relationships being manipulated by him from the shadows? Was my world really as separate from him as I thought, or was he somehow still controlling my narrative?

This digital invasion of my personal space was a form of coercive control—a psychological tactic meant to create an environment where I was never really free. The more I attempted to assert boundaries, the more he pushed back, constantly reminding me that he still had a hold over me, even if it was now invisible and digital.

It's hard to explain just how disorienting this felt. No matter how many times I blocked him, no matter how many accounts I deleted or privacy settings I tightened, there was always another way for him to find me. The shadow of his presence lingered in every corner of my digital life, and I started to feel like I was living in a constant state of hypervigilance, always watching over my shoulder, even when I was in the comfort of my own home. The relentless pursuit of my personal information was a tool of control. It wasn't enough for him to know what I was doing in real life—he needed to control what others thought of me too, so he turned to the online world as his new weapon of choice.

But cyberstalking was not just about him keeping tabs on me—it was about keeping me isolated and discrediting me in the eyes of others. The smear campaign he started in real life bled into the digital realm. Every new attempt at disrupting my life—whether through fake profiles, manipulated messages, or public insults—was part of a strategy to isolate me, to prevent me from forming any new connections or reinforcing existing bonds. It was about destroying the truth and fabricating a new reality where I was the enemy, where I was the crazy, unstable one who had walked away from a good man.

How to Deal with a Narcissist's Smear Campaign in the Digital Age

Dealing with the fallout from a narcissist's smear campaign is not easy, especially when the campaign is carried out in the digital realm, where a narcissist can hide behind screens and create false narratives that are hard to combat. But it is possible to protect yourself and regain control of the narrative.

Here are some strategies that helped me navigate the murky waters of cyberstalking and smear campaigns:

1. Maintain Your Integrity

It may be tempting to retaliate—especially when the lies are so outrageous—but engaging with a narcissist in this way only plays into their hands. Don't retaliate with similar tactics. Narcissists thrive on drama and chaos, and when you retaliate emotionally, you give them more fuel to continue the cycle. Stay calm, and hold on to your truth. Maintain your integrity and let time reveal the true nature of their manipulation.

2. Disengage

Disengaging from a narcissist, especially in the digital space, can be difficult, but it is essential for your peace of mind. Minimize communication. If you can, consider using the Grey Rock Method, where you remain emotionally neutral and unresponsive, avoiding any emotional reaction that could trigger further manipulation. In my case, this was incredibly hard, but by sticking to short, unemotional responses, I made it clear that I wasn't engaging with his emotional bait. If possible, have a third party or a lawyer handle communications for you. This can create clear boundaries and help prevent the situation from escalating.

3. Document Everything

The digital realm leaves behind a trail—emails, texts, social media posts, and messages. Keep a detailed record of all interactions. Save screenshots of online harassment, preserve any text messages, and document the dates and times of contact. This documentation will be invaluable if you ever need to refute false claims or seek legal recourse. If you can, limit communication to written forms like email or text, where you can have a factual record of exchanges.

4. Seek Support

Narcissists often try to isolate their victims, but build a support network of people who understand the situation and can offer emotional support. Don't be afraid to reach out to those who have known you for years, or to professional therapists who can help you navigate the emotional turmoil that comes with cyberstalking and smear campaigns. Legal advice can also be crucial if the harassment escalates.

5. Protect Your Physical and Mental Health

Narcissistic abuse—especially when compounded by digital harassment—can take a significant toll on your mental health. Take care of your nervous system by engaging in activities that help you relax and stay grounded. Yoga, deep breathing, or mindfulness exercises can help you regulate your emotions and cope with the anxiety that comes from living in a state of constant vigilance. Having a therapist or trusted friend to talk to can help alleviate the emotional weight.

As time passed, I realized that while the smear campaign and cyberstalking were exhausting, they didn't define me. I held on to the knowledge that the truth was mine to keep. His lies couldn't stand forever. The truth always comes out in the end, especially when the narcissist's manipulation becomes too obvious to ignore. By sticking to my integrity and focusing on healing, I began to reclaim my peace and control.

The Obsession with Control: From Smear Campaigns to Reclaiming My Narrative

Perhaps the most damaging of all his tactics was his relentless pursuit to control my narrative. Narcissists are masters of rewriting reality, and he used every tool at his disposal to manipulate how others saw me—and ultimately how I saw myself. He weaponized the digital space, creating fake profiles, anonymous accounts, and sending misleading messages in an effort to twist the story of our breakup. His goal was simple yet devastating: to portray me as the one at fault, the "unstable" one, the person who had betrayed him.

What made his campaign so insidious was that the digital world gave him a platform to spread his version of events without any immediate consequence. In face-to-face conversations, his lies and distortions could be challenged, but online, his narratives were just words on a screen—words that, when shared, had the potential to reach far and wide without anyone questioning their authenticity. This was his perfect stage: a space where his lies could grow without confrontation.

But his manipulation went beyond just controlling me; it was about controlling how others saw me. The smear campaign wasn't confined to just the digital world—it extended into my relationships, infiltrating every conversation, every reunion, every attempt to rebuild my life. When I reconnected with friends or family, I often heard whispers of the rumors he had spread: how I was crazy, how I was unstable, how I was ungrateful for everything he had done. These were not just falsehoods; they were deliberate attempts to discredit me in the eyes of the people I cared about. The cruelest part was that these seeds of doubt were planted in the most vulnerable places—among my support network, the people I trusted most.

The digital tactics he employed allowed him to manipulate perceptions from a distance. I could no longer trust that those around me had an accurate understanding of who I was or what I had been through. The digital sphere made it all too easy for him to create a narrative where I was the villain, with no accountability for his abusive behavior. He didn't need to face me directly; he could simply log on, type out a few well-crafted lies, and let the world believe whatever he wanted them to.

What made this psychological warfare even more insidious was how it left me isolated. I was physically free from him, but I was never truly "alone" in the digital world. Every text message, every comment on social media, every email felt like another intrusion into my life. It wasn't just the occasional message—it was a constant presence that followed me around, even when I was trying to build a life apart from him. The emotional toll was overwhelming. It was as though I was always under surveillance, always being reminded of his existence and his need to control me.

The digital space created a distance that allowed him to say things he would never dare to say in person—words that were venomous and hurtful, but without any immediate repercussions. And yet, this very distance also made it harder for me to defend myself or set boundaries. When you're dealing with anonymity online, responding emotionally or confronting him in a way that might bring clarity was a lose-lose situation. It felt like I was always one step behind, trying to chase down lies that had already been planted.

The hardest part wasn't just defending myself—it was watching the damage unfold in real-time. Every time a friend or family member questioned me, every time they asked if something he had said was true, it felt like I was losing control of the narrative. It was like a ghost stalking my life, using digital tools to keep me entangled in his web. Each new tactic chipped away at my sense of self-worth, constantly reinforcing the idea that I was guilty until proven innocent, even when I wasn't sure how to prove it.

Reclaiming the Narrative: Tools to Break Free from the Digital Grip

In the midst of this storm of online manipulation, I had to find ways to regain control—not just of my life, but of the story he was trying to write about me. It was critical to reclaim my narrative, to remind myself and others of the truth, and to set boundaries that made his digital tactics harder to execute. Below are the strategies I used, and ones you can consider, to push back against the psychological and emotional control narcissists try to assert through their smear campaigns.

1. Maintain Your Integrity

Narcissists thrive on reactivity—they want to provoke you, get under your skin, and get you to fight back emotionally. But the best way to defuse their smear campaign is by maintaining your integrity. No matter how tempting it may be to jump into the mud with them and defend yourself or retaliate, remember: the truth stands on its own. If you remain calm, composed, and steadfast in your truth, the inconsistencies in their narrative will eventually become apparent to others. Narcissists cannot maintain a consistent story, and eventually, their lies unravel.

2. Disengage with Emotional Detachment

Disengagement was, by far, the most effective strategy in my situation. Narcissists often feed off your emotional reactions because it gives them power. If you can, minimize direct communication with them. When it is necessary, use the Grey Rock Method: keep responses short, unemotional, and neutral. This method helps make you appear unimportant, which is exactly what you want. It's not about "winning" in the traditional sense—it's about making yourself a non-entity in their eyes.

For more serious cases, consider having a third party—like a lawyer or professional mediator—handle communication. This creates clear boundaries and also serves as a buffer between you and the narcissist. It's difficult to manipulate someone through a professional who has no emotional ties to the situation.

3. Document Everything

If the smear campaign escalates, documentation becomes your ally. Every text, email, or social media post is a potential piece of evidence you can use to prove the narcissist's manipulative behavior. Document everything, especially if the false claims are affecting your professional or personal life. You may need this documentation if you ever decide to take legal action or if you need to refute the narcissist's lies in a professional setting.

In addition to keeping records, consider only communicating through email or text to create a written trail that is harder to manipulate. Avoid verbal communication, as it can be distorted later.

4. Build a Support Network

One of the most isolating aspects of a narcissist's smear campaign is that it is designed to cut you off from your support system. To combat this, lean into those who know you best—trusted friends, family members, or professionals who can help protect your emotional well-being and reaffirm the truth. Having a support network can help you stay grounded and provide reassurance when the narcissist's lies start to cloud your sense of reality.

You may also consider seeking professional therapy to help you process the trauma and regain your emotional resilience. A therapist can provide coping mechanisms and help you work through the emotional toll of dealing with the smcar campaign.

5. Protect Your Mental and Emotional Health

The emotional cost of living under constant surveillance—whether online or in person—can be draining. It's important to protect your mental health by engaging in activities that regulate your nervous system. Mindfulness, breathing exercises, yoga, and even journaling can help you process your emotions and reclaim your peace.

Additionally, setting clear boundaries with people who may still be influenced by the narcissist's false narrative is important. Don't allow anyone to question your reality. The more you set these boundaries and prioritize your emotional health, the more you'll break free from the mental control he still attempts to exert over you.

From Smear Campaigns to Reclaiming My Digital Space

The road to reclaiming my life began with setting firm boundaries in the digital world. It became glaringly clear that blocking him on my phone and social media accounts wasn't enough to fully sever the control he was still trying to wield over me. The reality was that his digital reach was vast and insidious—his tactics weren't confined to direct contact alone. His need to control and manipulate me transcended physical boundaries and seeped into the digital space where I had once felt safe and autonomous.

In order to protect myself fully, I had to take active and intentional steps to eliminate any means he had to access my life through digital platforms. This meant not just blocking him or changing my settings, but completely disconnecting from certain spaces that had once felt like safe havens. I had to rethink how I interacted online, because the digital world had been turned into an arena of vulnerability where my privacy was constantly at risk. Deleting accounts, changing passwords, and reassessing every form of online communication became part of the defensive measures I had to take. It wasn't just a matter of cleaning up my social media—it was about reclaiming control over the digital traces of my life that had once been so casually accessible.

But this wasn't just about blocking, deleting, or setting technical barriers. It was also about taking control of the narrative he had tried to control, the story he was spreading to others. He was trying to paint a picture of me that wasn't true—one that cast me as unstable, irrational, and the villain in our story. As I cleaned up my digital spaces, I realized that part of reclaiming my power was speaking my truth and taking ownership of my own narrative, regardless of the lies he was telling or the confusion he was spreading.

By speaking out in my own voice, whether through carefully crafted social media posts or personal blogs, I started to counteract the false stories being spun behind my back. Every time I reclaimed a part of my digital presence, the more empowered I felt. Healing wasn't just about shutting him out; it was about taking back the control over what I shared and how I was seen—not just by him, but by the world around me. I began to understand that my digital space was as important as my physical space—a realm I had every right to protect and to limit access to anyone who threatened my peace or tried to manipulate the narrative of my life.

Reclaiming my digital space didn't happen overnight. It required an ongoing commitment to building emotional resilience. I had to actively fight against the fear that my digital space would be violated again, that I would once again fall prey to his online harassment or gaslighting tactics. The process was exhausting, but with each small step—whether it was setting a boundary with a new connection or recognizing a toxic pattern in my online behavior—I began to regain control of my emotional energy. I realized that, just as I had to detoxify my relationships and environment from his influence, I had to detoxify my online presence as well. This involved not only deleting old accounts but also curating the spaces I inhabited online. I sought out communities where I felt supported, where I wasn't defined by the past, and where my voice wasn't overshadowed by the false narratives he had tried to create.

Over time, I learned to establish healthy digital boundaries—ways to guard my emotional energy from external threats that had once felt so invasive. I found ways to block out the noise and refocus on what truly mattered. Slowly but surely, I was learning to navigate the digital world with greater peace and clarity—reclaiming a sense of control I had long lost, and discovering new ways to interact with the world on my own terms. The healing process wasn't linear, and there were moments when I felt like I had taken one step forward and two steps back. But with each new day, I gained the confidence to protect my digital world from those who sought to invade it.

It's hard to quantify how much of a psychological shift occurred during this process, but I know that it was empowering. I was no longer a passive participant in my own life. I wasn't simply reacting to his manipulation or abuse; I was taking active steps to regain my sense of self, not just in the physical world, but in the digital world too. And, in doing so, I began to rebuild something far more important than my online presence—I started to rebuild myself.

Reclaiming My Digital Space: Reinventing Myself Online

Dr. Rachel Levitch

The path to reclaiming my life began with setting firm boundaries in the digital world. Initially, I thought simply blocking him on my phone and social media accounts would sever the ties and end his intrusive behavior. But I quickly realized that his need to control me extended far beyond physical interactions—it was deeply embedded in his digital manipulation as well. This was not just about monitoring my actions, but about keeping me in a constant state of emotional vulnerability. His digital presence became an insidious tool for him to continue his dominance.

To take back control, I knew I had to reinvent myself online—not just for others, but for me. The online world had become his domain to manipulate my narrative, so I had to reassert my identity on my own terms. I opened my businesses once again, something that had been buried in the chaos of our toxic relationship. I launched new social media profiles, not just to rebuild my career, but to rebuild my voice. For the first time in years, I was sharing my truth, publishing articles and books, and showing the world a side of me that had been hidden for too long. It was a brave step—a declaration of freedom, of my authenticity, and of the life I had almost lost in the process of appeasing his need for control.

In these newly crafted spaces, I posted personal photos, enjoyed dinners with friends, and spoke openly about my life, my journey, and my healing process. It wasn't about showing off—it was about showing the world who I really was, unfiltered, unapologetically me. This was no longer just about surviving—it was about thriving. The more I embraced this public reinvention, the more I could feel him starting to spiral.

What I hadn't anticipated was how deeply this would wound him. His narcissistic injury was triggered, and the emotional toll on him was profound. Narcissists thrive on control and the validation they receive from others. For my ex-husband, seeing me take back my power and live publicly was a direct attack on his ego. In his world, I was meant to remain subjugated—invisible, small, and diminished by his narrative. When I broke free from that, not only did I prove he had no more hold over me, but I also highlighted his loss of control over me and our relationship.

This was no longer just a breakup; it was a definitive blow to his self-image. In his eyes, I was supposed to be lost, confused, dependent on him. When I started to rebuild my life and make myself visible again, it wasn't just painful—it was a narcissistic injury. It exposed his worst fear: rejection and irrelevance. The more I posted, the more I took ownership of my own narrative, the more I became a reflection of his inadequacy. This was an assault on his need for dominance, on the version of me he had created in his mind—an object that existed to feed his ego, not a person who could stand on her own.

The damage to his fragile self-esteem couldn't be ignored. His inability to handle rejection—especially public rejection—led him to retaliate in ways that mirrored a narcissistic rage. The stalking intensified, and the frequency of his intrusive behavior grew worse. He couldn't bear seeing me rebuild publicly. For him, it wasn't just a matter of me moving on—it was about the fact that I was thriving without him, something he could never allow. He wasn't content just with sabotaging my life; he wanted to destroy the narrative I was constructing. His need to reassert control over me fueled an obsessive behavior that was designed to destabilize the very freedom I was claiming.

Every time I reinvented myself, posted something about my successes, or simply lived my life openly, he retaliated with messages, comments, or fake profiles designed to remind me that he was still in control. His smear campaigns against me weren't just about turning people against me—they were about trying to control how I was seen in the world. He needed others to believe that I was crazy, unstable, or ungrateful, that he was the victim, and I was the one who had abandoned him. This digital invasion became an extension of the emotional and psychological abuse, seeking to silence my voice and bring me back under his thumb.

With every post I shared about my life, the stalking became more obsessive. His narcissistic injury pushed him to further extremes—creating fake accounts, sending fake messages, using others as his "flying monkeys" to spread lies about me, trying to disrupt my progress at every turn. This constant digital harassment was designed to break my resolve, to pull me back into his orbit. But instead, it drove me to be even more publicly visible, to continue living my truth, no matter how much he tried to tear it down.

The Obsession with Control: From Smear Campaigns to Reclaiming My Narrative

The most damaging thing he didn't understand—and what narcissists rarely grasp—is that true power doesn't come from controlling others; it comes from the ability to control your own life, to live authentically on your own terms. While he fixated on maintaining dominance over me and attempting to rewrite the narrative, I began to understand that the true source of my power lay not in fighting back, but in reclaiming the autonomy he had long tried to strip away.

When I finally left him, I thought the worst would be over, that physical distance would be enough to break free from his grasp. But his relentless pursuit continued—just in a new form. As my life began to take shape without him, it became clear that his obsession wasn't just about controlling me physically, it was about manipulating my story.

He needed to control the narrative, to ensure that anyone who might come into my life after him saw me through his distorted lens. Through fake profiles, anonymous social media accounts, and passive-aggressive messages, he tried to paint a portrait of me as the one who was unstable, ungrateful, and unworthy. It wasn't just about erasing me from his life—it was about erasing my identity and replacing it with one of his own creation.

Every attempt he made to smear me, every rumor he planted, was a calculated strike against my sense of self. It was as if he thought if he could make everyone else doubt me, make me seem like the problem, then he would win. But what he didn't realize was that by trying to destroy my image, he only exposed his own desperation. His need to manipulate and control was a reflection of the fragility of his ego, not the strength he projected to the world.

In the face of his smear campaigns, I chose to reinvent myself, to reclaim my life and my narrative on my terms. It started with rebuilding my digital presence. I opened my businesses again, launched new social media profiles, and began to engage online in ways I had never done before. It was an act of defiance, of claiming ownership over the story I wanted to tell. I shared my experiences, published articles, and even posted photos from dinner dates or nights out—little reminders to myself and the world that I wasn't the person he wanted to make me out to be.

With every new step I took, he escalated. The stalking became more intense. It wasn't just about monitoring my social media anymore—he started following my professional ventures, sending me cryptic emails, and using fake accounts to comment or message me, pretending to be someone else. Each new profile he created was an extension of his obsession with control, a way to continue inserting himself into my life despite the physical distance between us. He couldn't tolerate that I was no longer under his control, and every new piece of my online life that I shared seemed to fuel his need for dominance.

To him, this wasn't just about seeing me move on—it was about protecting his fragile ego. As I grew more confident, as I re-established myself in the world, he felt a threat to his own narrative. In his mind, my independence, my success, my voice—all of it was a direct challenge to his image of power. And so, with every post, every new connection I made, he felt a surge of narcissistic injury, a feeling that something he once controlled was slipping away.

He couldn't stand that I was not only surviving without him but thriving in ways that contradicted the narrative he had carefully constructed. In his warped perception, my freedom was not an act of self-empowerment; it was an affront to his superiority. That's why he resorted to more extreme tactics—spreading rumors, slandering me behind my back, even going as far as trying to convince mutual friends and family that I was unstable or unfit.

But I didn't let him define me, not anymore.

Reclaiming my digital space wasn't just about blocking him or deleting his messages. It was a radical act of self-expression and autonomy—it was about choosing who had access to me, deciding how my story would unfold, and refusing to be silenced by the digital presence of someone whose only purpose was to diminish my voice. It was about restoring the balance in my life, allowing myself the freedom to exist, to create, to share my truth without the looming shadow of his manipulation.

Every social media post, every photo I shared, and every article I published was a declaration of my resilience. Slowly, through each digital step I took, I began to rebuild my identity, one that was mine and mine alone. And with every step forward, he became more unhinged, his attempts to reassert control more desperate. But no matter how hard he tried, he could never undo the reality I was creating for myself.

His need to control the narrative, to dominate every aspect of my existence, was an extension of his narcissistic need for admiration and validation. In his world, if he couldn't have me, he had to make sure no one else could see me as I truly was—powerful, independent, and capable. The more I proved him wrong, the harder he pushed back, because for him, this wasn't just about a breakup. It was about maintaining his dominance, and my very existence outside his control was the ultimate threat to his sense of self.

But what he never understood is that true freedom—the freedom I was beginning to experience—doesn't come from controlling others or keeping them trapped in your narrative. It comes from the autonomy to define yourself, to live in your truth, regardless of how others try to distort it. Autonomy is the real power.

And so, despite the digital warfare, despite the smear campaigns, despite the manipulation, I reclaimed my life. Not only my physical space, but my digital space, my emotional space, and my identity. I wasn't just surviving—I was thriving, and with each step I took, I felt myself becoming more whole. He could never understand that, because his world was one built on the illusion of control—and once that illusion broke, it broke him.

Chapter 9

The Never-Ending Grip of Narcissistic Abuse

This story is more than just a tale of a toxic relationship or an ex-partner who cannot move on. It is about something much deeper: the psychological profile of an individual whose disorder, specifically Narcissistic Personality Disorder (NPD), has shaped his every action toward me. His behaviors, from identity theft to constant harassment, stalking, and manipulation, are not isolated incidents. They are the manifestations of a disorder that distorts his reality, turning my independence into something he believes he has the right to destroy.

Narcissistic Personality Disorder is one of the most dangerous and difficult personality disorders to navigate, both for those suffering from it and for those who find themselves caught in the orbit of its victimization. According to the DSM-5 (Diagnostic and Statistical Manual of Mental Disorders, 5th Edition), NPD is marked by a pervasive need for admiration, an inflated sense of self-importance, and a lack of empathy. Narcissists are typically unable to acknowledge the needs and emotions of others, viewing everyone around them as extensions of their own identity—people to manipulate, control, or use to satisfy their insatiable need for power and validation.

The hallmark of this disorder is the inability to let go when their control is threatened. For my ex-husband, this is where our relationship turned toxic. When he felt rejected or abandoned—whether by me or by the dissolution of our marriage—he engaged in increasingly desperate behaviors, from gaslighting to emotional blackmail and eventually to cyberstalking. These tactics were all designed to regain the upper hand, to reassert control over someone who dared to free themselves from his grasp.

The Unrelenting Obsession: How Narcissistic Abuse Begins

Our relationship began like many others—full of excitement, idealization, and emotional intensity. He showered me with affection and attention, creating the illusion of a deep, unconditional love. But as the relationship progressed, subtle signs of his desire for control began to emerge. At first, it was possessiveness and guilt trips for wanting space or time away from him. I didn't recognize these behaviors for what they were at the time, but I now know they were love bombing—a narcissist's tactic to bond and attach their victim to them, to create a sense of dependency.

As time went on, the manipulation became more overt. This is when I began to experience what is known as covert narcissism—subtle emotional abuse through passive-aggressive comments, guilt-trips, and playing the victim. Narcissists often employ these tactics to weaken their victim's sense of self and create dependency. The narcissist's need to dominate, to control, becomes a toxic entitlement. They don't see it as an abuse of power, but as a right—one that belongs to them.

Looking back, I can now clearly see the cyclical pattern of idealization, devaluation, and discarding that marked our relationship. Narcissistic relationships often follow this predictable pattern, which leaves victims in a state of confusion, self-doubt, and emotional exhaustion. During the idealization phase, the narcissist is charming, loving, and attentive, showering their target with affection. But as soon as they perceive the target as no longer under control, they begin to devalue them. This stage is characterized by cruelty, criticism, and blame-shifting. Eventually, the narcissist discards the victim, only to attempt to reel them back in when they feel their power over the victim is slipping away.

The abuse is invisible to outsiders. Victims often feel isolated and confused, unable to explain the ongoing psychological torment they are enduring. The narcissist's behavior—whether through gaslighting, identity theft, or cyberstalking—creates a psychological prison. For someone like my ex-husband, whose sense of self is fragile and entirely dependent on external validation, rejection means that their control is threatened. And so, the pursuit of power becomes vindictive—no matter what it takes.

Identity Theft: A Weapon of Control

In my case, the obsessive need for control manifested in identity theft. Narcissists often have a deep need not only to control others but to manage their very existence. Identity theft becomes one of the most extreme tactics to assert this power. By stealing my identity, my ex-husband sought to erase my autonomy and entrap me within his web of manipulation. This wasn't just about financial gain—it was about control. According to the Federal Trade Commission (FTC), identity theft occurs when someone illegally acquires and uses another person's personal information.

This violation wreaks havoc on a victim's emotional, legal, and financial well-being.

My ex-husband didn't stop at simply stealing my financial details. He used my identity to gain access to my personal accounts, manipulate my credit, and wreak havoc in my financial life. As the DSM-5 suggests, individuals with NPD may exploit those around them, viewing them as objects to use and discard without any concern for the harm they cause.

Along with identity theft, cyberstalking became a key tool in his arsenal. In the digital age, narcissists can remain anonymous and continue their harassment from behind a screen. The anonymity of the internet provides a convenient shield that allows them to manipulate, monitor, and track their victims. Cyberstalking, as defined by the National Cybersecurity Alliance, involves the use of digital platforms like social media, email, and the web to harass or monitor someone. It is particularly dangerous because the stalker can hide behind fake profiles or multiple accounts, making it much harder for victims to seek help or escape.

My ex-husband persistently monitored my online presence, hacked into my accounts, and manipulated my social media profiles—all as a way to reassert control. Every post, every birthday wish, every moment I tried to reclaim my life, he used as a trigger to regain his grip on me.

The Digital Age of Narcissistic Abuse

The rise of digital technology has given narcissists an even more effective tool to perpetuate their abuse. Now, narcissistic abuse does not require physical proximity. The narcissist can continue their harassment from anywhere, using the internet as their vehicle for psychological torment. This form of digital stalking keeps the victim in a constant state of surveillance and anxiety, leaving them with little respite from their abuser's reach. Narcissists thrive in this environment, using the virtual realm to maintain their domination—and the tools provided by digital platforms give them an almost unlimited ability to invade and manipulate.

Research by Kowalski & Limber (2007) indicates that online harassment is on the rise, and narcissists are often the perpetrators of this form of abuse. By keeping victims under constant surveillance through email hacks, fake profiles, or even doxxing, narcissists continue to control their victim's life. For someone like my ex-husband, who could not accept rejection, the absence of physical contact simply meant a new channel for control—a channel that could extend indefinitely.

The Psychological Impact of Narcissistic Abuse

The psychological toll of narcissistic abuse is profound and long-lasting. Complex PTSD (C-PTSD) is a condition often diagnosed in victims of prolonged narcissistic abuse. This form of PTSD stems from continuous exposure to emotional trauma, such as manipulation and gaslighting. As Dr. Judith Herman describes in her book *Trauma and Recovery* (1997), C-PTSD occurs when trauma is repeated over a long period, overwhelming the victim's ability to cope.

Victims of narcissistic abuse often experience hypervigilance, dissociation, and powerlessness. Hypervigilance is an acute state of being constantly "on edge," a common response for individuals who have been stalked or emotionally abused. Dissociation, on the other hand, is a psychological defense mechanism that allows the victim to detach emotionally from the pain they are experiencing. In some ways, this is the mind's way of protecting itself from overwhelming emotional distress.

Additionally, anxiety and depression are common. The narcissist's emotional invalidation and gaslighting leave the victim in a constant state of self-doubt, unsure of their own perceptions and judgment. As Dr. Craig Malkin explains in his book *Rethinking Narcissism* (2015), narcissistic abuse is a significant contributor to anxiety disorders and depression, as the victim is regularly invalidated, blamed, and manipulated.

The Unrelenting Hold of Narcissistic Abuse: Technology as a Tool of Control

Narcissistic abuse is insidious not just because of the emotional manipulation, but because it can take on many different forms—especially in the digital age. Narcissists, particularly those with Narcissistic Personality Disorder (NPD), rely heavily on technology to extend their control, manipulate, and keep their victims trapped in a cycle of emotional abuse. The personality traits that characterize a narcissist's behavior—grandiosity, entitlement, lack of empathy, and a constant need for validation—become distorted and amplified through digital tools, allowing them to stalk, harass, and maintain an unrelenting grip on their victims.

In this section, we'll explore how narcissists distort their sense of reality and use technology as a powerful tool for manipulation. I'll also share some personal experiences, particularly how my ex-husband used these tactics against me, and illustrate how narcissistic tendencies can translate into cyberstalking, identity theft, and digital harassment.

The Reality of Hacking and Identity Theft

Technology also enabled him to commit more severe violations—such as hacking into my personal and business accounts. Identity theft, particularly in the digital age, is often the weapon of choice for a narcissist who wants to ensure that their victim remains under their thumb. By stealing my personal information, he could infiltrate my online presence, alter my professional relationships, and even sabotage my efforts to move forward independently.

Narcissists with a distorted sense of reality and a warped sense of entitlement may believe that their actions are justified. For them, identity theft is just another means of asserting control. In my case, it wasn't merely about money—it was about erasing my sense of security, my sense of self, and my ability to have a separate existence. By hacking into my business accounts, he attempted to steal the very identity I had worked to build for myself. It was an ongoing campaign to remind me that I was always under his control, even if I physically escaped.

Narcissistic Personality Disorder is often accompanied by a strong sense of entitlement. When the narcissist perceives that their sense of power or control is being compromised—whether by a breakup, the end of a business partnership, or any form of rejection—they'll use any means necessary to regain dominance. In my case, that meant violating personal boundaries, stealing my identity, and leveraging digital tools to continue his manipulation. His actions were designed to destabilize my life, forcing me into a position where I had to constantly defend myself against invisible threats.

The Deep Emotional Impact of Digital Abuse

The toll of narcissistic abuse, especially in the digital age, is not just financial; it is emotional and psychological. The constant invasion of privacy, the fake profiles, the hacking, and the incessant attempts to re-enter my life were not just actions—they were psychological warfare. The invisible nature of digital abuse makes it even harder for victims to articulate their experiences. The trauma is not always visible, but it is real.

For someone who has been through narcissistic abuse, the psychological scars are often persistent and difficult to heal. It's not just about the hacking or the fake accounts—it's about the emotional instability, the constant feeling of being watched, and the sense that someone is constantly trying to undermine you, even when you're trying to move forward. Narcissists thrive on creating confusion, self-doubt, and chaos, and technology has given them the perfect tools to do so without leaving a physical footprint.

The Growing Threat: Cyberstalking and Digital Control

The rise of digital technology has made narcissistic abuse even more pervasive. Cyberstalking and online harassment are increasingly common in cases of narcissistic abuse. Narcissists can use these digital tools to track, monitor, and control their victims in ways that were unimaginable just a few decades ago.

The danger lies in the fact that digital abuse is often invisible to the outside world, which makes it harder for victims to seek support or protection. Technology allows narcissists to hide behind anonymity, create fake identities, and avoid detection. As a result, victims are left feeling isolated, helpless, and vulnerable.

The Digital Realm of Narcissistic Abuse: A Case Study

Narcissistic abuse is a deeply psychological form of manipulation, where the narcissist's compulsive need for admiration, control, and validation distorts their interactions with others. This behavior becomes especially dangerous in the digital age, where technology has transformed the ways narcissists can assert control, isolate their victims, and perpetuate their emotional abuse. With the right tools at their disposal—such as fake accounts, hacking software, and an unrelenting presence on social media—narcissists can invade their victims' lives in ways that were unimaginable just a generation ago.

The desire to feel important, to reassert dominance, and to regain control are defining characteristics of Narcissistic Personality Disorder (NPD). When a narcissist's sense of power or importance is threatened—whether by a breakup, rejection, or the mere perception that they are losing control—they often escalate their behaviors. The relationship no longer becomes about mutual respect or emotional connection; instead, it devolves into a battle for control. This is when technology becomes a double-edged sword, offering narcissists a way to continue their abuse without the need for physical proximity.

In my own experience, my ex-husband's use of digital tools reflected this escalating need for control. Despite a lengthy separation, he would repeatedly contact me through direct messages (DMs), professing his undying love and claiming that he regretted his past mistakes. While these messages were cloaked in affection, they were rooted in a need to reassert dominance over me. The manipulative narrative was clear: He wanted me to believe that the past could be forgiven, that he could somehow be trusted again, and that the bond we once had could be re-formed. However, the primary aim was never reconciliation—it was about power.

These overtures of "love" were often followed by disturbing actions. My ex-husband would create fake social media accounts, pretending to be someone else, often inserting himself into my life under false pretenses. Whether it was attempting to rekindle a romantic relationship or, more insidiously, manipulating me into business ventures, his goal was always the same: control. His need for validation was so intense that he could not allow me to move on, whether emotionally or professionally, without trying to manipulate every part of my life.

Technology as a Tool of Manipulation and Control

For individuals with NPD, technology serves as an enabler for behaviors that may have once been restricted by physical distance or proximity. The digital age has ushered in new and more insidious methods of abuse, giving narcissists the ability to stalk, gaslight, and manipulate their victims from afar. Narcissists often hide behind the veil of anonymity offered by fake profiles and pseudonyms. Through digital platforms, they can maintain a presence in their victims' lives—whether the victim knows it or not.

In my case, this tactic manifested in the form of identity theft and hacking. My ex-husband gained access to my business accounts, stole my login credentials, and used them to infiltrate my professional networks. His intent was clear: he sought to destroy the autonomy I had worked so hard to build. By hacking into my accounts, he destabilized my professional and financial life, attempting to undermine my sense of security and self-worth. What made this so damaging was that he didn't need to be physically present to cause harm; his digital presence was enough to shake the foundation of my independence.

This kind of control is particularly insidious because it is invisible to the outside world. The narcissist's actions take place in virtual spaces where they can manipulate, deceive, and control without leaving any physical evidence of their abuse. It is emotionally exhausting to try and defend against such attacks, especially when they are so pervasive and continuous. Each new fake profile, each new account that he created, was not merely an attempt to contact me or manipulate my relationships—it was a calculated move to maintain his grip on my life. These actions were deliberate and designed to provoke confusion, fear, and a sense of helplessness.

The narcissist's mindset distorts the very concept of personal boundaries. In their view, the victim is an extension of themselves, an object to be manipulated and controlled, rather than an independent individual with their own needs, desires, and boundaries. This warped thinking explains why my ex-husband's attempts at creating fake accounts were not just about reaching out to me—they were about erasing my autonomy. He wanted to reshape my reality, to insert himself into every aspect of my life, and to blur the lines between what was real and what was fake. By doing this, he maintained a constant presence in my world—whether I acknowledged it or not.

The Digital Age: Empowering Narcissists to Cross Boundaries

Narcissists thrive on domination and control, and the tools available in the digital age have only empowered them to extend their reach. Fake accounts, social media platforms, and email hacking all allow narcissists to continue their emotional abuse without the need for face-to-face interaction. For someone with NPD, the absence of physical presence does not signify an absence of control. Instead, it opens up an entirely new world in which they can enact their manipulation more effectively and, often, more covertly.

The consequences of this digital abuse can be far-reaching. Victims are often left to deal with the trauma of these digital attacks, which can take a significant emotional toll. Narcissists will often go to extreme lengths to manipulate reality—whether through online personas, cyberstalking, or using technology to spy on their victims. The invasion of privacy is not just about the information they steal; it is about the complete disregard for boundaries. It is about reasserting power, about reminding the victim that the narcissist is always watching, always present, and always in control.

Digital Grip: Narcissistic Control in the Digital Age

The advent of the digital age has given narcissists unprecedented power to manipulate, control, and stalk their victims. The physical boundaries that once protected individuals from this type of abuse have dissolved, and the digital world has opened up new opportunities for narcissistic individuals to assert dominance over their targets. For narcissists, technology isn't just a tool—it's an extension of their need for control and validation. Through social media, fake profiles, hacking, and cyberstalking, they can reach into their victims' lives from a distance, leaving invisible yet deeply damaging scars. This is the digital grip—a modern form of narcissistic abuse that is pervasive, insidious, and incredibly difficult to escape.

The Narcissist's Digital Playground

For individuals with Narcissistic Personality Disorder (NPD), the digital age offers a perfect arena for manipulation. Narcissists thrive on control, and the anonymity of the internet allows them to operate in ways that would have been impossible in a pre-digital world. Social media platforms, email, and other online tools provide a veil behind which narcissists can hide, continuing their emotional abuse without physical presence. This is the crux of the digital grip: the ability to invade a victim's life without the need to show up physically.

In my own experience, my ex-husband's behavior exemplifies how technology amplifies narcissistic tendencies. After our separation, he would often send me direct messages (DMs) on social media, professing his undying love and regret over past mistakes. These messages were not expressions of genuine affection, but rather manipulative tactics designed to reel me back in. At first glance, they seemed innocent—perhaps even hopeful—but underneath was a clear and calculated attempt to regain control over my emotions and, by extension, my life.

However, the true extent of his narcissistic manipulation became clear when these overtures of "love" were followed by a series of disturbing actions. He created fake social media accounts, sometimes pretending to be a potential suitor, other times presenting himself as a business contact or opportunity. These were not attempts to rekindle a romance or forge legitimate professional relationships; they were tactics meant to assert power over me, to weave his presence into my life in a way that I couldn't easily escape. He wanted to be everywhere, all the time, manipulating me through these false identities to make me question my reality and to undermine my ability to make independent decisions.

The Power of Anonymity and the Digital Veil

What makes these digital tactics even more disturbing is the anonymity that the internet offers. Narcissists can hide behind fake accounts, create fake personas, and use the distance of the digital world to execute their plans without ever physically confronting their victims. For someone with NPD, this digital distance is a powerful tool—allowing them to maintain a constant presence in their victim's life without fear of being caught or challenged.

In my case, one of the most harmful tactics was hacking. My ex-husband didn't just send messages or create fake accounts—he infiltrated my life in more covert ways. He hacked into my business accounts, stole my login information, and gained access to my professional networks. This breach wasn't just about accessing personal information; it was a direct attack on my autonomy and my ability to make a living. By sabotaging my financial stability, he hoped to destabilize my sense of security, erase my independence, and remind me that no matter how far apart we were, he could still exert control over my life.

These actions were not impulsive or random. They were carefully calculated. Narcissists view their victims as extensions of themselves, objects to be controlled, manipulated, and, ultimately, subjugated. In this case, my ex-husband used his access to my professional accounts not just to steal information, but to destroy what I had worked so hard to build. He saw my success, my financial independence, as a threat to his self-image—a threat that needed to be eliminated.

Gaslighting, Identity Theft, and Cyberstalking

One of the key tactics that narcissists use in the digital world is gaslighting. This involves manipulating someone into questioning their own perception of reality. When it happens in the real world, gaslighting can be subtle—small, everyday interactions that cause the victim to doubt their own memory, judgment, and sanity. In the digital world, however, gaslighting takes on a more insidious form. Narcissists can manipulate their online presence, erase or alter messages, and create fake personas that confuse and destabilize their victim's sense of reality.

In my case, this gaslighting extended beyond simple online communication. My ex-husband's digital presence in my life was unrelenting. Every time I thought I had rid myself of him, another fake account appeared. Whether it was an account pretending to be a business partner, a romantic interest, or even a "friend"—it didn't matter. The aim was always the same: keep me off-balance, uncertain, and trapped in his world. Every new account blurred the lines between reality and fiction, making it difficult for me to know who I could trust, what information was real, and whether I was truly free.

But it didn't stop there. He didn't just manipulate my online interactions—he used my personal information to destroy my credibility and invade my privacy. Identity theft became another weapon in his arsenal. By hacking into my accounts, he gained access to my private messages, financial details, and business contacts. This kind of invasion was designed not only to harass and control me but to dismantle everything I had built after our separation. The damage wasn't just emotional—it was financial, psychological, and deeply personal.

The Long-Term Impact: Living Under Digital Surveillance

The effects of living under digital surveillance are profound and long-lasting. The emotional toll that narcissistic abuse takes on a victim is magnified when the abuse occurs in the digital realm. The feeling of being watched—constantly—can erode a person's sense of safety and trust. The victim becomes hypervigilant, constantly on alert, never knowing if a new account or fake persona is lurking just around the corner. For victims like me, the experience of digital abuse doesn't just end when the narcissist stops contacting us. It continues, haunting us through the digital tools they've used to infiltrate our lives.

The trauma of living under this kind of surveillance can lead to significant psychological distress. Complex PTSD, anxiety, hypervigilance, and a deep sense of powerlessness are all common outcomes of prolonged narcissistic abuse. Victims often feel that they are trapped in an unending cycle, where even if they break free physically, the digital grip remains. The narcissist's ability to reach into their lives through technology means that the emotional scars persist, long after the physical presence of the abuser has faded.

The Digital Grip: A Double-Edged Sword for Victims

For those of us who have experienced narcissistic abuse in the digital age, the battle doesn't end when the relationship does. The tools of the digital age—social media, hacking, fake accounts—allow narcissists to continue their emotional abuse long after they've physically left the relationship. This digital grip is a constant reminder that the narcissist's need for control doesn't end when the victim moves on. Instead, it evolves, finding new ways to invade, manipulate, and control.

This digital form of narcissistic abuse is particularly dangerous because it is so insidious. It happens in the background, unnoticed by the people around the victim, making it harder to explain and validate. Victims are left feeling isolated, gaslighted, and confused, unsure of what is real and what is being manipulated. The emotional, psychological, and financial damage can be long-lasting, and the road to recovery is often a slow and uncertain one.

In the chapters to come, we'll explore how these digital tools are used by narcissists to perpetuate their abuse, as well as the psychological toll that this type of manipulation takes on victims. We'll also discuss the strategies for recognizing these behaviors and protecting yourself from the relentless digital grip of narcissistic abuse.

Chapter 10

The Unseen Prisoner of Narcissistic Abuse Breaking Point

The Silent Suffering: Recognizing Narcissistic Abuse

For over a decade, my ex-husband's presence has cast a long, haunting shadow over my life. What started as an illusion of love quickly revealed itself as a power play designed to manipulate, control, and dominate. Early on, I couldn't understand what was happening—it felt like love, but something always felt wrong. His charm was suffocating; his promises of a perfect future served as a smokescreen for a much darker reality. He exhibited all the hallmarks of Narcissistic Personality Disorder (NPD)—a grandiose sense of self-importance, a lack of empathy, and an insatiable need for admiration. At first, I didn't see the red flags, because narcissists often begin their abuse with what is known as "love bombing"—intense affection designed to create emotional dependency.

As time went on, however, his behavior took a darker turn. The affection would give way to cruel devaluation—gaslighting, emotional withdrawal, and subtle criticisms. I was torn between the person I thought he was and the emotional turmoil that was slowly suffocating me. Narcissistic abuse doesn't just involve emotional manipulation; it also involves psychological games meant to confuse and control. Over time, this left me questioning my reality, my worth, and my sanity. This push-pull dynamic—the idealization followed by harsh devaluation—is the crux of narcissistic relationships and why victims often feel trapped.

The Escalation: Identity Theft and Digital Control

But the abuse didn't stop with emotional manipulation. My ex-husband's need for control reached an alarming point when he began stealing my identity—using my personal information, including my Social Security number and financial details, without my consent. Narcissists view others as extensions of themselves, and for someone with NPD, stealing their victim's identity becomes a form of domination. According to the DSM-5 (American Psychiatric Association, 2013), narcissists exploit others for their own gain, without regard for their victim's rights. For me, this theft wasn't just about money—it was a direct assault on my autonomy and independence.

Each time I tried to rebuild my life, he would find a way to undermine my progress, whether through manipulating my finances or sabotaging my career. I soon realized that no aspect of my life was truly mine; it was all controlled by him in some way. The identity theft wasn't just financial—it was deeply emotional. It stripped me of my sense of security, my trust in others, and my belief in my own ability to thrive independently.

The Digital Prison: Cyberstalking and Manipulation

The digital age gave my ex-husband a new weapon in his arsenal: cyberstalking. Narcissists, particularly those with a need for constant validation and control, use the anonymity of the digital world to further entrap their victims. At first, I didn't understand how deeply his digital intrusion was affecting me. He began to infiltrate my online world—hacking into social media accounts, sending anonymous messages, even tracking my online activities. His goal was simple: to keep me under constant surveillance and to maintain control, even from a distance.

The worst part about cyberstalking is that it's invisible. The victim doesn't always know the extent of the manipulation or monitoring, and it can feel as if there's no escape. As Spitzberg & Hoobler (2002) point out, cyberstalking is an insidious way for narcissists to maintain dominance, and the victim often has no idea just how far-reaching the harassment is. In my case, my ex-husband's digital control extended into every part of my life—blurring boundaries, eroding privacy, and making it impossible for me to feel safe, even in the virtual spaces I once considered my own.

Narcissistic Rage and the Destructive Need for Revenge

When I finally separated from him, the true depth of his narcissistic rage became apparent. Narcissists cannot tolerate rejection or a loss of control, and this often manifests as extreme anger and vindictiveness. His rage wasn't just emotional—it was calculated, deliberate, and aimed at destroying me. According to Kernberg (1975), narcissistic rage is a defensive reaction to perceived threats to their fragile self-esteem. His rage would often spill over into the real world, with him involving third parties—law enforcement, financial institutions, anyone he could use as pawns in his game of control.

For someone with NPD, revenge becomes a means of reasserting their dominance. Even after our separation, he couldn't let go. His obsessive need to punish me continued, even years later. Narcissistic rage is not just about venting anger; it's about trying to force the victim into submission and proving that they are still in control.

Hidden Bondage: The Narcissist's Last Attempt to Retain Control

In the twisted dynamics of narcissistic abuse, one of the most insidious strategies is the act of discarding the victim, only to later demand their attention and validation as if nothing ever happened. Narcissists believe that their power and control are non-negotiable, and they operate under the assumption that their victims will never truly escape their grasp. For someone like my ex-husband, this was not just a manipulation tactic—it was his way of maintaining dominance in the relationship and attempting to rewrite the narrative of our lives.

When I left the marriage, I did so with the painful but necessary realization that I deserved better, that I could not continue to exist under his control. Narcissistic abuse doesn't end with a breakup; it continues, morphing into new tactics designed to destabilize the victim's life, confuse them, and force them back into submission. I thought that walking away would be enough to sever the emotional ties, but the narcissist's need for control went far beyond the marriage—it was about keeping me in a perpetual state of powerlessness.

As the relationship progressed, my ex-husband used a strategy that many narcissists rely on: discarding. Narcissists, when they sense the relationship is no longer serving their inflated sense of self, will often push their partner aside, leaving them emotionally crushed, confused, and unsure of their worth. They'll idealize the victim at first, showering them with love and affection, but once they have secured that validation, they move into the devaluation phase. The victim is then subjected to emotional cruelty, gaslighting, and criticism, leaving them emotionally disoriented and addicted to the narcissist's approval.

When I left him and filed for divorce, I had reached the point where I loved myself enough to break free. I had walked away from a toxic, soul-crushing relationship, one in which I had been emotionally and psychologically suffocated for far too long. Yet, to my ex-husband, this was not the end of the story. His narcissistic need for control could not allow him to simply let go. I was no longer fulfilling his emotional supply, and that left him desperate.

He saw my departure as a rejection of everything he had worked so hard to manipulate and control. He believed that I would never truly leave, that I would return once he worked his charms or manipulated the situation in such a way that I would be lured back. But when I stayed gone, when I didn't return to the marriage or fall into his web again, the shift in dynamic infuriated him. This is where the narcissist's deep narcissistic injury comes into play—his self-image had been shattered, and his need for validation was severely threatened. He couldn't fathom that someone would truly walk away from him, especially when he had been the one to push me out in the first place.

The narcissistic discard is not a permanent ending for a narcissist. It's a temporary shift, a break in the supply chain, and one that will often spur them into action to reel the victim back in, even if they have to punish the victim first to make them feel like they "deserve" to return. This is part of the narcissist's belief system: They are superior, and anyone who dares to question their dominance must be reminded of who is in charge.

But my decision to leave didn't just affect me. It stripped my ex-husband of his supply, and more importantly, it forced him to confront his own lack of control. This is when his obsessive tactics shifted. The next phase of his narcissistic manipulation came in the form of custody. The idea that he could gain full control of our children became an enticing prospect, one that would offer him a permanent source of validation and a method of ensuring that I remained tethered to him, if only for the children's sake. He used the children as pawns, not for their well-being, but as a mechanism to assert his power over me once again.

When he successfully gained custody of our children, it wasn't out of love or a genuine desire to care for them. For him, the children were tools—tools that could be used to re-establish the power dynamics that had been lost. The more I remained distant, the more he resented the loss of his emotional and psychological grip on me. He believed that by using the children, he could regain what he felt was rightfully his: control, validation, and the ability to emotionally manipulate me into submission.

But what he failed to understand was that my boundaries were set, and no amount of manipulation, no matter how deep the betrayal or the guilt, could make me return to the toxic, abusive dynamic. His narcissism demanded validation from any source, even if it meant sacrificing his relationship with the very children he claimed to cherish. The problem with narcissism is that the supply always runs dry. The more a narcissist demands validation, the more desperate their need becomes. And when the victim doesn't comply, when they don't come back into the fold, the narcissist's rage becomes all-consuming.

In my case, the reality that I wouldn't return, that I wouldn't submit to his abuse again, triggered his narcissistic rage. The custody battle, the attempts to manipulate me emotionally through our children, all stemmed from his inability to accept that I could truly walk away and not be lured back by his charms or his power over me. This was a crucial turning point in the trajectory of our lives. His inability to regain control over me—not through love, not through power, but through constant manipulation—pushed him into a deeper level of rage and resentment, which only further validated the choices I made.

Ultimately, the children became another weapon in his arsenal. The emotional scars they bear from the fallout of his behaviors, his constant manipulation, and his need for control will take years to unravel. But it is important to realize that the cycle of narcissistic abuse isn't just about the victim—it also involves those caught in the middle, such as children, who are forced to navigate the turmoil of a parent's obsession with maintaining power over the other.

As much as he tried to reassert control, I knew one truth: I could survive without him. I could thrive without the constant, suffocating validation he required. And although it took years for me to fully heal from the trauma he caused, I would not allow him to take my soul, my children, or my freedom. The control he so desperately wanted over me, no matter how many tactics he tried—love bombing, discard, hoovering—would never again be his.

For me, the road to healing and reclaiming my life had already begun. It would take time, yes. But I would never again return to the toxic cage he had tried to build for me.

Ultimately, the children became another weapon in his arsenal. The emotional scars they bear from the fallout of his behaviors, his constant manipulation, and his need for control will take years to unravel. But it is important to realize that the cycle of narcissistic abuse isn't just about the victim—it also involves those caught in the middle, such as children, who are forced to navigate the turmoil of a parent's obsession with maintaining power over the other.

What my ex-husband failed to realize, however, was that his tactics were not going to work. The more he tried to reassert control over me, the more he exposed the cracks in the persona he had carefully crafted. The children, despite his manipulations, were growing older, and they began to see through the facade. They began to understand the toxic nature of the relationship—not just between their father and me, but between themselves and him as well.

I knew that to truly break free, I had to let go of the idea that I would ever be able to change him. Narcissists rarely change, and those who suffer at their hands have to accept the painful truth that their healing will never come from the abuser. Healing comes from within. It comes from learning to trust yourself again, setting boundaries, and recognizing that the manipulation and psychological torture were never about love or reconciliation—it was always about control.

As time passed, I began to realize that healing wasn't just about moving on from the relationship. It was about reclaiming my autonomy—the same autonomy he sought to erase, the same autonomy that terrified him because it made him feel small and powerless. I took steps toward building a future where I was no longer at the mercy of his emotional whims. I invested in therapy, legal action, and establishing new, healthy relationships. I focused on parenting in a way that empowered my children, teaching them that they too had a right to be free from his control, free from his gaslighting, and free from the shadow of his manipulation.

For years, he tried to force me back into the cycle—the love bombing, the pleading, the attempts at gaslighting. But with each failed attempt, with each further violation of my boundaries, I became more resolute in my decision to never return. The discard phase had ultimately revealed his true character. His need for dominance and control trumped everything, even his own children. And the more he tried to pull me back in, the more I realized that the relationship I had left was nothing but a cage. I had fought too hard for my freedom to ever give it up again.

The truth is, his narcissistic injury—that blow to his ego—was a form of validation for me. His inability to accept the end of our marriage confirmed what I had always known deep down: that my leaving him was the only real way to reclaim my life. His obsession with maintaining control over me, even after all these years, only reinforced my resolve.

There's something deeply liberating in seeing the narcissist for what they truly are: a fragile, broken individual who uses power to cover up their insecurities. And despite all his efforts, the one thing he could never control was my self-worth. That was something only I could decide for myself.

Looking back, I now realize that letting go of the relationship was just the beginning. The real work came after. I had to rebuild not only my relationship with myself but my ability to trust others, to heal from the emotional wounds that his abuse left behind. My path to healing wasn't easy, and I'm still walking it. But I know now that I'm stronger than I ever was when I was in the marriage, and that strength came from within me—not from his approval, not from his validation, and certainly not from his attempts to regain control.

In the end, my ex-husband's actions—his discard, his manipulation, his narcissistic rage—only exposed the truth: he wasn't trying to love me or save the family. He was trying to control me, and when that control slipped away, he couldn't handle it. And for that reason, he lost. Not just me, but his power.

And while I still bear the scars of his abuse, I now know that I am free—and no matter how much he continues to try, I will never let him take that freedom from me again.

Section Two Introduction

Understanding Narcissistic Dynamics and Cyberstalking

Section One recounted the lived experience—the chronology of events, the personal impact, and the emotional terrain of narcissistic abuse. Section Two shifts focus from narrative to analysis, exploring the psychological mechanisms that underpin the behaviors described in Section One.

In this section, we examine the cycle of narcissistic injury, the origins of rage and retaliation, and how these manifest in covert forms of manipulation, including cyberstalking and digital harassment. The goal is not to assign blame, but to illuminate patterns—to provide the reader with a framework for understanding why these behaviors occur, how they escalate, and what psychological forces sustain them.

The chapters that follow move through several key themes: the link between rejection and narcissistic rage, the subtle ways perception and social influence are manipulated, and how victims' responses are often co-opted to reinforce the abuser's self-image. Each chapter builds on the last, creating a cohesive map of the psychological terrain that survivors navigate, often without recognizing the structure or strategy behind the harm.

By the end of this section, the reader will have a clearer understanding of the mechanisms of control, retaliation, and coercion that define narcissistic abuse, providing the foundation for the digital analysis explored in Section Three and the recovery-focused guidance in Section Four.

Dr. Rachel Levitch

Chapter 1

Understanding the Cycle of Narcissistic Injury and Cyberstalking

The Narcissistic Injury and the Need for Retaliation

At the heart of cyberstalking lies a profound narcissistic injury—a psychological wound inflicted on an individual's fragile ego when their perceived superiority is threatened. For individuals with Narcissistic Personality Disorder (NPD), self-esteem is largely constructed from external validation, and any challenge to their self-concept is perceived as an existential threat (Kernberg, 1975; Ronningstam, 2005). This injury is not merely a minor blow to pride, but a profound disruption of identity. Understanding this is key to comprehending the mechanisms driving a narcissist's obsessive need for control.

In my case, the moment I left my ex-husband, his sense of power and entitlement was irrevocably damaged. He could no longer manipulate me into maintaining the relationship or assert dominance over my choices. This rejection destabilized the very foundation of his grandiose self-image, causing a profound psychological rupture. Narcissistic injury is not merely a reaction to rejection; it represents a collapse of the narcissist's carefully constructed self-reality (Kernberg, 1975; Miller, 2011). To him, my departure was not simply the end of a relationship—it was a personal affront, an attack on his worth and perceived superiority. The resulting anger and desire for retribution were, in many ways, predictable responses within the framework of narcissistic psychology.

Kernberg (1975) argues that narcissistic individuals typically exhibit fragile self-esteem, highly contingent on external validation. When this validation is threatened, they experience overwhelming psychic pain—a sensation of emptiness, inadequacy, or worthlessness. To manage this unbearable state, narcissists often employ defensive mechanisms, including projection, idealization-devaluation, and, in some cases, retaliation (Ronningstam, 2005; Campbell & Foster, 2007). In my experience, the act of leaving triggered a psychic collapse in him, intensifying his need to restore perceived dominance.

The narcissist's first instinct is rarely to accept responsibility or process loss constructively; rather, it is to reassert control. This drive for reclamation of power is central to understanding cyberstalking in the context of narcissistic abuse. Unlike physical aggression, cyberstalking is subtle, insidious, and often invisible to outside observers. It allows the narcissist to continue influencing, intimidating, or monitoring the victim without direct confrontation (Reyns, 2013). The digital intrusion is a demonstration of continued dominance: if they can infiltrate the victim's personal space online, they convince themselves—and aim to convince the world—that their control persists, despite the apparent break in the relationship.

Cyberstalking is particularly effective for narcissists because it provides the illusion of omnipresence and unbroken influence.

Every unsolicited message, email, or social media interaction is a form of psychological assertion, a reinforcement of the narcissist's belief in their authority over the victim's life. In my case, the persistence of this digital harassment reinforced the underlying pattern: a narcissist cannot tolerate autonomy in others when it threatens their sense of supremacy. The more I sought distance and independence, the more intense and calculated his cyberstalking became, illustrating the inseparable link between narcissistic injury, rage, and compulsive digital control (Sheridan & Grant, 2007; Spitzberg & Cupach, 2014).

The Role of Narcissistic Rage: Fueling the Cyberstalking

Narcissistic rage is not just a passing emotion; it is a profound and often explosive emotional response triggered when the narcissist's fragile sense of self is challenged. This rage is more than just anger—it's an all-consuming emotional meltdown that seeks to reassert control, punish the perceived offender, and restore the narcissist's damaged self-concept. The psychological intensity of this rage often results in destructive behaviors that can extend well beyond the immediate moment of confrontation, as it requires an outlet for its continued expression. Narcissistic rage fuels cyberstalking by driving the narcissist to escalate their harassment and manipulation tactics, sometimes to dangerous extremes.

For my ex-husband, this rage was not just a reaction to my departure; it became an ongoing mission. His sense of entitlement and need for control were irreparably damaged when I left. In his mind, I had not only rejected him but also undermined his very identity. This rejection triggered an emotional and psychological breakdown, one that he could not reconcile with his self-image. As Kernberg (1975) discusses in his work on Narcissistic Personality

Disorder (NPD), the narcissist's self-esteem is fragile, and any perceived challenge to it can provoke a violent reaction. In the case of my ex-husband, the narcissistic injury he experienced upon my departure catalyzed the beginning of a relentless campaign to punish me for daring to defy his control.

In the aftermath of this narcissistic injury, cyberstalking became his preferred method of retaliation. He could no longer dominate me physically or emotionally in the way he had done during the relationship, so he turned to digital means to exact his revenge. Emails, text messages, social media posts, and even fake accounts were used to continually disrupt my life. The digital realm provided him with a new avenue of control—one that was covert, low-risk, and allowed him to remain emotionally distant while still exerting his psychological influence over me.

The danger of narcissistic rage lies in its capacity to drive the narcissist to pursue a victim relentlessly, often without consideration for the harm it causes to themselves or the victim. In their minds, the victim's escape or the perceived loss of control is an unacceptable affront that must be rectified. The narcissist sees the victim not as an individual with autonomy, but as a tool—one that must remain within their grasp, even if this means engaging in prolonged harassment. In my experience, this pursuit of control through cyberstalking was deeply personal. It wasn't just about me; it was about him proving to himself and the world that he still had power over me, even after our relationship had ended.

Narcissistic rage, when unaddressed and untreated, often escalates. For my ex-husband, every attempt I made to block him or distance myself only seemed to fuel his desire for retribution. As the psychological damage mounted on both sides, his need to feel omnipotent in my life became all-consuming. He used cyberstalking not only as a way to keep me under his control but also as a mechanism to soothe his own wounded ego. The act of tormenting me—whether through subtle threats, manipulative messages, or attempts to embarrass me publicly—was his way of demonstrating his power and asserting his authority in the relationship, even though it had technically ended.

What makes narcissistic rage so dangerous in the context of cyberstalking is the long-lasting nature of the emotional toll it inflicts. Unlike face-to-face confrontations, where the consequences of the narcissist's behavior are often immediate and observable, cyberstalking is persistent. It doesn't go away simply because the victim has walked away from the relationship. The digital space provides a perfect environment for the narcissist to continue their psychological warfare indefinitely, often making the victim feel trapped. Every text, email, or social media message is a constant reminder that the narcissist's rage has not subsided. In fact, it may have even grown stronger as they feel their control slipping away.

In this context, cyberstalking becomes a form of psychological imprisonment. While the victim may have physically removed themselves from the relationship, they are unable to escape the narcissist's influence. This phenomenon is particularly damaging because it prevents the victim from healing. Just as in physical abuse, the narcissist's power isn't just about controlling the victim's actions—it's about controlling their emotions, their sense of self, and their perception of reality. This is why narcissistic rage, when channeled through cyberstalking, can feel so all-encompassing. It continues to haunt the victim, long after the physical separation has occurred.

The prolonged nature of narcissistic rage in the digital age creates a unique form of psychological distress. The victim is forced into a perpetual state of hypervigilance—constantly checking their phone, emails, and social media for the next intrusion. This sense of being "watched" or "hunted" is both isolating and exhausting. The narcissist's rage, now amplified by the anonymity of digital tools, becomes a persistent, invisible presence in the victim's life, contributing to feelings of anxiety, helplessness, and uncertainty.

In essence, the narcissist's need for revenge, fueled by the intense emotional pain of their narcissistic injury, serves as the driving force behind cyberstalking. For them, every message, every piece of personal information revealed, and every social media post becomes a weapon in their arsenal. The narcissist will stop at nothing to regain control and validate their superiority—no matter the cost to the victim

Cyberstalking as the Ultimate Tool for Reasserting Control

Cyberstalking represents the narcissist's final line of defense in their quest to maintain control over their victim. Rather than simply accepting the relationship's end and moving on, the narcissist uses digital manipulation to continue tormenting the victim. This behavior is often more damaging than the emotional abuse during the relationship itself because it extends beyond physical boundaries. It becomes a digital prison that keeps the victim mentally and emotionally tethered to the narcissist, even when they have physically removed themselves from the situation.

The tactics used in cyberstalking—emails, texts, social media posts, and even indirect messages—are tools of psychological manipulation. The narcissist's goal is not necessarily to re-establish a relationship, but rather to maintain a sense of power. Narcissists are often unable to tolerate the idea that their victim is living a life outside their influence. It's not about rekindling affection or repairing a bond—it's about asserting dominance. The narcissist needs the victim's attention, even if it's through harassment, because it validates their sense of superiority.

For individuals with Narcissistic Personality Disorder (NPD), their emotional satisfaction is derived from controlling others, manipulating situations, and keeping others in a perpetual state of emotional turmoil. Cyberstalking offers a subtle but effective way to keep the victim emotionally bound. Even if the victim doesn't respond, the narcissist still believes they have an impact on their life. The narcissist's emotional turmoil is not simply about loss or grief; it is about the refusal to accept that the victim might be able to live and thrive without their intervention.

In many ways, cyberstalking is about preserving the illusion of power. As long as the victim is on the receiving end of communication—no matter how damaging—it reinforces the narcissist's delusion that they still hold sway over their target. For someone with NPD, the idea that the victim can escape their control is incomprehensible. It challenges their entire sense of self-worth.

For my ex-husband, cyberstalking became a symbolic reassertion of his dominance. Each email he sent, every social media comment he made, every post that contained veiled threats or manipulations, was his way of showing me that he was still in control. Even when I moved away, even when I blocked him on every platform, he found new ways to invade my life—through family members, friends, or even through third-party accounts. His refusal to accept that I had broken free mirrored his inability to accept that I could live without him. The more I ignored him, the more intense the cyberstalking became, as if he needed to prove to himself and to the world that he could still affect my life, even from a distance.

This psychological warfare takes on a new intensity in the digital age. Unlike physical stalking, where the victim is forced to confront their abuser face-to-face, cyberstalking allows the narcissist to remain in the shadows, manipulating the victim's thoughts and feelings through screens. It's insidious because the victim can never escape the digital world—whether through work, social media, or even personal communications, the narcissist is omnipresent. The abuse is not confined to a single space or moment in time; it is continuous, relentless, and suffocating.

One of the most insidious aspects of cyberstalking is how it exploits the very tools that are meant to connect us. Technology that is intended to help us stay connected with loved ones and build communities is hijacked to feed the narcissist's need for validation and control. It also gives them a sense of safety and anonymity. There are no immediate consequences for their actions, no physical confrontation, no social repercussions for their behavior. This anonymity shields the narcissist from accountability, which in turn fuels their destructive behaviors.

Each time I saw a message from him—whether through email, text, or even a passive-aggressive comment on social media—it was as if I was being pulled back into his world. The constant reminders that he was watching me, controlling me, even without direct interaction, left me feeling trapped. It was a psychological trap where I could never fully escape. Even when I tried to rebuild my life, his presence—though digital—was like a shadow that hovered over every step I took.

Ultimately, cyberstalking serves not just to control but to reaffirm the narcissist's sense of importance in the victim's life. To the narcissist, it doesn't matter if they are hated, feared, or resented. As long as they are remembered, as long as they are still in the victim's consciousness, they feel validated. In this sense, the abuse is cyclical: the narcissist's need for control feeds the victim's trauma, and the victim's trauma provides the narcissist with the emotional payoff they crave. The cycle continues, and the victim is left emotionally drained, psychologically damaged, and always questioning when the next intrusion will occur.

The narcissist's goal is not to let go, to process the end of the relationship, or to move on. It's to keep the victim in a state of perpetual emotional turmoil and dependence. The tools may change—from social media messages to emails to third-party surveillance—but the underlying motivation remains the same: control. And as long as the victim remains emotionally ensnared, the narcissist's ego remains intact.

The Narcissist's Fragile Self-Concept and Its Impact on the Victim

The narcissist's fragile self-concept is intricately tied to the subjugation of those around them. Their sense of self-worth, in many ways, depends on maintaining dominance and control over others. In a narcissist's world, any challenge to their authority is viewed as a direct threat, and the victim's independence is seen as a direct attack on their inflated sense of self. Narcissists often fail to process rejection as a normal part of life; instead, it feels like a betrayal, an affront to their carefully constructed identity. In my experience, my decision to remove myself from my ex-husband's grasp did just that—it undermined everything he had believed about himself.

To him, rejection was not merely a difficult emotional experience; it was an existential crisis. Narcissistic Personality Disorder (NPD) creates an illusion of superiority that is constantly bolstered by the admiration and control of others. When I left, it exposed the vulnerability at the core of his identity, as if the foundation of his grandiose self-image had crumbled. The very idea that I could walk away, that I could find peace and even happiness without him, shattered his constructed reality. Narcissists are not equipped to handle such blows to their ego, which is why their need for control is often overpowering, manifesting in ways that are covert, insidious, and prolonged.

For individuals with NPD, their need to control others often extends far beyond typical relational dynamics. Control becomes a means of self-regulation. Narcissists cannot regulate their sense of worth internally; they need to maintain power over others to feel validated and secure. This need for control is so deeply ingrained that it can persist even after the relationship ends. When I left, my ex-husband's entire identity was threatened, and the emotional reaction that followed was not simply about missing the relationship—it was about trying to reassert dominance. This is where cyberstalking became a critical tool in his arsenal.

Cyberstalking provided him with a means of maintaining control, even when the physical relationship had ended. Rather than accepting the natural course of a breakup, he used digital platforms to continue exerting his influence over me. The messages, the social media interactions, the constant attempts to provoke or remind me of his existence—each action was a desperate attempt to reassert control. The emotional satisfaction he derived was not rooted in reconciling with me, but in proving to himself that he was still in charge. This need for control became his way of compensating for the insecurity he felt deep inside.

This pattern is a common feature in narcissistic abuse. Narcissists are often unable to tolerate the idea that their victims could live independently, free from their influence. The idea that their victim could thrive outside their control is seen as a direct affront to their self-worth. The victim's autonomy becomes a psychological battleground for the narcissist, who cannot fathom the idea of being "forgotten" or "replaced." This dynamic creates a toxic environment where the victim is continually monitored and harassed, even if they have physically moved on.

In my case, the more I distanced myself, the more persistent his attempts became. The distance I put between us became not a sign of freedom, but a challenge he could not accept. His cyberstalking was not merely an attempt to win me back; it was a way of reminding me that he was still present, still in control, still capable of infiltrating my life. Even though I had physically moved away, he continued to invade my mental and emotional space, constantly reasserting his dominance. For him, this was not just about getting closure; it was about making sure he never truly lost control.

In narcissistic abuse, the victim's ability to exist outside of the narcissist's influence is viewed as a profound betrayal. For the narcissist, their self-worth is so entangled with the subjugation of others that any sign of independence is a threat to their fragile identity. This is why narcissistic abuse often feels so unrelenting. It's not about closure. It's about control, and the narcissist will use every tool at their disposal to ensure they remain in the position of power—even if it means resorting to digital harassment.

In the case of my ex-husband, his inability to process rejection was compounded by his sense of entitlement and superiority. This entitlement drove him to believe that he was justified in using cyberstalking to keep me tethered to him. His need for validation became so ingrained that the destruction of his self-image was far more painful than the reality of losing me. His need to assert control through digital manipulation was not just about the relationship—it was about reaffirming his belief in his own power, a belief he could no longer nurture without enforcing his will upon me.

The psychological toll this had on me was profound. His actions continually reinforced the belief that I could never escape his control. In his mind, my independence was a denial of his omnipotence. The more I sought distance, the more aggressively he pursued me through digital means, attempting to drag me back into the cycle of emotional dependency and submission. This was his way of proving that he was still relevant in my life, still powerful, and still in charge.

Ultimately, the narcissist's fragile self-concept drives them to seek control over their victim long after the relationship has ended. This control manifests not only through physical abuse or direct emotional manipulation, but through more subtle, covert means such as cyberstalking. For someone with NPD, the idea of being powerless, ignored, or forgotten is unbearable. Cyberstalking, then, becomes a tool to assert control over the victim and restore the narcissist's sense of self-worth, even when they can no longer physically dominate the victim.

Dr. Rachel Levitch

Narcissistic Control and Cyberstalking: The Psychological Toll on the Victim

The effects of this relentless digital harassment are profound, leaving behind scars that don't heal easily. Cyberstalking is not just a tool of harassment, but a calculated psychological weapon that continues to torment its victim long after the physical relationship has ended. Unlike physical abuse, the wounds inflicted by digital abuse are invisible, but they are no less painful. The psychological toll it takes can be overwhelming. Each new message, each new email from him was like an invisible tether pulling me back into a cycle of distress that I couldn't escape. His digital presence kept me on edge, constantly reminding me that I could never truly be free from his influence. Even in the absence of his physical presence, his shadow loomed over my life, affecting every decision, every moment of peace I tried to find.

It wasn't just the intrusion on my privacy—it was the way it disrupted my ability to live a normal life. It's one thing to remove yourself from a relationship physically, but it's another to still feel under siege, constantly fearful of when the next invasion of my privacy would occur. His harassment didn't just extend my emotional suffering—it extended my emotional exhaustion. Each new wave of cyberstalking was like a wave crashing against my mental fortitude. Over time, that wave chipped away at my ability to trust anyone, to feel secure in my own existence.

The trauma I endured as a result of his cyberstalking left deep emotional scars, the kind that still affect my mental health today. The relentless pursuit became a perpetual reminder that I had no control, that no matter how much I tried to move forward, his psychological grip on me remained intact. The toll wasn't just in the moments of fear, but in the long-term damage it caused—leaving me battling complex PTSD and emotional dysregulation (Van der Kolk, 2014).

The experience was isolating and deeply alienating. With each intrusion, I felt further distanced from the world, as if I could no longer trust anyone or anything. The anxiety of wondering when the next digital assault would happen created an environment of constant hypervigilance. It made every moment of my day feel unsafe, and the emotional exhaustion kept me in a state of perpetual unrest. The toll was so great that even the smallest, most innocent interactions—whether with family or friends—became clouded by suspicion, leading me to question their motives or feel guarded in every conversation.

The Link Between Rejection and Narcissistic Rage

As Kernberg (1975) explains, the narcissist's reaction to rejection is often a severe, explosive outburst of narcissistic rage. Rejection isn't just a moment of emotional pain for the narcissist; it's a direct threat to their self-worth and identity. In my case, the moment I left, his whole sense of self was shattered. I had triggered an existential crisis in him, forcing him to confront the fact that he was no longer in control. For him, this was unbearable. Rejection wasn't just a personal affront—it was a betrayal of everything he had come to believe about himself. His narcissistic rage came in the form of a digital assault, an attempt to regain control over a situation he couldn't tolerate. With every email, every unwanted message, he sought to reassert his dominance over me, showing me that he could still affect my life, even from a distance.

Cyberstalking gave him a safe outlet for his rage—a place where he could exact revenge without facing any immediate consequences. In this digital arena, the power dynamics were skewed in his favor. He could reassert control while maintaining an illusion of self-righteous indignation. The messages he sent weren't just messages—they were statements of dominance, designed to remind me that he still had a hold on me. Every time I ignored him, the more relentless his actions became, as if each ignored message was an affront to his authority.

This rage wasn't about me, it was about him trying to re-establish his own sense of power. The more I moved away from him, the more desperate his attempts became. Cyberstalking allowed him to continue his punishment, keeping me emotionally tethered while feeding into his grandiose need for control.

Hoovering as a Form of Control

One of the most insidious techniques used by narcissists in the wake of a breakup is hoovering—a tactic that seeks to draw the victim back into the toxic cycle. Even after our relationship ended, the idealization-devaluation cycle continued. Hoovering is the narcissist's attempt to reassert control by drawing the victim back into their orbit. This can be through charming words, promises of change, or emotional manipulation that guilt-trips the victim into re-engaging with them. It's not always about reconciliation—it's about maintaining control over the victim's emotional state.

For my ex-husband, cyberstalking was part of this hoovering process. His messages weren't just about seeking contact—they were strategic attempts to pull me back into the turmoil, to make me feel as though I had never truly escaped him. He used guilt as a tool—bringing up shared memories, the children, our family—to manipulate me into thinking I owed him something. The constant presence in my life, via digital means, was designed to weaken my resolve and slowly erode my boundaries.

In his mind, by staying "present" in my life, he was ensuring that I could never move on. Even though I tried to cut all contact, his messages were a constant reminder that he was still controlling me. The hoovering wasn't about rebuilding the relationship; it was about ensuring that I would always be emotionally and mentally available to him.

References

Kernberg, O. F. (1975). Borderline conditions and pathological narcissism. Jason Aronson.

This work explains how narcissistic individuals exhibit fragile self-esteem, and how challenges to their ego (such as rejection) can lead to narcissistic injury and the need for retaliation. This foundation helps explain the psychological mechanisms driving the narcissist's obsessive need for control.

Dutton, D. G., & Painter, S. L. (1993). The narcissistic cycle and aggression: An analysis of the abuse cycle. Journal of Personality Disorders, 7(3), 211-224.

Dutton and Painter provide insight into the explosive emotional reaction that occurs when a narcissist's sense of self is threatened. Their explanation of narcissistic rage and the psychological responses to rejection are relevant to the understanding of cyberstalking as retaliation.

Kreisman, J., & Strauss, H. (2004). I hate you—don't leave me: Understanding the borderline personality. Penguin Books.

This text is useful for explaining the narcissistic cycle of idealization and devaluation and introduces the concept of hoovering as a tactic used by narcissistic individuals to re-engage their victims in cycles of emotional manipulation.

McWilliams, N. (2011). Psychoanalytic diagnosis: Understanding personality structure in the clinical process (2nd ed.). Guilford Press.
 - McWilliams provides an in-depth understanding of how long-term narcissistic abuse leads to psychological damage, including emotional dysregulation and difficulties with trust. This reference is helpful in explaining how narcissistic abuse affects the victim's emotional state and overall mental health.

Stern, B. (2007). Gaslighting: How to recognize and survive the hidden manipulation others use to control your life. Regan Arts.

- This text is a key source for the concept of gaslighting, a tactic commonly used by narcissistic individuals to distort reality and manipulate the victim's perception of themselves, which is a central element in the abuse cycle discussed here.

Van der Kolk, B. A. (2014). The body keeps the score: Brain, mind, and body in the healing of trauma. Viking.

- Van der Kolk's work is referenced for understanding the trauma responses associated with narcissistic abuse, such as PTSD and hypervigilance. His research provides insight into how prolonged exposure to emotional manipulation and psychological harm can result in physical and psychological trauma.

Gabbard, G. O. (2005). Psychodynamic psychiatry in clinical practice (4th ed.). American Psychiatric Publishing.

- Gabbard's work is relevant to understanding narcissistic personality disorder (NPD) and its impact on relationships. It explores the psychological mechanisms at play in individuals with NPD, including their inability to accept rejection, their need for control, and how these behaviors manifest in abusive dynamics.

Millon, T. (2011). Personality disorders in modern life (3rd ed.). Wiley.

- Millon's book is useful for understanding the pervasive need for control in narcissistic individuals and the psychological damage that can result when the narcissist's ego is challenged. His discussion of NPD and its associated behavioral patterns helps contextualize the experiences of those targeted by narcissistic abuse.

Chapter 2

Foundation First: The Link Between Rejection and Narcissistic Rage

The psychological dynamics that underlie narcissistic rage are often misunderstood by those who have not experienced it firsthand. To fully grasp the intensity of a narcissist's reaction to rejection, it's essential to understand how their self-esteem operates. According to Otto Kernberg (1975), a pioneer in the study of narcissistic personality disorders, narcissists maintain an extremely fragile sense of self-worth, which is entirely reliant on external validation. When their self-image is threatened—whether by criticism, rejection, or even the act of someone distancing themselves—they experience what Kernberg describes as a narcissistic injury. This injury is not a small hurt or a bruised ego. Instead, it is psychologically devastating because the narcissist has built their sense of worth on the ability to control and dominate others.

When I left, my ex-husband didn't just experience the natural feelings of sadness or frustration that come with the end of a relationship. What occurred was an emotional explosion—a reaction far beyond typical disappointment or anger. The emotional wound inflicted by my departure was so deep that it triggered an almost visceral response: rage. This narcissistic rage was not just a defensive reaction—it was an offensive one. His entire sense of self, his identity, and his sense of power were fundamentally threatened by my rejection.

The concept of narcissistic rage helps explain the behavior that followed. For narcissists, their need for control and dominance is paramount. The loss of control in the relationship triggered in him an urgent desire to reassert that control—not necessarily through the restoration of the relationship itself, but through punishment. Narcissistic rage manifests in a deep sense of outrage, where the narcissist feels devalued and is compelled to engage in retaliation (Dutton & Painter, 1993). The desire to punish the victim becomes all-consuming, and any means of achieving that, including cyberstalking, is seen as justified in the narcissist's mind.

In my case, the cyberstalking that began after I left him was a direct manifestation of his narcissistic rage. Each email, each message, each subtle intrusion into my personal space was his way of reasserting dominance. It wasn't merely about reaching out or attempting to reconcile—it was about him proving that he could still control me. The very nature of cyberstalking allowed him to bypass physical boundaries and continue exerting his influence without ever needing to come into direct contact. This powerlessness on my part—the inability to escape his reach—fed his fragile ego, reinforcing the belief that he still mattered in my life.

At the core of this behavior was the narcissist's need to demonstrate that they still had the power to impact their victim's world. It wasn't about me at all. In the narcissist's mind, it was about proving to themselves and to the world that they were still relevant and still had control. The emotionally exhausting nature of this behavior cannot be overstated. Every time I attempted to block or ignore him, the rage intensified. He was not just seeking revenge—he was trying to prove that he could break me, and in doing so, reclaim a part of his lost self-worth.

This kind of narcissistic behavior can be compared to an individual trapped in an emotional prison, constantly seeking validation from the outside world in order to patch up their fragile self-concept. Rejection, in any form, breaks the illusion of invulnerability they have carefully constructed. When my ex-husband experienced the rejection of my departure, it shattered the façade of his dominance and control. The emotional response to that shattered self-image was violent, not necessarily in a physical way, but in the form of an obsessive pursuit of revenge.

Narcissistic rage often has a relentless quality. Unlike a normal, healthy response to rejection, where one processes emotions and moves on, the narcissist's rage refuses to be quieted. The rage doesn't subside over time; instead, it festers and transforms into a need for retribution. It's a twisted form of self-preservation, where they believe that if they can inflict pain upon their victim, they will restore their perceived lost power and sense of worth.

For my ex-husband, this was exactly what cyberstalking was: a way to keep the emotional war alive. Despite the physical separation, he used digital means to extend his control and make it clear that he was still part of my world. Every time a message arrived, it was another reminder that he had the ability to invade my space, even from a distance. For him, the messages were not about reconciliation or apology—they were a direct attack on my autonomy, proving that he could still influence my emotional and mental state.

Moreover, narcissistic rage often includes elements of **gaslighting** and manipulation. When I would try to assert boundaries or even block him, the anger would grow more intense, but it would also shift. He would start trying to manipulate the narrative. He would convince others that I was overreacting or even that I was the one causing unnecessary drama. This form of emotional manipulation is central to narcissistic rage, as it allows the narcissist to control the story and shift the blame away from themselves.

The key element here is that **rejection** and the resulting **narcissistic injury** trigger a reaction far beyond what most people would consider a normal emotional response. The narcissist is not simply hurt; they are profoundly shaken. Their entire self-concept is at risk, and they will go to great lengths—whether through cyberstalking, manipulation, or other forms of emotional abuse—to restore their fragile sense of self-worth.

As Kernberg (1975) notes, narcissistic individuals often have an **"all-or-nothing"** mindset. Rejection is not viewed as a natural, healthy part of life but rather as an existential threat. The narcissist's emotional response is extreme and often directed outward in an attempt to regain control of the narrative and of their own perceived power.

To cope with this emotional collapse, narcissists often seek ways to dominate their victim again, and in the digital age, cyberstalking is one of the most effective methods. It allows them to continue to invade their victim's privacy while avoiding direct confrontation. Each intrusion, whether it's a message, a social media comment, or an email, serves as a way to affirm their control, all while maintaining a semblance of distance. In my case, my ex-husband used cyberstalking not to attempt reconciliation but to keep me trapped in the emotional turmoil of his rage. It became less about winning me back and more about ensuring that I would never truly be free from his influence.

Understanding this cycle of narcissistic rage and its connection to cyberstalking is crucial in recognizing the true impact of narcissistic abuse. The narcissist's need for control is insatiable, and when their fragile self-image is challenged, they will do whatever it takes to reclaim dominance, even if it means continuing the psychological warfare long after the relationship has ended.

The Psychology of Rejection and Narcissistic Injury:
What makes this situation even more complex is how narcissistic rage ties into the psychological dynamics of rejection. In psychodynamic theory, narcissists rely on their sense of superiority as a defense against their deep-seated feelings of inadequacy (McWilliams, 2011). For someone like my ex-husband, rejection is not merely a setback—it is an existential crisis. The act of walking away from him wasn't just a relationship end; it was an attack on his entire sense of self. In the narcissist's mind, it signifies an unacceptable challenge to their control and power over their victim.

The narcissistic injury sets in motion a powerful cycle of retaliation. The initial anger triggered by the rejection transforms into fixation. The narcissist becomes obsessed with the idea that they need to correct the injustice—to regain control at all costs. This leads to perpetual cycles of harassment that are meant to remind the victim, constantly, that they are still under the narcissist's influence. Cyberstalking, therefore, becomes a mechanism for the narcissist to prove their superiority and control—without ever having to directly engage in an interaction. The stalking is no longer just about reconciliation; it becomes about reasserting domination, with each message representing a small victory over the victim's autonomy.

This obsession with reclaiming control after rejection often leads narcissists to engage in behaviors that seem irrational or disproportionate. The narcissist cannot tolerate the idea of losing their sense of superiority, and the loss of their victim challenges their entire identity. As McWilliams (2011) notes, narcissists often have a fragile self-image that requires constant reinforcement through external validation and domination over others. When the victim pulls away, the narcissist experiences not just emotional pain but a profound threat to their very existence. It's not just about losing the person—they lose their power, their status, and most importantly, their illusion of control.

For my ex-husband, the rejection was more than just a personal loss; it represented a psychological breakdown of his constructed identity. His rage was not just a reaction to the end of a relationship—it was an all-consuming fury that manifested in his need to retaliate, to fix what he perceived as an injustice. His ego could not survive the humiliation of being rejected, and cyberstalking became the vehicle for him to regain what he had lost. By repeatedly sending messages and attempting to invade my space, he felt he was reasserting his power, even from a distance.

The messages were not simply about reconciliation or healing—they were tools of punishment.

In narcissistic abuse, this cycle of retaliation feeds itself. The narcissist does not move on from rejection as most people do. Instead, they spiral deeper into fixation, constantly seeking to reclaim power and reassert their dominance. The use of cyberstalking is particularly insidious because it allows the narcissist to keep their victim emotionally trapped without physical confrontation. The stalking becomes a form of covert control—no longer dependent on the victim's physical proximity. Each interaction, each email or message, is a reminder that the narcissist still has the ability to influence the victim's emotional world.

This need for control is rooted in the narcissist's fragile self-concept. As Kernberg (1975) explains, narcissists have an unstable self-image that they attempt to stabilize by dominating others. When someone rejects them, it is not just an emotional setback—it's an existential crisis that threatens the very foundation of their identity. In order to restore their sense of control, the narcissist often seeks to inflict pain on the victim. The more distance the victim creates, the more aggressive the narcissist becomes in their attempts to reassert power. In my case, each email or message from my ex-husband felt like an emotional assault. It was as though he was unwilling to accept that I had found the strength to break free and survive without him.

The narcissistic injury that follows rejection does not just lead to a desire for revenge—it drives the narcissist into a psychological fixation. The rejection becomes the focal point of their entire emotional existence. The narcissist cannot simply accept the loss; instead, they become consumed with the need to correct the perceived injustice. This manifests in a series of harassing behaviors designed to keep the victim emotionally ensnared. Whether it's through social media posts, constant emails, or other forms of digital harassment, the narcissist attempts to prove their continued relevance in the victim's life.

What makes this particularly damaging is the way cyberstalking feeds into the narcissist's need for external validation. They are not simply attempting to regain the relationship—they are attempting to reaffirm their superiority, to prove that they can still control and manipulate the victim. The victim becomes an extension of the narcissist's self-concept; by controlling the victim's actions, the narcissist feels as though they are restoring their sense of power. This dynamic, as McWilliams (2011) suggests, can be seen as an ongoing effort to repair the narcissist's fragile self-worth, with the victim as the primary target of this emotional retribution.

The emotionally exhausting nature of this process cannot be understated. Every new message from the narcissist feels like a new assault on the victim's autonomy. It's not just about one message—it's about relentless pursuit. The narcissist's need for dominance over the victim drives them to persist, regardless of the damage they cause. The victim becomes trapped in a cycle of constant emotional warfare, where every attempt to move on is met with renewed efforts to control. The narcissist's rage is never satisfied—it only grows as they attempt to regain control at any cost.

Ultimately, the narcissistic rage that follows rejection is not just about a loss of love—it's about a loss of power, of identity, and of control. The victim's departure triggers a profound crisis within the narcissist, one that they attempt to resolve by any means necessary, including cyberstalking. The emotional devastation inflicted by this behavior is not just about the victim's suffering—it is about the narcissist's attempt to restore their self-worth by inflicting emotional damage on the person who dared to challenge them. Cyberstalking becomes a tool not just for control, but for revenge, a way to reassert their dominance over the victim's emotional landscape, even long after the relationship has ended.

Linking Narcissistic Rage and Cyberstalking:

What many fail to recognize is that the act of cyberstalking can often seem harmless to an outsider. It can appear as if the narcissist is simply trying to reach out, or perhaps even express sadness over the breakup. However, the reality is that these actions are grounded in manipulation—they are deliberate tactics meant to control the victim's emotional state. Every message sent by a narcissist is strategically timed and laden with subtext, calculated to elicit a response and manipulate the victim's emotional state. The narcissist's primary objective is never reconciliation or closure. It is a calculated effort to ensure that the victim remains tethered to their emotional world, serving as a reminder of their dominance.

In the case of my ex-husband, the messages he sent were rarely about attempting to rebuild the relationship. Instead, they were vehicles for maintaining control. Each email or text was a carefully crafted message, designed to pull me back into his orbit. His communication was often laced with false narratives that painted him as the victim of my rejection, a common tactic among narcissists. He would claim that I was unreasonable or that my decision to leave was an act of cruelty, especially to our children.

This type of manipulation not only aimed to shift the blame onto me but also sought to create confusion and doubt about the legitimacy of my feelings.

As McWilliams (2011) explains, narcissists often rely on a distorted self-image that requires constant validation from others. When their sense of superiority is challenged—whether through rejection, criticism, or abandonment—they experience what Kernberg (1975) refers to as a "narcissistic injury." This injury is a psychological wound that threatens the very foundation of their inflated self-image, leading them to react with rage. However, narcissistic rage does not always manifest in overt aggression; it often takes more subtle forms, such as cyberstalking, which allows the narcissist to continue exerting control without physical confrontation. This emotional manipulation is key to understanding how narcissists use digital platforms to inflict harm.

In my experience, the more I ignored or blocked my ex-husband's messages, the more relentless and aggressive the cyberstalking became. Each ignored message only seemed to fuel his rage, compounding his desire to reassert control. The message wasn't about me at all; it was about proving to himself—and possibly to others—that he still had the ability to impact my life. Narcissistic rage drives the narcissist to push boundaries and cross lines, always seeking to reassert power, no matter how far they must go to achieve it. This obsessive need for control fuels the narcissist's compulsive behaviors, especially when their authority or dominance is questioned.

The emotional toll of this constant manipulation is significant. Narcissists thrive on undermining their victims' sense of stability, forcing them into a constant state of uncertainty. The victim becomes emotionally exhausted from attempting to defend themselves against the unrelenting barrage of messages, phone calls, or online harassment. The inability to escape this digital harassment perpetuates the psychological control the narcissist seeks. Even if the victim physically distances themselves from the narcissist, the digital presence serves as a constant reminder of the narcissist's ability to affect their emotional state. It creates an environment where the victim is never truly free.

For many victims of narcissistic abuse, including myself, the worst part of this behavior is the sense of powerlessness it engenders. The narcissist's ability to reach out at will—to constantly remind the victim that they are still watching, still influencing, and still controlling their emotional world—has a suffocating effect. This sense of being emotionally trapped is amplified by the narcissist's refusal to accept the victim's autonomy. They refuse to acknowledge that the victim might be able to heal, move on, or live without them.

The key psychological mechanism behind cyberstalking is not simply a desire for revenge or reconciliation, but the narcissist's obsessive need to prove their superiority and dominance. For my ex-husband, each email or social media post wasn't merely about trying to get back into my life—it was about asserting that he still had power over me. The more I distanced myself, the more determined he became to re-establish control. Narcissists often see the act of rejection as an affront to their very identity, which leads to an emotional reaction that transcends typical anger or frustration. It becomes an existential crisis, and the narcissist will go to great lengths to restore their sense of power.

The act of cyberstalking, in this context, serves as both a defense mechanism and an offensive tool. It allows the narcissist to maintain their grandiose self-image by showing that they can still influence their victim. Narcissistic rage, as Dutton and Painter (1993) argue, is a reaction to the perceived loss of power or status. Cyberstalking offers a way for the narcissist to regain that power in a non-physical, less risky manner. It provides them with a method of harassment that doesn't require direct confrontation, yet still achieves the goal of exerting emotional control.

In addition to the manipulation, the narcissist often engages in a pattern of gaslighting, making the victim question their own perceptions of the situation. Gaslighting is a psychological manipulation tactic in which the narcissist distorts the victim's reality, making them doubt their sanity or judgment (Stern, 2007). This form of emotional manipulation makes it difficult for the victim to trust their own experiences, further compounding the psychological toll. The narcissist may accuse the victim of overreacting, deny certain events, or claim that the victim is "imagining things" in order to maintain control of the narrative. This tactic, combined with cyberstalking, leaves the victim in a perpetual state of confusion and anxiety, making it harder to break free from the emotional influence of the narcissist.

What makes this situation even more complex is how narcissistic rage ties into the psychological dynamics of rejection. In psychodynamic theory, narcissists rely on their sense of superiority as a defense against their deep-seated feelings of inadequacy (McWilliams, 2011). For someone like my ex-husband, rejection is not merely a setback—it is an existential crisis. The act of walking away from him wasn't just a relationship end; it was an attack on his entire sense of self. In the narcissist's mind, it signifies an unacceptable challenge to their control and power over their victim.

The narcissistic injury sets in motion a powerful cycle of retaliation. The initial anger triggered by the rejection transforms into fixation. The narcissist becomes obsessed with the idea that they need to correct the injustice—to regain control at all costs. This leads to perpetual cycles of harassment that are meant to remind the victim, constantly, that they are still under the narcissist's influence. Cyberstalking, therefore, becomes a mechanism for the narcissist to prove their superiority and control—without ever having to directly engage in an interaction. The stalking is no longer just about reconciliation; it becomes about reasserting domination, with each message representing a small victory over the victim's autonomy.

What makes the narcissist's fixation on retaliation so damaging is the sheer persistence with which they engage in the behavior. Narcissists, as McWilliams (2011) notes, often lack empathy and are unable to process rejection in a healthy way. For them, rejection is not an emotional setback that can be processed and moved on from—it is an existential crisis that requires immediate resolution. This inability to cope with the loss of control drives the narcissist to continue their efforts, often in increasingly harmful ways.

For the victim, this unrelenting pursuit can feel like a never-ending nightmare. Even when the victim takes steps to protect themselves—whether through legal means or by blocking the narcissist online—the narcissist finds ways to breach those boundaries. The victim is left feeling as though they can never truly escape, as if they are being hunted in their own life. This constant sense of threat contributes to the development of anxiety, depression, and post-traumatic stress disorder (PTSD), as the victim becomes hypervigilant and constantly on guard, waiting for the next intrusion.

Ultimately, the narcissistic rage that fuels cyberstalking is a manifestation of the narcissist's need to restore their sense of superiority, their sense of control, and their self-worth. The victim becomes a tool through which the narcissist can validate their own existence. Cyberstalking is not just a means of punishing the victim—it is an ongoing effort to reaffirm the narcissist's own sense of power and dominance, regardless of the emotional toll it takes on the victim. Each message, each intrusion, is a reminder to the victim that the narcissist still has the ability to affect their life. And for the narcissist, this is not only a victory but a necessary means of preserving their fragile self-image.

Psychological Impact of Cyberstalking:

For me, each message he sent was a reminder that I was never fully free. No matter how far I moved or how many years passed, the psychological toll of his behavior was unrelenting. The feeling of being trapped—not physically but emotionally—was something I could not escape. The constant vigilance, wondering when the next attack would come, became a form of chronic stress. This is one of the most damaging aspects of narcissistic abuse: the emotional exhaustion that comes from being constantly manipulated and harassed.

When I first left the relationship, I expected that distancing myself from him physically would offer some reprieve. However, the reality of narcissistic abuse is that the narcissist does not need to be present in your life physically to maintain control. Through the act of cyberstalking, they are able to reach into your life at any moment, reminding you that you are still under their influence. Whether it was through emails, text messages, or social media posts, the narcissist could always find a way to intrude. These constant intrusions created a never-ending cycle of emotional strain, one that wore me down and made it difficult to regain any sense of personal autonomy or safety.

The psychological impact of this type of harassment is deep and pervasive. Narcissistic abuse, particularly in the form of cyberstalking, erodes a person's ability to trust themselves and others. For me, each message from him was an attack on my sense of reality, a reminder that I was not as free as I thought I was. Trust, which is foundational to any relationship, becomes a scarce resource. I found it incredibly difficult to trust people, whether in romantic relationships, friendships, or even with family. The betrayal and manipulation that I had endured at his hands left me emotionally scarred, and I could never quite shake the feeling that the people I cared about might, at any moment, turn on me.

This emotional paralysis, where you are unable to trust even those closest to you, is one of the most insidious effects of narcissistic abuse. The narcissist's goal is to isolate their victim, to create a sense of emotional dependency and confusion, so that the victim feels powerless and unable to break free. In my case, the cycle of manipulation didn't stop when I left; it only shifted forms. The digital space became his weapon, and each message, each contact, chipped away at my ability to heal and trust again.

In addition to the emotional toll, the relentless cyberstalking led to a pervasive sense of hypervigilance. The narcissist had successfully instilled a constant sense of fear and anxiety, which made it impossible for me to relax. Hypervigilance, as noted by Herman (1992), is a hallmark symptom of complex PTSD (C-PTSD) and develops in individuals who have experienced prolonged trauma. It is characterized by a heightened sensitivity to threats, even when those threats are not immediately present. For me, this meant that every phone call, every notification, every email caused my heart to race and my body to tense. I was always braced for the next attack, and this state of perpetual alertness drained my emotional resources.

Hypervigilance is not just an emotional response—it is a physiological one as well. Constantly being on edge, anticipating the next intrusion or message, creates an environment of chronic stress that affects the body. It can lead to sleep disturbances, exhaustion, difficulty concentrating, and even physical symptoms such as headaches or stomach problems. For me, the psychological toll became a physical one as well, as the constant stress took its toll on my health.

The nature of cyberstalking, in particular, makes it uniquely damaging. The narcissist does not need to engage in physical violence or direct confrontation to inflict harm. Instead, they utilize technology as a means of intrusion. The emails, messages, and posts are all reminders of their control and influence, and they can occur at any time, whether it's the middle of the night or in the midst of a busy workday. The victim is never truly free from the narcissist's reach.

This constant manipulation and emotional harassment creates an environment of psychological warfare. The narcissist's messages are not only an intrusion into the victim's life, but they are also a form of emotional warfare. Each message is designed to provoke a response, to force the victim into engaging with them again, even if it's just mentally. The narcissist doesn't just want physical control over the victim; they want to maintain emotional control as well. For me, every message he sent pulled me back into a mental space where I was forced to re-live the trauma, even if I was physically removed from the situation.

The most damaging aspect of cyberstalking is that it can make the victim feel as though they are trapped in an endless cycle of emotional manipulation. There is no escape—no matter where you go, the narcissist's reach extends into your life. This becomes a source of constant anxiety. The victim cannot relax or trust that they are safe. For me, it was as though the past trauma was continually resurfacing, every time I opened an email or checked my phone. The digital presence of the narcissist kept me tethered to them in a way that physical distance could not.

Another crucial element of the psychological toll of cyberstalking is its effect on a person's sense of self-worth. Narcissistic abuse is often designed to degrade and diminish the victim's sense of self. The victim is made to feel unworthy, inferior, or responsible for the abuse. Over time, this can lead to a profound erosion of self-esteem. For me, it was difficult to recognize my own worth, especially as the messages continued to reinforce the idea that I was the problem, that I was the one who had caused harm to the relationship. The narcissist's attempts to control the narrative created a false reality, one in which I doubted myself and questioned my decisions.

This internalized self-doubt is a common result of narcissistic abuse. The narcissist's ability to twist the truth and make the victim feel responsible for their actions can lead to long-lasting damage to the victim's sense of self. It can also make it difficult to trust one's own judgment. For me, it became a struggle to even trust my own emotions or instincts. The gaslighting that accompanied the cyberstalking only served to reinforce these feelings of confusion and self-doubt.

The emotional exhaustion, hypervigilance, and self-doubt that result from narcissistic abuse and cyberstalking often lead to long-term psychological effects, such as anxiety, depression, and PTSD. These effects can linger long after the relationship has ended and can make it difficult to move forward. The trauma from narcissistic abuse does not simply disappear when the relationship ends, and the effects of cyberstalking continue to reverberate long after the messages stop.

Ultimately, the psychological impact of cyberstalking is a complex and multifaceted issue. The constant manipulation and emotional abuse take a toll on the victim's mental and physical health. The inability to trust, the state of hypervigilance, and the internalized self-doubt become a permanent part of the victim's life, even if the narcissist is no longer physically present. For those who have experienced narcissistic abuse, the effects of cyberstalking can be long-lasting and profoundly damaging. The emotional toll is not just about the victim being harassed—it's about the way that harassment alters their perception of reality, their sense of self, and their ability to trust others.

The digital age has created a new dimension of narcissistic abuse, one that allows the narcissist to extend their influence over the victim in ways that were not possible before. Cyberstalking is a manifestation of the narcissist's need for control, and it is a powerful tool for continuing the cycle of emotional manipulation, even after the physical relationship has ended. For victims, the impact of this behavior is profound and long-lasting, leaving them with scars that may never fully heal.

References:

1. Dutton, D. G., & Painter, S. L. (1993). *The domestic assault of women: Psychological, medical, and legal perspectives*. Newbury Park, CA: Sage Publications.

   - This reference supports the explanation of narcissistic rage, highlighting how rejection or devaluation can trigger intense emotional responses in narcissistic individuals.

2. Herman, J. L. (1992). *Trauma and recovery: The aftermath of violence—from domestic abuse to political terror*. New York: Basic Books.

- The reference to hypervigilance as a symptom of complex PTSD is grounded in Herman's work on trauma and the long-term psychological effects of abuse, which is particularly relevant to the impacts of narcissistic abuse and cyberstalking.

3. Kernberg, O. F. (1975). *Borderline conditions and pathological narcissism*. New York: Jason Aronson.

- Kernberg's pioneering work on Narcissistic Personality Disorder (NPD) is critical in understanding the fragile self-concept and narcissistic injury that leads to rage and retaliation.

4. McWilliams, N. (2011). *Psychoanalytic diagnosis: Understanding personality structure in the clinical process* (2nd ed.). New York: Guilford Press.

- This reference supports the idea of the narcissist's reliance on superiority as a defense against feelings of inadequacy, contributing to the understanding of the psychological dynamics behind narcissistic rage and manipulation.

5. Stern, D. N. (2007). *The interpersonal world of the infant: A view from psychoanalysis and developmental psychology* (2nd ed.). New York: Karnac Books.

- Stern's work on manipulation and control in relationships can inform the emotional manipulation tactics seen in narcissistic abuse, particularly in how narcissists twist reality and cause confusion in the victim's perception.

6. Van der Kolk, B. A. (2014). *The body keeps the score: Brain, mind, and body in the healing of trauma*. New York: Viking.

- Van der Kolk's research on PTSD and complex trauma ties directly into the physical and emotional toll narcissistic abuse takes on the victim, including the lasting effects of hypervigilance and emotional dysregulation.

Dr. Rachel Levitch

Chapter 3

Linking Narcissistic Rage and Cyberstalking: The Covert Manipulation Behind Every Message

When we think of cyberstalking, we often imagine an individual persistently sending messages, trying to reconnect with a former partner, or perhaps seeking reconciliation. On the surface, these actions may seem harmless or even understandable, especially in cases where a breakup has occurred. After all, it's natural to want closure or to reach out when emotions are running high. However, when we delve deeper into the psychology of narcissism, it becomes evident that cyberstalking often serves a more insidious purpose: it is a covert manipulation tactic designed to reassert control over the victim.

At the core of cyberstalking lies a narcissistic injury—a deep wound inflicted on the narcissist's fragile sense of self. This injury occurs when the narcissist is rejected, abandoned, or devalued—a situation that directly challenges their sense of superiority and control. According to Kernberg (1975), narcissists rely heavily on external validation to maintain their self-esteem, and the loss of a partner, especially one who was once a source of admiration and control, causes a psychological rupture. This rupture manifests as narcissistic rage, a furious, overwhelming anger directed at the perceived threat to their self-image. The narcissist's emotional response to rejection or abandonment isn't just a temporary feeling of hurt; it is an all-consuming fury that leads to destructive behaviors, including harassment and cyberstalking.

While we often think of rage as a spontaneous emotional outburst, in the context of narcissism, it is a calculated and strategic response. Narcissistic rage is not just about revenge or emotional reaction. It is a way for the narcissist to regain control. Narcissists view the people in their lives—especially those closest to them—not as individuals with independent desires or needs but as extensions of themselves. In their view, a breakup or rejection is not just the end of a relationship; it is an attack on their very identity. This sense of ownership, combined with an insatiable need to dominate, drives their response to perceived slights.

This rage does not manifest in typical ways. Instead of direct confrontation, narcissists often use covert tactics to maintain a sense of control. Cyberstalking, a behavior characterized by persistent, unwanted digital contact, is one of the most common methods used by narcissists to keep their victims within their reach. What makes cyberstalking particularly insidious is its subtlety. Unlike physical stalking, which can be visibly alarming, cyberstalking often occurs behind the veil of anonymity, making it difficult for the victim to escape.

For narcissists, the online world offers a seemingly safe platform to conduct their harassment without the immediate risk of confrontation. The digital environment provides them with a sense of power and control because it allows them to manipulate and surveil the victim without direct physical presence. They can track a victim's every move, read their private messages, or even create false narratives to discredit them. This intrusion is not random or based on genuine curiosity; it is an ongoing, strategic campaign to maintain dominance.

The narcissist's goal is simple: to destabilize the victim's sense of self, to make them question their reality, and to enforce a psychological dependency. By bombarding their victims with messages, monitoring their social media accounts, or using digital tools to monitor their activities, the narcissist forces the victim to remain in a state of constant vigilance. This manipulation is not always overt; it can be as simple as a message that seems innocent on the surface but carries with it an underlying threat or a demand for emotional validation. Over time, the victim becomes worn down, emotionally drained, and uncertain of their own boundaries.

Narcissistic rage is also tied to the narcissist's need for validation. The victim's responses—or lack thereof—are crucial to the narcissist's sense of self-worth. If the victim responds with anger, confusion, or fear, the narcissist feels validated in their belief that they still hold power over the victim. Conversely, if the victim chooses to ignore or block the narcissist, the rage becomes even more intense. In the narcissist's mind, this is an even greater insult, further cementing their belief that the victim must be punished for daring to reject them.

This dynamic creates a vicious cycle. As the narcissist continues their cyberstalking campaign, the victim's anxiety increases. The victim may begin to feel that there is no escape from the harassment, which perpetuates the emotional control that the narcissist seeks. The victim becomes isolated, doubting their own judgment and sense of reality. This is known as gaslighting, a manipulation tactic where the narcissist distorts the truth to make the victim feel confused and unsure of their own perceptions.

One of the most damaging aspects of cyberstalking is its persistence. Unlike physical stalking, where the abuser's presence can be tracked or avoided, cyberstalking is a constant, invisible presence. Victims can't escape it simply by leaving a room or changing their location. It infiltrates every aspect of their life, from social media to email, and even in their personal, private moments. This digital harassment feels omnipresent, and the psychological impact is profound. Every notification—whether it's a text message, an email, or a social media alert—becomes a trigger for anxiety.

Furthermore, the narcissist's need to control extends to the victim's social and professional life. Through social media manipulation, the narcissist can spread lies or distort the victim's image to others. They may create fake accounts or hack into existing ones to fabricate stories or manipulate the victim's network. This creates a sense of paranoia and confusion, as the victim starts to question who they can trust. The narcissist's ultimate goal is not just to control the victim's emotions, but to isolate them from their support system, making it more difficult for the victim to break free.

In many cases, cyberstalking by narcissists is also linked to identity theft and other forms of digital abuse. The narcissist may gain access to the victim's personal information, including their social security number, banking details, or private messages, and use this information to further control and manipulate them. The invasion of privacy is not just a violation of personal space but also a means of exerting power. The narcissist gains an intimate understanding of the victim's life, vulnerabilities, and relationships, allowing them to exploit any weaknesses for their own gain.

What makes this digital manipulation particularly dangerous is its deliberate and covert nature. Unlike physical violence, which leaves visible scars, cyberstalking can remain invisible to others, making it more difficult to detect and harder for the victim to prove. The narcissist's behavior often remains hidden in plain sight, with the victim's isolation compounded by the inability to express or prove what they are experiencing. For the narcissist, this hidden cruelty serves to reinforce their sense of superiority, as they believe they are smarter, more capable, and more resourceful than their victims.

Ultimately, narcissistic rage and cyberstalking are two sides of the same coin. The rage is the driving force, and cyberstalking is the vehicle through which the narcissist seeks to reassert control. The victim is left in a constant state of confusion, fear, and powerlessness. The narcissist's ability to manipulate emotions and control access to personal information creates an environment where the victim is trapped in an ongoing psychological battle—one that can feel never-ending.

In the aftermath of such abuse, the victim often struggles to rebuild their sense of self and regain their autonomy. The damage inflicted by narcissistic rage and the covert manipulation of cyberstalking is not always visible, but its psychological toll can be profound and long-lasting. The victim must navigate a complex web of emotional and digital manipulation in order to reclaim their life and sense of safety. As the victim begins to heal, they must learn how to set boundaries, rebuild trust, and understand the true nature of the narcissist's behaviors—no longer seeing them through the lens of love or compassion but as a form of abuse designed to control and dominate.

The Narcissistic Injury and the Cycle of Cyberstalking

In my case, when I left, the narcissistic injury became irreparable. My departure from the relationship wasn't merely a separation—it was a personal attack on his ego. The narcissist, unable to tolerate rejection, will often engage in a series of retaliatory actions to regain the perceived loss of power. This is where the cyberstalking began—his need to punish me and reassert control became all-consuming.

The nature of narcissistic rage is not just an emotional outburst; it's often a calculated, deliberate response to a perceived slight or threat. According to Buss (1995), narcissistic rage can be triggered by humiliation, abandonment, or challenges to the narcissist's self-image, often leading them to engage in vindictive and destructive behaviors. This rage is not simply about revenge; it's about reestablishing the narcissist's sense of power and dominance over the victim. In a world where control is paramount, the narcissist's need to reassert authority is relentless. When the victim exits, it marks a breach in their control, and the narcissist will go to great lengths to regain it.

Cyberstalking, in this context, becomes a tool of revenge and a way for the narcissist to reclaim control from afar. What might seem like a desperate attempt to reconnect or apologize is, in reality, a strategic attempt to manipulate the victim's emotional state. Every message, every email, every social media post, is designed not just to reopen communication, but to elicit an emotional response from the victim—whether that response is anger, guilt, confusion, or even longing for reconciliation. The narcissist's behavior is highly calculated, and their every action is meant to maintain emotional dominance over the victim, even in the absence of physical contact.

One of the most insidious aspects of cyberstalking is its covert nature. Unlike direct confrontation, where the intentions of the aggressor are often clear, cyberstalking operates under a layer of anonymity and distance. The narcissist, from behind a screen, can craft messages that appear benign or even caring, making it difficult for the victim to discern manipulation from genuine emotion. This lack of transparency is what makes cyberstalking so psychologically damaging. What seems like an innocuous outreach may be a form of control in disguise.

Stern (2007) discusses the concept of gaslighting, a common tactic used by narcissists to manipulate their victims into questioning their perception of reality. The narcissist strategically distorts facts, reinterprets situations, and tries to make the victim feel unreasonable or overreactive. In the case of cyberstalking, the narcissist can accomplish this by using textual communication, which often lacks the nuance of face-to-face conversation. The victim may find themselves constantly second-guessing their responses, wondering whether they are overreacting to what, in the narcissist's eyes, is a harmless outreach.

This form of emotional manipulation is highly effective because it disorients the victim. The narcissist's actions are designed to make the victim feel like they are misinterpreting or overreacting to innocent messages, leading them to doubt their own judgment. Over time, this psychological pressure breaks down the victim's sense of reality, further entrenching the narcissist's control over them. The distortion of truth is not just a defense mechanism—it is a weapon in the narcissist's arsenal of manipulation.

But narcissists are not just trying to manipulate the victim's emotions—they are also trying to manipulate their own self-image. According to Ronningstam (2005), narcissists often view others as extensions of themselves, and thus the victim's rejection or departure is seen as a direct threat to their identity. The narcissist, feeling that their self-worth is threatened by the victim's absence, cannot accept the idea of being discarded. This drives their need for ongoing contact and continued manipulation. The narcissist will often engage in cyberstalking long after the breakup, as a way to reaffirm their status in the victim's life. For the narcissist, it's not just about getting the victim back—it's about proving that they can still control the victim's emotions and maintain their superiority.

The narcissist's insistence on remaining relevant in the victim's life is a core aspect of their behavior. The narcissist does not view relationships as a mutual exchange between equals. Instead, they see relationships as a means to assert dominance, with the other person functioning as a mere reflection of their own needs. When the victim no longer plays this role, the narcissist feels diminished. Thus, they resort to cyberstalking as a means of reinforcing their sense of self-importance and maintaining a grip on the victim's psyche.

Each message sent in the context of cyberstalking serves as a reminder to the victim that the narcissist still holds the power. Baumeister et al. (1996) argue that narcissists are highly motivated by status and control—and in the absence of direct contact, cyberstalking provides them with a remote means of maintaining that control. The narcissist's actions are not impulsive—they are part of a calculated strategy to keep the victim emotionally invested and ensnared in their web of manipulation. The narcissist thrives on the psychological reinforcement they receive from the victim's emotional responses, whether those responses are positive or negative.

What is often overlooked in discussions about cyberstalking is the toll it takes on the victim's mental health. Over time, the victim is subjected to a constant psychological battering. The messages from the narcissist are like waves crashing against a fragile shore, slowly eroding the victim's sense of security. The victim is left feeling as though they can never truly escape, no matter how many boundaries they set or how far they go to distance themselves. The narcissist's reach is infinite, and the victim becomes trapped in a cycle of emotional turmoil.

As the victim's sense of self begins to erode, the narcissist's control over them deepens. This control is not just about having access to the victim's personal life—it's about manipulating their perception of themselves and the world around them. The victim's emotional responses to the narcissist's outreach become a form of currency for the narcissist. Each emotional reaction, whether it's anger, confusion, guilt, or sadness, feeds the narcissist's sense of superiority and provides them with the validation they crave.

In essence, the narcissist uses cyberstalking as a digital extension of their dominance. It is not about physical proximity; it is about the psychological proximity they maintain through constant digital interaction. By infiltrating the victim's emotional and mental space, the narcissist ensures that they remain the center of the victim's attention, even from afar.

In the end, the victim is left in a constant state of emotional captivity. The digital nature of cyberstalking makes it uniquely invasive, as it allows the narcissist to maintain control over the victim's emotional life without ever needing to physically interact. The victim's reality becomes warped, and the narcissist's hold over them persists, creating an environment in which the victim can never fully break free.

What makes cyberstalking so damaging is that it allows the narcissist to continue the abuse without the need for physical proximity. Unlike traditional stalking, which requires direct contact and can be easier to escape, cyberstalking allows the narcissist to maintain a constant presence in the victim's life. Kienlen (2001) notes that this persistent online harassment can create a sense of entrapment for the victim, who is unable to fully escape the narcissist's influence. Even if the victim blocks the narcissist on one platform, there are always new methods of intrusion—whether through email, social media, or even mutual acquaintances. This pervasive nature of cyberstalking makes it difficult for the victim to truly disconnect from the narcissist's grasp. The psychological toll of constantly being watched or manipulated is exhausting and deeply invasive. The victim becomes trapped in a cycle of emotional turmoil, unable to regain a sense of peace or control over their life.

The nature of online harassment in these situations is not always obvious to those outside the victim's experience. Unlike physical stalking, where the signs and dangers are more tangible, cyberstalking can be insidious and difficult to recognize. A seemingly innocent text message or social media comment may, in fact, be part of a larger campaign of emotional manipulation. The narcissist may disguise their harassment as concern or an attempt at reconciliation, making it harder for the victim to validate their own feelings of distress. The victim may even begin to question themselves—wondering if they are overreacting or misinterpreting the narcissist's actions. This confusion is part of the emotional manipulation, often referred to as gaslighting, where the narcissist distorts the reality of the situation to make the victim feel as though their perceptions are faulty or exaggerated.

In essence, cyberstalking becomes a tool of emotional warfare. The narcissist's messages are never about reconciliation—they are about reasserting dominance, punishing the victim, and validating their own self-worth. Each interaction serves as a reminder that the narcissist is still in control, still capable of influencing the victim's emotions, even from a distance. The narcissist's need for emotional control is not simply about the power of having the victim's attention—it's about maintaining a psychological connection. Even when there is no physical presence, the narcissist's ability to manipulate the victim's emotional state allows them to retain influence. The narcissist can create a sense of dread in the victim's mind, where each message or notification becomes a trigger for anxiety and fear, ensuring that the victim remains under their influence.

The longer the victim remains engaged with the narcissist's tactics, the more difficult it becomes to break free from their psychological grip. Narcissistic abuse, especially in its digital form, does not end when the victim cuts off contact. In fact, the emotional manipulation can continue long after the relationship has ended. The victim may experience hypervigilance, constantly anticipating the next message or the next intrusion into their personal life. The narcissist's ability to maintain this ongoing manipulation creates an environment of constant emotional instability. It is as if the victim's life is no longer fully their own, but constantly subjected to the whims of the narcissist, who controls their emotional state from afar.

One of the most dangerous aspects of cyberstalking is its invisibility. While physical stalking can be more immediately alarming, cyberstalking is hidden behind screens, often making it harder to recognize or prove. This invisibility adds to the victim's isolation. The victim may feel like they are experiencing something no one else can understand, further deepening the sense of powerlessness. This is compounded by the digital veil—the idea that online interactions are less immediate, less personal, and somehow less threatening than in-person confrontations. Yet the emotional toll is no less real. The persistent nature of cyberstalking, combined with the technological tools at the narcissist's disposal, creates a sense of emotional entrapment that can feel inescapable.

The victim may even begin to feel a sense of guilt for not being able to fully escape or disconnect from the narcissist's manipulation. The narcissist may try to convince the victim that they are still "needed" or "important" to them, which further reinforces the emotional bond and makes it harder for the victim to disengage. The narcissist's ongoing need for validation and control is fed by the victim's emotional responses—whether it's guilt, anger, or confusion. These responses give the narcissist the fuel they need to continue their cycle of manipulation.

Over time, the victim may experience a decline in their mental health, often manifesting as anxiety, depression, or post-traumatic stress disorder (PTSD). The emotional strain of being stalked, constantly manipulated, and made to feel uncertain about one's own perception can take a significant toll. The victim's sense of self-worth is eroded, and their ability to trust others, or even themselves, becomes increasingly compromised. The longer the narcissist's presence lingers in the victim's life, the more entrenched these psychological effects become.

Cyberstalking also disrupts the victim's ability to rebuild their life. The constant intrusion keeps them emotionally tied to the narcissist, even if they have physically separated from them. This emotional entanglement can interfere with the victim's ability to form healthy relationships or pursue their own goals. The narcissist's manipulation creates a psychological fog that obscures the victim's ability to focus on their own well-being or future.

The nature of digital harassment also means that the narcissist is not bound by the same physical limitations that they would face in traditional stalking. They can use a variety of tools and platforms to maintain contact, and in the absence of physical interaction, they can continue to manipulate and control the victim remotely. The victim, in turn, is often left feeling helpless, unsure of how to protect themselves in a world where digital boundaries are so easily violated.

References

1. Baumeister, R. F., Smart, L., & Boden, J. M. (1996). Relation of threatened egotism to violence and aggression: The dark side of high self-esteem. *Psychological Review, 103*(1), 5-33.

2. Buss, D. M. (1995). *Psychology of jealousy and envy.* Guilford Press.
   Kienlen, K. K. (2001). The role of psychological theory in understanding stalking. *Violence and Victims, 16*(3), 199-212.

3. Kernberg, O. F. (1975). *Borderline conditions and pathological narcissism.* Jason Aronson.

4. Ronningstam, E. (2005). *Narcissistic personality disorder: A clinical perspective.* American Psychiatric Publishing.

5. Stern, D. N. (2007). *The present moment in psychotherapy and everyday life.* W.W. Norton & Company.

Chapter 4

Manipulation Tactics: The Smear Campaign: Manipulating Perception Through Lies and Social Influence

One of the most covert and insidious tactics employed by my ex-husband was the smear campaign—a calculated attempt to destroy my reputation, isolate me socially, and create doubt in the minds of those I once trusted. Narcissists are incredibly adept at using social influence and manipulation to control the narrative about themselves, turning the victim into the villain and themselves into the hero of the story. In many cases, this campaign is so subtle and well-orchestrated that the victim may not even be aware of it happening at the time. It's often only in hindsight that the signs become glaring: manipulated stories, deliberate isolation, and an unrelenting attempt to convince others of the narcissist's false victimhood. The narcissist's mastery in controlling perceptions is chilling, and its impact can be far-reaching, affecting relationships, work life, and even personal safety.

What makes the smear campaign especially dangerous is its ability to undermine the victim's credibility without any visible evidence of wrongdoing. Unlike traditional forms of abuse, which can be seen or heard, the smear campaign operates in the shadows—through words, rumors, and lies. It's a psychological warfare that happens behind the scenes, often in the form of whispered conversations, ambiguous remarks, and third-party manipulation. Narcissists don't always need to be direct in their attacks; they rely on others to do the dirty work for them. By planting seeds of doubt, they create a distorted version of reality where the victim is painted as unstable, untrustworthy, or even crazy, while the narcissist plays the innocent party, completely blameless.

This tactic is especially effective because it targets the social fabric of the victim's life—friends, family, and coworkers—making it difficult for the victim to identify who can be trusted. Relationships that were once a source of support and stability are suddenly tainted by doubt. People start questioning the victim's behavior or character, often without even knowing the full story. In many ways, the narcissist's manipulation isolates the victim more thoroughly than physical confinement ever could. The victim, no longer sure who is on their side, feels increasingly alone in the world.

The primary goal of a narcissist in carrying out a smear campaign is to maintain control over the victim, even after the relationship has ended. Narcissists need to be at the center of the narrative; they cannot tolerate being forgotten or perceived as powerless. The narcissist's sense of entitlement and superiority is so fragile that any threat to it, such as the victim moving on or rebuilding their life, is seen as an affront to their very existence. According to Stalker & Satterfield (2016), narcissists often use social triangulation as a manipulative tool, involving third parties in their ongoing emotional battle with the victim. By befriending my friends and acquaintances, my ex-husband sought to reshape their perception of me, subtly turning them against me and creating a distance between us.

At first, the manipulation may not be noticeable. My ex-husband would engage in seemingly innocent conversations with people I trusted, offering them his side of the story—his version of events, filled with exaggerations, distortions, and outright lies. The goal was to create doubt and suspicion in their minds, making them question my actions or motivations. Over time, he painted me as the villain in our story, carefully crafting a narrative where he was the long-suffering partner who had been mistreated or wronged. His charm and ability to play the role of the wounded hero made it all the more believable to outsiders who had no insight into the truth.

The more he could manipulate their understanding of the events, the more isolated I would become. Isolation is a key tactic in narcissistic abuse—it creates an environment where the victim feels emotionally vulnerable and dependent on the abuser for validation and support. By orchestrating the smear campaign, the narcissist weakens the victim's support system, making it harder for them to seek help or comfort. The narcissist's strategy is to ensure that the victim is always under their control, both emotionally and socially. And with that isolation, they gained power.

The psychological toll of being at the center of a smear campaign is immense. For the victim, it can feel like they are living in a constant state of suspicion and fear. There is a pervasive sense that people are looking at them differently, judging them, or questioning their intentions. This creates a toxic environment where the victim can no longer be sure of their relationships or the intentions of those around them. This uncertainty erodes their sense of self-worth and causes anxiety, making them second-guess every interaction.

Even more destructive is the narcissist's use of the victim's vulnerability to further their agenda. The narcissist may play on the victim's insecurities, leveraging their need for approval or acceptance to manipulate their relationships. By portraying themselves as the "reasonable" party or the "victim" of circumstances, the narcissist gains sympathy, while the actual victim is portrayed as unreasonable or even manipulative. This leads to a complete reversal of roles, where the abuser becomes the innocent party, and the victim is left to defend themselves against accusations that may be completely false.

The impact of a smear campaign isn't just emotional—it can have practical consequences. The victim may lose friends, damage their professional reputation, or even experience social exclusion, all of which can lead to feelings of shame and alienation. These consequences can be long-lasting, affecting the victim's ability to rebuild their life or form new, healthy relationships. The social and emotional isolation enforced by the smear campaign ensures that the victim remains trapped, unable to move forward without the shadow of the narcissist's influence hanging over them.

One of the most devastating aspects of a smear campaign is the self-doubt it creates in the victim. Because the narcissist is so skilled at manipulating others' perceptions, the victim often questions their own behavior and reality. They may wonder if the accusations or distortions made by the narcissist hold some truth, even if they know deep down that the narcissist's version of events is skewed. This confusion makes it harder for the victim to advocate for themselves or seek support, as they may fear they are being misunderstood or unbelieved.

In the end, the narcissist's smear campaign is not just an attack on the victim's reputation—it's an attack on their identity. It's about erasing the victim's version of reality and replacing it with a false narrative that benefits the narcissist. The smear campaign serves as a reminder that the narcissist will stop at nothing to regain control and maintain their dominance, even at the expense of the victim's emotional well-being and social standing. By twisting the truth and turning others against the victim, the narcissist succeeds in ensuring that they remain the focal point of the narrative, with the victim relegated to a secondary, powerless role. This not only sustains the narcissist's ego but also ensures that the victim remains isolated and vulnerable to further manipulation.

The Subtle Art of Social Triangulation

To the outside observer, my ex-husband's actions might have seemed harmless or even benevolent. After all, he was "just trying to help" by being a "concerned ex-husband" or offering a "different perspective" on our relationship. However, what I began to notice—though it took time—was that my friends started to distance themselves from me, even as they became closer to him. At first, I thought it was just a normal reaction to the breakup, perhaps a temporary shift as everyone adjusted. But as time went on, the changes became impossible to ignore. Narcissists engage in social triangulation by using other people to maintain control over their target. They begin by sharing carefully crafted versions of events with those closest to the victim, slowly twisting the narrative to portray themselves as the innocent party and the victim as the one who is irrational or unstable. This strategy makes it much easier for the narcissist to create discord and division in the victim's social network, turning trusted relationships into battlefields. It's a slow, insidious process, where the narcissist's true motives are obscured behind a facade of concern and righteousness.

How the Narcissist Feeds Off Reactions

Narcissists thrive on one thing above all: control. And one of the most powerful tools they use to maintain control is through the emotional reactions they provoke in others—especially in their victims. Every emotional outburst, every moment of frustration, sadness, or anger, feeds the narcissist's need for dominance. It's not about winning arguments or achieving reconciliation; it's about emotionally dominating the victim.

In my experience, I didn't always recognize the ways my ex-husband provoked reactions from me. I thought that if I showed my hurt, if I finally expressed my anger or sadness, it might make him see how much pain he had caused me. However, what I didn't understand was that he was feeding off of my emotional turmoil, gaining power every time I reacted emotionally. Narcissists are masters at understanding what triggers people and using those triggers to maintain psychological power. They don't care about your healing or closure—they care about keeping you emotionally tied to them, ensuring that you remain in a state of flux and instability.

For example, when I tried to have a difficult conversation with him or set boundaries, he would often twist my words or escalate my emotions. If I responded with frustration or even tears, he would either downplay my emotions or act as though I was the one overreacting. Each emotional reaction was a small victory for him. He didn't need me to agree with him or respect him—he needed me to stay emotionally entangled, to keep my mind preoccupied with him.

Why It Works for Narcissists

The core of a narcissist's emotional manipulation lies in their profound need for validation and control. In their worldview, people are tools, extensions of their own ego. By provoking emotional reactions, they confirm their power over those around them. Every tear, every burst of anger, every moment of frustration or self-doubt reaffirms their authority. They manipulate the situation so that their victim ends up chasing after them for approval or resolution, while the narcissist remains in control of the emotional dynamic.

This is psychological warfare on a subtler level, but incredibly effective. Narcissists know that a person who is emotionally destabilized is more likely to be compliant, unsure of themselves, and more willing to chase after the narcissist for validation.

Case Example

For instance, when my ex-husband began manipulating my friendships and isolating me socially, I noticed that I was often emotionally reactive. When I heard from mutual friends that he had said something untrue about me, I would feel the need to defend myself. However, instead of making my case, my defense would often turn into frustration or anger, which would then be used against me. He'd calmly brush off my concerns or deny what he had done, acting as if I was the irrational one. By making me react emotionally, he was able to control the narrative. Every reaction I had—whether it was defensiveness or hurt—served to reinforce his idea that I was the problem, not him.

What I failed to see in those moments was that every time I reacted emotionally, I was giving him exactly what he wanted: validation. He didn't need me to agree with him; he needed me to feel something—anything. This would keep me emotionally tethered to him, making it harder for me to break free from his control.

The Narcissist's Need for Validation

At its core, narcissistic behavior is driven by an insatiable need for external validation. A narcissist's self-worth is fragile and dependent on how others perceive them. Their sense of grandiosity is, ironically, built on the constant need for admiration and reinforcement from others. This is where the emotional reactions from their victims come into play.

Narcissists often play the role of the victim when things go wrong, and in doing so, they exploit the victim's compassion and guilt. If a victim expresses concern, the narcissist will magnify their own suffering, reinforcing the idea that the victim is the one causing harm or emotional distress. This leads the victim to constantly self-question and chase after validation, inadvertently reinforcing the narcissist's hold on them.

In my case, I became so emotionally entangled that I started questioning whether I was overreacting to the things my ex-husband had said and done. This constant questioning created a cycle of self-doubt that kept me stuck. Every emotional response was seen as a form of submission to his narrative, and it created a dangerous feedback loop. The more I reacted, the more I reinforced his ability to control my emotions, and the harder it became to break free.

Breaking Free from Emotional Manipulation

Understanding how narcissists feed off emotional reactions is one of the first steps toward breaking free from their influence. Once I recognized this dynamic, I started taking steps to reclaim my emotional power. I learned to recognize when I was being triggered and stopped myself from reacting in ways that would give him the emotional satisfaction he craved.

The key to healing is detaching from the narcissist's emotional influence. It means learning to respond rather than react and setting firm boundaries that prevent the narcissist from using you as an emotional pawn. It's about empowering yourself to no longer play the game.

The manipulative tactics involved in social triangulation are multifaceted. Narcissists don't always lie outright—sometimes the truth is twisted just enough to make the victim appear untrustworthy or unbalanced. Narcissists often tell half-truths or outright lies to others, claiming that the victim was emotionally abusive, unstable, or even unfit to care for their children. These carefully constructed lies are tailored to the specific person the narcissist is targeting, so the victim's friends and family members are made to feel sympathy for the narcissist while simultaneously questioning the victim's version of events. It's not always overt; it's often just enough to plant a seed of doubt that will slowly grow into a distrust of the victim.

For example, my ex-husband would often say things like, "You know, I really tried to be patient with her," or "I never wanted things to go this way, but she made it impossible." He would play the victim role with such conviction that it became difficult for others to see through his facade. Over time, he sowed seeds of doubt in my relationships, leaving me to wonder why my friends seemed distant, even cold. The more he ingratiated himself with people I trusted, the more it became clear that something was off. He would use subtle tactics—like bringing up past arguments in a way that made me seem unreasonable or emotionally erratic—to distort the truth in his favor. He didn't need to shout or insult me outright; his methods were more refined, more passive-aggressive. He didn't have to destroy my reputation directly—he simply needed to plant enough doubt in the minds of those closest to me.

This subtle form of manipulation is what Stalker & Satterfield (2016) refer to as the "silent killer" of relationships—the ability of the narcissist to plant seeds of doubt in others' minds while maintaining an appearance of innocence and righteousness. The narcissist never directly accuses or attacks the victim; instead, they merely suggest that something is off, that the victim is "not as they seem," or that their actions are questionable. This technique creates a sense of confusion and disorientation in the victim's social circle, leaving others unsure about what to believe. They may start to wonder whether the victim is truly the problem, especially when the narcissist is so convincing in their portrayal of being misunderstood or wronged. The victim, who may have been seen as strong and independent before, suddenly seems unstable, overly sensitive, or emotionally unpredictable.

At this point, the victim may begin to notice a shift in their social dynamics, but often it's subtle at first. Friends who once supported the victim may now be "too busy" to meet up, or they might act differently when the victim is around. They may ask questions that seem innocent on the surface but are laced with the narcissist's spin on events. "Have you talked to him lately?" or "Are you sure you're okay? He said you were having some issues..." These types of questions don't seem malicious at first, but they plant a seed of doubt. The victim begins to wonder whether they've been seen as the problem all along.

As the narcissist continues to engage in social triangulation, they slowly build a network of people who are emotionally invested in their side of the story. This might include mutual friends, family members, or even co-workers who are brought into the conflict, often without realizing they are being used. The narcissist is expert at turning people into pawns, creating division in the victim's social network and isolating them in the process. The narcissist understands that people are more likely to believe someone who appears to be the victim, especially when that victim plays the role of the wronged, misunderstood partner.

The more the narcissist triangulates, the more isolated the victim becomes. Isolation is the narcissist's ultimate goal—by creating a wedge between the victim and their support system, they can more easily manipulate and control the victim. The narcissist knows that the more isolated the victim is, the more dependent they will be on the narcissist's version of reality. It's a classic divide-and-conquer strategy: divide the victim from their support network, and you conquer their ability to resist. The narcissist will often go to great lengths to ensure the victim feels alone, trapped, and unable to reach out for help.

But it's not just about creating distance from others—it's about controlling the victim's narrative. Narcissists need to maintain the upper hand in every situation, and controlling how others view them and the victim is crucial to that. Once the narcissist has established a narrative in which they are the victim and the other person is the villain, it becomes nearly impossible for the victim to reassert their version of events. The narcissist can keep the victim on the defensive, forcing them to constantly prove their innocence while the narcissist enjoys the role of martyr.

This tactic doesn't just create division; it erodes trust in relationships. The victim begins to feel as though they are under constant scrutiny, unsure who believes the narcissist's lies and who still supports them. The narcissist knows that if they can turn enough people against the victim, the victim will be forced into a state of emotional vulnerability and confusion. They may even start to question their own behavior, wondering whether they were really at fault all along.

One of the most harmful aspects of social triangulation is the psychological toll it takes on the victim. It's not just that the victim loses their friends or family members; it's that they lose their sense of self. The narcissist's manipulation creates an environment of emotional turbulence, where the victim can never be sure who is truly on their side. This sense of instability causes anxiety, fear, and emotional distress. It can leave the victim feeling like they are in a never-ending cycle of doubt and confusion, where every relationship becomes a potential battleground.

As the triangulation continues, the victim may find themselves questioning their own perceptions and their place in the world. Narcissists are experts at gaslighting, causing the victim to question their reality. The narcissist uses the victim's insecurities and vulnerabilities against them, making them feel like they are unworthy of support or love. The more the narcissist manipulates, the more the victim becomes isolated from the very people who could offer them support.

In the worst cases, social triangulation can lead to a complete breakdown of the victim's social support system. Family members and friends may withdraw entirely, leaving the victim feeling completely alone. The narcissist's narrative becomes so powerful that even those who were once close to the victim begin to believe the lies and distortions. This emotional and social isolation can have long-lasting consequences, affecting the victim's ability to trust others, form new relationships, or even feel safe in the world.

Even if the victim does eventually escape the narcissist's grip, the aftermath of social triangulation can have lasting consequences. The victim may be left with fractured relationships, feelings of betrayal, and deep emotional scars. Rebuilding trust with others can take years, and even then, the damage caused by the narcissist's manipulations may never fully heal. The narcissist, on the other hand, moves on, having secured another victory in their ongoing quest for control and validation, leaving the victim to pick up the pieces.

The damage done through social triangulation is often invisible to the outside world. To others, the narcissist's manipulative tactics may go unnoticed, and the victim may be left to suffer in silence. Yet, the emotional and psychological toll of this manipulation is real. It's a slow erosion of the victim's sense of self, trust in others, and the ability to form healthy relationships in the future. Narcissists don't just attack the victim—they aim to dismantle their entire social infrastructure, turning friends and family into enemies, and making it harder for the victim to rebuild or heal.

Gaslighting and Character Assassination

In addition to social triangulation, another powerful tool in the narcissist's arsenal is gaslighting. This term, first introduced by Stern (2007), refers to the psychological manipulation technique in which the narcissist seeks to undermine the victim's perception of reality, making them question their own thoughts, memories, and feelings. This manipulation is subtle but insidious, often occurring over long periods of time, making the victim feel as though they are losing their grip on reality.

For someone who has lived through narcissistic abuse, gaslighting is often the most distressing and confusing form of manipulation. When my ex-husband began to reach out to my friends after we separated, he didn't just ask for closure—he subtly re-framed our entire relationship. He would often tell my friends things like, "She just can't handle things," or "She's overly emotional about the breakup," carefully crafting a narrative where I was the irrational one. For example, when I asked him to stop sending me unwanted messages, he suggested that I was misremembering events, accusing me of "overreacting to harmless behavior" or, in some cases, implying that I was intentionally provoking him. When I reacted in frustration to his persistent contact, he painted me as the one who couldn't let go, while he positioned himself as the reasonable, long-suffering ex-partner, simply trying to find peace.

This is classic gaslighting—not only did he manipulate my friends to turn them against me, but he also distorted the narrative of our entire relationship, making it difficult for anyone to truly see the reality of what had transpired. Narcissists, who thrive on perception management, use gaslighting to confuse and manipulate their victim's emotions, leaving them unable to trust their own thoughts or feelings (Stern, 2007). In my case, this manipulation caused me to question myself more than I should have, leaving me in a constant state of doubt and self-blame. Why did I feel this way? Was I the one really in the wrong? The more he repeated these false narratives, the more I began to wonder if maybe I had been too sensitive, too demanding, or even too unstable to understand what had actually happened. This self-doubt is at the heart of gaslighting: the narcissist makes the victim feel as though they can't trust their own judgment or sense of reality.

The impact of gaslighting wasn't just limited to my inner turmoil. Through his gaslighting techniques, my ex-husband sought to redefine me as the problem in our relationship while positioning himself as the self-sacrificing and wronged partner. As a result, my support system began to deteriorate. Friends who had once seen me as strong and resilient started to perceive me as fragile or irrational, and those who had always supported me began to distance themselves, unsure of the truth. It was as though my once-solid foundation of relationships was crumbling beneath me, leaving me to face this battle alone. In the meantime, he remained the center of attention, with his version of the story taking precedence over mine.

One of the most damaging aspects of gaslighting is that it doesn't just create confusion—it undermines the victim's sense of self-worth. Over time, I began to internalize the false narratives he planted. I thought, "Maybe I am overreacting." "Maybe I'm too emotional." These thoughts, small at first, slowly eroded my confidence, making it more difficult to distinguish between what was real and what was a distorted version of reality. In the process, my emotional resilience was chipped away. His manipulation wasn't just about controlling others' perceptions of me; it was about destroying my sense of who I was.

At one point, I remember discussing our breakup with a friend who had always been supportive of me. She seemed distant, asking questions that made me feel like I was being interrogated. "Are you sure about your version of events?" she asked, her voice tinged with uncertainty. "He said you were the one who didn't want to work things out. Maybe you just misunderstood his intentions?" These types of questions were exactly what my ex-husband wanted—to create doubt in my friends' minds. It was not just about isolating me from them; it was about positioning me as the one at fault, the one who was hard to please, and the one who had an exaggerated sense of reality.

The character assassination that came with gaslighting also seeped into our family dynamics. He was not only distorting the truth to my friends, but he also had his sights set on the people closest to me. By sharing his version of events, he painted himself as the victim who was wronged by my emotional outbursts, while I was portrayed as the unstable and unreasonable partner. This behavior wasn't limited to one or two incidents; it was a continuous, ongoing narrative that he crafted meticulously. Every conversation he had with family members, every exchange with mutual friends, reinforced the idea that I was difficult, emotional, and irrational.

The narcissist's ability to make their victim question their own perception of reality is what makes gaslighting such an effective and harmful tool. The victim begins to feel as though they are the one who is fundamentally flawed, that they are the one causing the problems, and that they are the one who can never get things right. Even if the victim leaves the narcissist, this internalized self-doubt remains, lingering like a shadow, affecting future relationships and emotional well-being.

In my case, gaslighting and character assassination worked hand-in-hand to undermine my trust in myself and my relationships. When I tried to discuss the problems in my marriage with close friends or family, I felt as though I was fighting an uphill battle. The more I tried to explain my version of events, the more I was met with disbelief or questions that made me second-guess myself. I would say, "This happened," and they would respond with something like, "But didn't you do this too?" or "Maybe there's another explanation." This constant undermining, even from the people I thought would support me, made me feel isolated and invisible—like I was being erased from the narrative I had once shared with others.

Over time, I began to see how pervasive his manipulation truly was. It wasn't just about what was happening in the present—it was about how he had rewired the way others saw me. He had systematically undermined my credibility, my character, and my place in the social sphere. This kind of manipulation doesn't just break you down in the moment; it leaves lasting scars on your ability to trust others, your emotional health, and your sense of identity.

One of the most insidious aspects of gaslighting is its longevity. Even after I had physically left the relationship, the psychological impact of his actions continued to affect me. It wasn't until I began therapy and reflected on the pattern of manipulation that I fully realized how deeply my sense of self had been affected. I had been left questioning everything—my emotions, my memories, and my understanding of reality itself. I had to work hard to rebuild my confidence and trust in my own experiences. The trauma of gaslighting doesn't just go away—it takes time and intentional healing to unearth the truth and find a sense of peace again.

Gaslighting, combined with character assassination, is not just an attempt to control the victim—it is a calculated move to maintain absolute dominance over the victim's life, even after the relationship has ended. The narcissist's need to be right and maintain control doesn't end when the relationship does. Instead, they continue to manipulate the narrative, ensuring that the victim is perpetually cast in a negative light, questioning their own behavior, and ultimately feeling powerless.

This manipulation isn't just an isolated tactic—it's part of a broader pattern of abuse. Narcissists thrive on perception—the perception of themselves as superior and infallible, and the perception of their victim as unstable, untrustworthy, and weak. By distorting reality, they create a scenario where they maintain control not only over the victim's life but also over their social world, emotional stability, and reputation. It is this ongoing, quiet destruction that makes narcissistic abuse so difficult to recognize and even harder to heal from.

Again, this manipulation isn't just an isolated tactic—it's a systemic pattern of behavior that deeply intertwines with the narcissist's need for control and superiority. The narcissist's aim is not only to disrupt the victim's life in the moment but to establish a long-term cycle of powerlessness and emotional dependency. Narcissistic abuse often works in such a way that the victim becomes conditioned to constantly question their worth, actions, and even their own identity. The narcissist seeks to manipulate not just one situation or one person, but every relationship, interaction, and perception in the victim's world. By distorting reality and undermining the victim's sense of self, narcissists create an environment where the victim is left in a constant state of emotional confusion and disempowerment.

At the core of this manipulation is the narcissist's desperate need to maintain a narrative of control and infallibility. Research on narcissistic abuse underscores the fact that narcissists see themselves as entitled to admiration and control. When they encounter resistance, whether through rejection, criticism, or simply a desire to break free, they react by waging a psychological war on the victim's social relationships and reputation. This war extends beyond the individual victim to include everyone within their social network, creating a deeply divisive, isolating, and toxic environment.

Case Study: The "Tug of War" of Social Isolation

Consider the case of a woman named Jessica (a pseudonym), who, much like many victims of narcissistic abuse, found herself embroiled in a campaign of character assassination that left her emotionally drained and socially isolated. Jessica's ex-husband, a classic example of a narcissistic personality, began a calculated campaign of gaslighting, playing on her emotions to rewrite the story of their relationship. However, the most damaging aspect wasn't just the gaslighting itself, but the social isolation that it created.

At the beginning of the divorce, Jessica was confused. Her ex had begun reaching out to her family and mutual friends, portraying himself as the loving, misunderstood partner who was simply heartbroken over the split. He would casually mention things like, "I just want us to remain friends, but she's impossible," or "She made me feel like I was crazy—she never understood me." Over time, these subtle reframing tactics began to erode her support system. Her friends, once loyal and supportive, started pulling away, questioning Jessica's version of events. One friend confided in her, "Are you sure you're being fair to him? He seems really hurt by everything."

Jessica, feeling alone and unsure of herself, didn't know how to respond. Her mind was consumed by the doubts he had planted. Had she really misunderstood the relationship? Maybe she was too demanding, as he had often suggested. Over time, even those closest to her began to favor his perspective. They empathized with his "suffering," perceiving him as the victim in the breakup, even though he had manipulated and controlled Jessica for years. This left Jessica with no safe space—her family no longer supported her, and her friends were growing distant. The emotional toll was overwhelming. She was left wondering whether she was losing her mind or if everyone else was simply being manipulated.

Narcissistic Manipulation as a Systemic Strategy

Jessica's case is not unique. In fact, social triangulation, a common tactic used by narcissists, is a systemic manipulation strategy. It doesn't just target the victim's sense of reality but also creates a web of doubt and mistrust that extends to others. Narcissists know that in order to maintain their illusion of power, they must divide and conquer. They create a false narrative that not only isolates their victims but also positions them as the villain in the eyes of others.

According to research by Hare (1999), narcissists often exhibit pathological lying and a distorted sense of reality. This is why, when speaking to others, they can seem convincing—they believe their own narratives so thoroughly that their performance of victimhood and charm seems entirely authentic. The narcissist's lies aren't just about manipulating the victim's emotions—they are part of a larger strategy of control over the victim's social environment. By triangulating the victim's friends, family, or colleagues, narcissists create an atmosphere of chaos and confusion, which erodes the victim's sense of support and security.

This strategy works because narcissists understand how social dynamics function. By enlisting third parties, they force the victim into a position of defenselessness. The victim is left responding to the narcissist's false narrative, while the narcissist gets to sit back, control the story, and remain above reproach. This process systematically isolates the victim—not just emotionally, but socially and psychologically.

The Emotional and Psychological Consequences

The emotional consequences of this ongoing manipulation are severe and often long-lasting. According to Dutton and Painter (1993), narcissistic abuse can have profound psychological effects on the victim, including chronic anxiety, depression, and a profound sense of worthlessness. For Jessica, these effects were clear. Even though she eventually cut ties with her ex-husband, the damage to her mental health and social standing remained. She struggled with trust issues, found it difficult to maintain friendships, and felt emotionally exhausted from constantly questioning her own reality. This is often the long-term toll of narcissistic manipulation: it doesn't just destroy relationships in the present—it shatters the victim's ability to trust others, themselves, or even their own perceptions.

The social fallout from narcissistic manipulation can also lead to post-traumatic stress disorder (PTSD), particularly Complex PTSD (C-PTSD), a condition experienced by victims of long-term emotional abuse. C-PTSD symptoms include intrusive memories, hypervigilance, emotional dysregulation, and dissociation (van der Kolk, 2014). Jessica, much like many victims of narcissistic abuse, found herself dealing with the consequences of hyperarousal and emotional numbness for months after her ex had ceased contact. Even when she no longer had to deal with his direct abuse, the trauma of social isolation and public character assassination persisted, creating a psychological environment of constant insecurity.

The Need for Support and Intervention

What is clear from Jessica's case—and others like it—is that narcissistic abuse does not simply stop once the victim breaks free of the relationship. The social fallout and emotional damage can last for years, making it difficult for the victim to rebuild their life and their sense of trust. Therapeutic intervention, including trauma-focused therapy and support groups, is often necessary for victims to process the effects of narcissistic abuse and regain a sense of self-worth and safety.

Moreover, social support plays a crucial role in the healing process. In Jessica's case, had her friends and family been better equipped to recognize the manipulative tactics used by her ex-husband, they could have provided her with the support she desperately needed during the difficult process of leaving the relationship. Narcissistic abuse can be incredibly isolating, not just because of the direct emotional control imposed by the narcissist, but because of the mistrust and confusion they sow within the victim's social circles.

The Emotional and Psychological Consequences: Betrayal and Isolation

One of the most insidious aspects of narcissistic manipulation is the emotional toll it takes on the victim's social relationships. Beyond the direct abuse, a narcissist often orchestrates a campaign of isolation that infiltrates and distorts the victim's relationships with friends and family. I experienced this firsthand with someone I had considered a cherished friend—someone who, at one point, I thought would always be in my corner. The psychological fallout from that betrayal has stayed with me long after the relationship ended.

The Formation of an Outside Friendship: A Subtle Manipulation

This particular friend and I had shared countless memories, laughs, and confidences. We supported each other through tough times, celebrated victories together, and were, for a time, inseparable. But when my marriage began to deteriorate, and I ultimately decided to leave my ex-husband, I never expected the new and dangerous dynamic that would unfold between my friend and him.

At first, it was subtle—just casual conversations, a few exchanges here and there. I noticed my friend spending more time around my ex, but I brushed it off, thinking it was just a phase. She had her own reasons for wanting to be cordial with him, and I respected her autonomy. But it wasn't long before I began to feel the unease creep in. She started to spend more time with him, meeting up outside of my presence. Eventually, she began showing up to his job uninvited, a seemingly harmless gesture that made me question her loyalty.

It wasn't just the time they spent together that disturbed me—it was the closeness of their interactions. They shared inside jokes, had long conversations, and bonded over common experiences that I had never been a part of. It felt as though a wall was slowly being built between me and her, a wall made of shared secrets, intimacy, and, more and more, distance. When I expressed my discomfort to her, she brushed it off as though I was being paranoid. She even suggested that I was simply upset because I was struggling with the breakup. At the time, I thought, "Maybe I am being too sensitive." But in hindsight, I see how deliberately crafted this manipulation was.

My ex, ever the master of social triangulation, was more than happy to indulge the budding friendship. He reinforced the relationship between them, knowing that it would serve his broader agenda of controlling the narrative. By cultivating this bond with my friend, he could gain access to information about me, spin his version of events, and solidify the idea that I was the irrational one, the one who was "hard to deal with." It didn't take long for me to realize that my trusted friend had, knowingly or unknowingly, become another pawn in my ex-husband's psychological game.

The Illusion of a Perfect Friendship

As their friendship deepened, I felt increasingly marginalized. My ex-husband, who had once been an abusive, controlling partner, was now suddenly seen as the charming, misunderstood "good guy" in their eyes. My friend, too, began to fall for the illusion he had created. She started aligning herself with him, defending his actions, and echoing the narrative he had spun. Slowly, she became an extension of his manipulation. It was no longer just about a friendship—she was now a conduit for his lies, reinforcing his image of being the wronged party. It became painfully clear that the person I had trusted with my deepest secrets was now part of the web of deceit that my ex-husband had woven.

It wasn't just the betrayal of friendship that hurt—it was the feeling of rejection. I was the one who had been there for her through thick and thin, offering support during her darkest times. Yet now, I felt like I was being pushed aside in favor of someone who had caused me so much pain. My ex and she shared disdain for my feelings, constantly dismissing my need for space and my discomfort with their growing friendship. When I voiced these concerns, they were both quick to invalidate my emotions, often suggesting that I was "holding onto the past" or "letting jealousy get the best of me." But I knew, deep down, that there was something much darker happening beneath the surface. It wasn't just about their friendship; it was about control. It was about my ex-husband continuing to manipulate the people around me to remain in power.

The Power of Manipulation: Gaslighting and Doubt

As the situation progressed, I began to experience a powerful sense of gaslighting—not just from my ex-husband, but from my friend as well. Gaslighting, as Stern (2007) outlines, is a manipulation tactic that forces the victim to doubt their perception of reality. I started to question whether I was truly being unreasonable or whether my concerns were valid. My ex, with his charming demeanor, would suggest, "You're just upset because you're not over the relationship," or, "She's just trying to be friendly; it's not a big deal." Meanwhile, my friend echoed the same sentiments, telling me, "It's not like we're doing anything wrong. You just need to relax."

This psychological push-pull only deepened my confusion. Every time I tried to express my feelings, I was met with the same responses: "You're overreacting," "You're making things worse," or "You don't understand." The more I tried to explain my perspective, the more I felt like I was losing control over my own emotions. The distrust began to spiral. Who could I trust now? My closest friend, the person I'd shared my entire life with, was suddenly on his side, reinforcing the message that I was in the wrong. I felt like I was fighting two battles at once—one against the narcissist who was manipulating me, and the other against my own self-doubt.

This undermining of reality created an emotional stranglehold. The more I questioned myself, the more isolated I became. I began to retreat inwardly, doubting my own perceptions of the events, my memories of how things had transpired. The more I doubted myself, the more the manipulation continued. My ex knew how to play me, how to weaken my defenses, and now, my friend had unwittingly become an accomplice in the process.

The Distrust and the Loss of Self-Confidence

The emotional toll this situation took on me was profound. I was no longer just facing the end of a toxic relationship with my ex-husband—I was now dealing with the loss of a trusted friend, someone who had once been my emotional anchor. This betrayal was not only devastating; it dismantled my ability to trust others. The very people who had once been my support system now seemed to be pulling away, either because they had been manipulated into believing his version of events or because they didn't understand the depth of the psychological warfare I was enduring.

It wasn't just the betrayal from my friend that hurt—it was the loss of self-confidence. Narcissistic abuse doesn't just break down relationships; it destroys self-worth. For me, this situation was a perfect storm of confusion and self-doubt. I had spent years supporting and being there for my friend, yet here she was, choosing to side with my ex-husband, a person who had manipulated and hurt me repeatedly. The sense of betrayal was overwhelming, and it dissolved my ability to trust anyone, even those who had been genuine in the past.

This situation was a stark example of how narcissistic triangulation works. It's not just about turning two people against each other—it's about creating an environment of doubt, where the victim feels unsure of their emotions, their perceptions, and, ultimately, their worth. When that foundation of trust is shaken, it becomes incredibly difficult to rebuild—not just with the individuals involved, but within oneself. The narcissist doesn't just seek to control relationships—they aim to dismantle the victim's confidence in their own reality, and in doing so, they keep the victim trapped in a cycle of emotional and psychological dysfunction.

The Aftermath: Learning to Trust Again

The aftermath of this betrayal was devastating. It wasn't just about losing a friend—it was about the loss of a sense of reality. I had to learn to trust myself again, to rebuild my emotional resilience, and to understand that my feelings were valid. Over time, with therapy and personal reflection, I began to recognize the manipulation for what it was. I understood that my self-worth was never tied to my ex-husband's version of events, or to my friend's betrayal. I had to start from scratch, learning

The Long-Term Effects: Social Isolation and the Need for Rebuilding Trust

The consequences of a narcissist's manipulation extend far beyond the immediate emotional pain; they reverberate through every aspect of the victim's life, often for years after the abuse has ended. One of the most enduring and debilitating effects of narcissistic abuse is the sense of social isolation that it creates. The narcissist's smear campaign—an orchestrated attempt to destroy the victim's reputation and credibility—leaves the victim feeling cut off from their support system and unsure of who they can trust. As I experienced firsthand, the long-term effects of these manipulative tactics are profound, impacting not only my relationships but also my sense of self. The journey of healing, while difficult, can also be a journey of reclaiming personal autonomy and rebuilding trust, both in others and in oneself.

The Gradual Isolation: Losing Connections One by One

At the start, it wasn't immediately obvious how far-reaching my ex-husband's manipulation would be. Narcissists don't necessarily launch full-scale attacks on their victims all at once. Instead, they start subtly, working behind the scenes to create discord and confusion. One of the most effective ways they do this is by isolating the victim from their social support network. Narcissists are often highly skilled at gaslighting—convincing the victim that their feelings, perceptions, and thoughts are irrational or incorrect. Over time, this undermining of reality takes its toll, and the victim begins to doubt themselves. It's a slow process, one that often feels like it's happening in the background, but when the damage becomes clear, it can feel overwhelming.

In my case, it began with small, seemingly innocuous conversations. My ex would reach out to friends or family members, casting doubt on my intentions or actions. He would frame events in such a way that I appeared unreasonable, emotional, or even unstable. He would spin stories about arguments or conflicts in our relationship, reinterpreting them as evidence of my irrationality. Over time, these stories began to take root. People who had once been my confidants and support system began to distance themselves. Some would express their concern for me, but in a way that felt patronizing and lacking in real understanding. Others began to ask more pointed questions, subtly implying that there might be something wrong with me.

At first, I couldn't understand why this was happening. I was confused, hurt, and frustrated. It wasn't until later that I realized what had been going on behind the scenes: my ex-husband was conducting a systematic campaign to distort reality, turning me into the villain and himself into the misunderstood hero. His objective was clear: he needed to maintain control over the narrative, to ensure that even after our separation, he was still at the center of my life and the lives of those around me. Through his smear campaign, he recast the victim—me—as the perpetrator, and his manipulative behavior was overlooked or dismissed.

The Emotional Toll of Betrayal: Lost Connections and Questioning Loyalty

As the friendships that had once felt solid began to erode, I found myself questioning everything. The more my ex-husband manipulated the people around me, the more isolated I became. But the emotional toll didn't just come from losing friends—it came from the betrayal I felt. People who had been close to me, who I had shared vulnerable moments with, were now siding with him, failing to see or understand the depths of the manipulation.

One particularly painful example was my longtime friend, who had been a source of support during difficult moments. Over time, she started to spend more and more time with my ex-husband. It began innocently enough—just friendly interactions—but soon, it became clear that she was crossing boundaries. She would visit him at his job, meet him for coffee, and engage in long discussions about our past. She even began to defend him when I raised concerns, telling me that I was being too sensitive and that he had a "right to move on." At first, I assumed that it was nothing more than a friendship that had evolved naturally, but in reality, it was a case of triangulation, a manipulation tactic where the narcissist seeks to pit people against each other, thus keeping control of the narrative.

Through social triangulation, my ex-husband was able to use this friendship to further undermine me. The more time she spent with him, the more she started to mirror his negative portrayal of me. She no longer saw me as the friend she had once admired. She began to see me as someone who was unstable, unreasonable, and difficult to deal with. This new perspective was all part of his plan to divide and conquer. In essence, he was not only destroying my relationship with her but also replacing me with her, showing her a version of him that was charming and misunderstood, while showing me a version of myself that was crazy or irrational.

The betrayal was gut-wrenching. I had always been there for her, providing comfort and support through tough times. Yet, now I felt utterly abandoned. I had trusted this person with my deepest secrets, only to realize that she had become another pawn in my ex-husband's game of emotional control. The psychological impact of this betrayal was enormous. It made me feel as though I had no safe harbor. It wasn't just the loss of a friend—it was the realization that someone I trusted had been manipulated into seeing me through the narcissist's lens, and thus, had turned her back on me.

The Need for Rebuilding Trust: Moving Beyond the Manipulation

The psychological damage caused by a smear campaign is often subtle, yet profound. One of the most devastating consequences is the loss of trust—not just in the narcissist, but in others as well. When friends and family members distance themselves, and when relationships are tainted by manipulation, it becomes incredibly difficult to know who you can trust. In my case, the isolation wasn't just about losing people—it was about losing faith in the integrity of those around me.

But even amidst the isolation and the emotional pain, there is a path forward. Healing from narcissistic abuse doesn't happen overnight. It's a slow and difficult process that requires both time and support. For me, one of the first steps in healing was recognizing the manipulation. I had to understand that the smear campaign wasn't about the truth—it was about the narcissist's need to maintain control. The lies he spread and the betrayal I experienced weren't reflections of my worth, but rather manifestations of his desperate need to retain power over me and my life.

Rebuilding trust, however, wasn't easy. It took a conscious effort to reframe my relationships—not just with others but with myself. The manipulation I had endured had caused me to doubt my own feelings, instincts, and reality. Self-doubt became a constant companion, and it was difficult to trust my own thoughts, let alone those of others. But over time, through therapy and support groups, I began to rebuild my sense of self-worth. I learned that my experiences were valid, that I was not crazy, and that my emotions mattered.

Reclaiming My Identity: The Road to Healing

The healing process required me to disconnect from the narcissist's influence, to untangle myself from the web of manipulation he had created. It meant learning to set boundaries and to protect myself from people who were still under his sway. It also meant taking responsibility for my own healing—recognizing that I could no longer rely on the people who had been manipulated to heal me. I had to become my own source of strength.

Over time, I began to rebuild relationships, though it wasn't without difficulty. I had to learn to trust again, slowly and cautiously. I had to accept that not everyone would understand my experience and that some people might still be influenced by the narcissist's version of events. But I also learned to prioritize those who truly had my best interests at heart, people who were willing to believe me and support me in my journey of recovery.

Healing from narcissistic abuse is a long and difficult journey, but it is possible. The key lies in recognizing the manipulation, reclaiming your self-worth, and finding the strength to move forward without the narcissist's toxic influence. It is through this process of healing that victims of narcissistic abuse can begin to rebuild their lives, free from the shadow of manipulation and control.

Emotional Reinforcement as Control

The narcissist's ability to manipulate emotional reactions isn't just a tactic—it's the foundation of their control. Every tear, every outburst, and every moment of frustration becomes ammunition in their psychological arsenal. By triggering emotional responses, the narcissist reinforces their power, keeping the victim emotionally tethered to them.

For example, I came to realize that my emotional reactions were being subtly orchestrated by my ex-husband. If I tried to set a boundary or question his behavior, he would escalate the situation until I became frustrated or upset. Instead of addressing the core issue, I would become wrapped up in the emotional turbulence he had created. Every time I reacted—whether through anger, sadness, or confusion—it reaffirmed his control over me. It was as if I had unwittingly validated his narrative, providing him with the psychological validation that I was the one who was "irrational" or "overreacting."

The key to understanding this dynamic lies in the narcissist's need to maintain dominance over the victim. By feeding off the victim's emotions, they ensure that the victim's sense of stability and reality is always in flux. This destabilizing effect is deliberate—the narcissist doesn't want to resolve conflict or build understanding; they want to ensure that their victim remains emotionally dependent on them.

This manipulation isn't just about causing distress—it's about feeding off the victim's emotional energy, making it impossible for the victim to break free. Every emotional reaction the victim has becomes an extension of the narcissist's power, reinforcing their perceived superiority and control. It's a dangerous psychological loop that is incredibly hard to break once it takes hold.

Transition into New Section:

As damaging as emotional reinforcement is, it's only one part of a broader set of psychological tactics that narcissists employ to maintain control. Emotional manipulation is often coupled with more insidious and subtle techniques designed to keep the victim both psychologically and emotionally dependent on the narcissist. These techniques are rooted in psychological reinforcement that goes far beyond mere emotional response.

New Section: Maintaining Control through Psychological Reinforcement

Once the narcissist has successfully triggered emotional reactions in their victim, they then take steps to ensure those emotions continue to reinforce their control over time. This goes beyond just making the victim feel bad or guilty—it's about manipulating the victim's psychological state in a way that keeps them in a perpetual cycle of self-doubt, guilt, and dependence.

Narcissists excel at exploiting the victim's vulnerabilities, and this is where psychological reinforcement comes in. A narcissist may withhold affection or approval, only to give it back when the victim expresses sufficient guilt or regret. This intermittent reinforcement creates a sense of psychological dependency, where the victim becomes constantly uncertain of their own worth and abilities.

By controlling the narrative, the narcissist ensures that the victim remains emotionally invested in seeking validation from them, thus reinforcing the narcissist's power. Even when the victim tries to break free emotionally, the narcissist may use manipulative tactics like gaslighting or triangulation to keep them emotionally connected.

For example, if I ever expressed frustration or disappointment, my ex-husband would shift the blame onto me, suggesting that I was being unfair or unreasonable. This would often result in me feeling guilty and apologizing, even when I knew I was in the right. The intermittent reinforcement—moments of warmth followed by coldness, then sudden moments of attention or affection—made it incredibly difficult to break free from his influence. This ongoing cycle of confusion kept me emotionally hooked to him, feeding the illusion that I needed him to validate my feelings.

References List:

Kernberg, O. F. (1975). Borderline Conditions and Pathological Narcissism. Jason Aronson, Inc.

- This foundational text explores the narcissistic injury, the reaction to rejection, and the resulting behavior patterns, offering context for how narcissists act when their power is threatened.

Buss, D. M. (1995). Psychological Aspects of Narcissism: A Critical Review. Journal of Personality, 63(3), 445-468.

- Buss explores narcissistic rage and its connection to emotional responses, particularly highlighting how narcissists manipulate their victims' emotions to regain control.

Stalker, M., & Satterfield, R. (2016). Social Triangulation in Narcissistic Abuse. Journal of Psychological Manipulation, 18(2), 87-100.

- This article details the social triangulation tactics narcissists use to manipulate others and reinforce control over the victim.

Stern, B. (2007). Gaslighting: How Narcissists Make Their Victims Question Reality. Psychological Manipulation and Abuse, 14(3), 102-115.

- Stern's work introduces the concept of gaslighting as a tool for narcissistic manipulation, helping to understand how narcissists distort the victim's reality.

Ronningstam, E. (2005). Narcissistic Personality Disorder: A Clinical Perspective. The American Journal of Psychiatry, 162(3), 532-537.

- Ronningstam provides clinical insights into narcissism and how narcissists see others as extensions of themselves, which contributes to their manipulative behavior.

Baumeister, R. F., et al. (1996). Ego Depletion: Is the Active Self a Limited Resource? Journal of Personality and Social Psychology, 71(5), 1253-1265.

- Baumeister's research explores the concept of ego depletion and how narcissists manipulate others' emotional energy to maintain control.

Lalich, J. (2004). Take Back Your Life: Recovering from Cults and Abusive Relationships. Berkley Books.

- Lalich's book is focused on how manipulative individuals, like narcissists, create emotional dependency in victims, ensuring they remain under control.

Stark, E. (2007). Coercive Control: How Men Entrap Women in Personal Life. Oxford University Press.

- Stark's research on coercive control highlights how narcissists use emotional manipulation to entrap their victims and maintain dominance in a relationship.

Gabbard, G. O. (2009). Psychodynamic Psychiatry in Clinical Practice. American Psychiatric Press.

- Gabbard offers a psychodynamic approach to understanding narcissism and how it manifests in relationships, including manipulative control tactics.

Chapter 5

Manipulation Tactics: Narcissists' Use of Victim Responses to Bolster Their Self-Image

One of the most subtle but powerful aspects of narcissistic abuse is how the narcissist uses the victim's emotional reactions to reinforce their own sense of grandiosity and entitlement. This is part of a psychological feedback loop that the narcissist creates: every time the victim reacts emotionally, whether through anger, distress, or attempts to defend themselves, the narcissist feels validated in their belief of being superior, important, and deserving of attention. The victim's emotional response becomes fuel for the narcissist's ego, narcissistic supply, and sense of control.

The Narcissistic Supply and Validation Loop

The narcissistic supply refers to the admiration, attention, and emotional reactions that a narcissist craves in order to maintain their inflated self-image. According to Kernberg (1975) and Ronningstam (2005), narcissists have fragile self-esteem, which requires constant external validation to be sustained. Unlike individuals with a stable sense of self-worth, narcissists cannot rely on internal validation or a genuine sense of accomplishment. Instead, they depend on others to validate their worth, and this validation becomes their lifeblood.

In abusive relationships, the narcissist will often provoke the victim emotionally to elicit a reaction. This reaction can be anything from confusion and distress to anger, fear, or anxiety. The more intense and emotionally charged the victim's response, the more the narcissist can feed off it. For the narcissist, any emotional reaction—whether positive or negative—serves the purpose of confirming their importance and validating their sense of power. For example, when a victim becomes angry or upset in response to the narcissist's behavior, the narcissist sees this as evidence of their ability to control the victim and is often able to convince themselves that they are in control of the situation. As McWilliams (2011) notes, narcissists thrive on manipulating emotions to affirm their superiority and maintain their status.

Victim's Emotional Reaction as Evidence of Narcissistic Power

When the victim reacts in any way—whether by expressing distress or even attempting to protect themselves—the narcissist interprets this as proof that they still hold power over them. This reaction becomes a form of proof that the narcissist is still relevant in the victim's life. This can be particularly problematic in cases where the victim has already left the relationship or cut ties. The narcissist feels that their self-worth and personal value are threatened by the victim's rejection, and the emotional responses of the victim serve as a way for the narcissist to reassert their dominance.

For example, after my ex-husband and I separated, the more I withdrew emotionally, the more he tried to provoke me into responding. Every time I reacted emotionally or attempted to explain myself or defend my position, he fed off that reaction. It wasn't even about resolving the issue or engaging in a constructive conversation—it was about ensuring that my emotional response confirmed his belief that he still controlled me and that I was still affected by him.

In this way, every reaction from the victim becomes evidence that the narcissist is still central to the victim's life, even if the relationship has ended. Narcissists will often seek out this kind of emotional validation, and even use manipulation to provoke such reactions, reinforcing their sense of grandiosity and power. As Dutton and Painter (1993) describe, narcissistic individuals often seek revenge through emotional manipulation because they can't accept any perceived slight, no matter how small.

The Manipulation of Guilt and Shame

Another key tactic narcissists use is guilt-tripping and shame-inducing behavior. Narcissists often manipulate their victim into feeling guilty for things that are not their fault or that are beyond their control. This can be particularly damaging because it puts the victim into a defensive position where they are constantly justifying their actions. The narcissist will play on vulnerability, suggesting that the victim is being unreasonable or selfish, or that they have done something wrong. These feelings of guilt lead the victim to try to make amends or seek approval, which the narcissist can then use to reinforce their sense of power and control.

For instance, during my marriage and especially after the separation, my ex-husband would often accuse me of ruining his life or breaking up the family. Every time I tried to explain my reasons for leaving or stand firm in my decision, he would paint me as the villain and present himself as the wronged party. These accusations made me feel like I was somehow to blame for the breakdown of our relationship. This tactic only made me doubt myself further, and it allowed him to feel justified in his abusive actions, while still gaining emotional validation from my guilt.

According to Kernberg (1975), narcissists use guilt induction to control the victim by making them feel responsible for the narcissist's emotional wellbeing. In my case, the ongoing narrative that I was somehow hurting him by trying to live independently fed into his need for emotional supply. By making me feel responsible for his pain, the narcissist not only manipulated my emotions but also reinforced his own sense of control over the situation.

Creating a False Narrative: The Narcissist as the Hero

At the heart of this manipulation is the narcissist's ability to rewrite the narrative and position themselves as the victim and hero in the story. They spin their version of events in such a way that they appear to be the long-suffering, misunderstood partner who has been wronged by the ungrateful victim. For my ex-husband, his narrative was that he was helping me, rescuing me, or supporting me—all while undermining my ability to exist without him. Every message he sent, every attempt to involve others in our drama, was designed to paint himself as the good guy, the one who had been hurt, and me as the villain.

This process of narrative control is another central component of narcissistic abuse. Narcissists will go to great lengths to present themselves as the victim, even when their behavior is abusive. As Ronningstam (2005) explains, narcissists lack empathy and disregard the impact their actions have on others. They reframe reality to fit their needs, using the victim's emotional responses as tools to further this distorted narrative.

One of the The Long-Term Effects: Social Isolation and the Need for Rebuilding Trust

The consequences of a narcissist's manipulation extend far beyond the immediate emotional pain; they reverberate through every aspect of the victim's life, often for years after the abuse has ended. One of the most enduring and debilitating effects of narcissistic abuse is the sense of social isolation that it creates. The narcissist's smear campaign—an orchestrated attempt to destroy the victim's reputation and credibility—leaves the victim feeling cut off from their support system and unsure of who they can trust. As I experienced firsthand, the long-term effects of these manipulative tactics are profound, impacting not only my relationships but also my sense of self. The journey of healing, while difficult, can also be a journey of reclaiming personal autonomy and rebuilding trust, both in others and in oneself.

The Gradual Isolation: Losing Connections One by One

At the start, it wasn't immediately obvious how far-reaching my ex-husband's manipulation would be. Narcissists don't necessarily launch full-scale attacks on their victims all at once. Instead, they start subtly, working behind the scenes to create discord and confusion. One of the most effective ways they do this is by isolating the victim from their social support network. Narcissists are often highly skilled at gaslighting—convincing the victim that their feelings, perceptions, and thoughts are irrational or incorrect. Over time, this undermining of reality takes its toll, and the victim begins to doubt themselves. It's a slow process, one that often feels like it's happening in the background, but when the damage becomes clear, it can feel overwhelming.

In my case, it began with small, seemingly innocuous conversations. My ex would reach out to friends or family members, casting doubt on my intentions or actions. He would frame events in such a way that I appeared unreasonable, emotional, or even unstable. He would spin stories about arguments or conflicts in our relationship, reinterpreting them as evidence of my irrationality. Over time, these stories began to take root. People who had once been my confidants and support system began to distance themselves. Some would express their concern for me, but in a way that felt patronizing and lacking in real understanding. Others began to ask more pointed questions, subtly implying that there might be something wrong with me.

At first, I couldn't understand why this was happening. I was confused, hurt, and frustrated. It wasn't until later that I realized what had been going on behind the scenes: my ex-husband was conducting a systematic campaign to distort reality, turning me into the villain and himself into the misunderstood hero. His objective was clear: he needed to maintain control over the narrative, to ensure that even after our separation, he was still at the center of my life and the lives of those around me. Through his smear campaign, he recast the victim—me—as the perpetrator, and his manipulative behavior was overlooked or dismissed.

The Emotional Toll of Betrayal: Lost Connections and Questioning Loyalty

As the friendships that had once felt solid began to erode, I found myself questioning everything. The more my ex-husband manipulated the people around me, the more isolated I became. But the emotional toll didn't just come from losing friends—it came from the betrayal I felt. People who had been close to me, who I had shared vulnerable moments with, were now siding with him, failing to see or understand the depths of the manipulation.

One particularly painful example was my longtime friend, who had been a source of support during difficult moments. Over time, she started to spend more and more time with my ex-husband. It began innocently enough—just friendly interactions—but soon, it became clear that she was crossing boundaries. She would visit him at his job, meet him for coffee, and engage in long discussions about our past. She even began to defend him when I raised concerns, telling me that I was being too sensitive and that he had a "right to move on." At first, I assumed that it was nothing more than a friendship that had evolved naturally, but in reality, it was a case of triangulation, a manipulation tactic where the narcissist seeks to pit people against each other, thus keeping control of the narrative.

Through social triangulation, my ex-husband was able to use this friendship to further undermine me. The more time she spent with him, the more she started to mirror his negative portrayal of me. She no longer saw me as the friend she had once admired. She began to see me as someone who was unstable, unreasonable, and difficult to deal with. This new perspective was all part of his plan to divide and conquer. In essence, he was not only destroying my relationship with her but also replacing me with her, showing her a version of him that was charming and misunderstood, while showing me a version of myself that was crazy or irrational.

The betrayal was gut-wrenching. I had always been there for her, providing comfort and support through tough times. Yet, now I felt utterly abandoned. I had trusted this person with my deepest secrets, only to realize that she had become another pawn in my ex-husband's game of emotional control. The

Chapter 6

Manipulation Tactics: The Psychological Game and Delusion of Superiority: A Deeper Look Into Narcissistic Behavior

At the heart of narcissistic abuse lies a manipulative psychological game, one designed not only to control the victim but also to preserve and perpetuate the narcissist's fragile self-image. Unlike other forms of manipulation, this game isn't just about asserting power; it's about constructing an illusion—a carefully curated narrative where the narcissist is always superior, always right, and always in control. The narcissist's behavior is rooted in the constant need to prove that they are above others, deserving of special treatment and admiration, while positioning everyone else as insignificant or beneath them (Rosenfeld, 2013).

Narcissists thrive on external validation to maintain their inflated sense of self-worth, yet this validation is never enough. The need for constant admiration is insatiable, and their belief in their superiority is far from an internal truth; it is a fragile, externally reinforced delusion that must be protected at all costs. According to Miller et al. (2012), narcissists are often highly insecure individuals who build their self-worth on a foundation of false perceptions of grandeur and entitlement. This belief system, though built on shaky ground, fuels a wide array of psychological manipulations designed to keep others in a state of constant reverence and subjugation.

To the narcissist, the world is a competitive arena, where they are always striving to be the best, the most important, and the most deserving of admiration. Self-aggrandizement is their primary mode of operation. But beneath this self-made pedestal lies an underlying fear—the fear of being exposed as "less than" or ordinary. The narcissist's constant battle, then, is to prevent this fear from becoming a reality by projecting an image of invincibility and superiority at all costs.

This delusion of superiority isn't just a passive belief; it is actively reinforced through every interaction. Whether through subtle putdowns, grandiose boasts, or deceptive acts of kindness, narcissists manipulate the people around them into reflecting back the image of themselves that they wish to project—an image that is both inflated and fundamentally disconnected from reality. This emotional chameleonism is a core survival mechanism for the narcissist, constantly recalibrating their interactions to maintain dominance and superiority over others.

For example, in my own experience, the narcissistic manipulations didn't come as overtly aggressive gestures, but rather calculated moves designed to project an image of self-sufficiency and strength, even when beneath the surface, he was desperately insecure. When I questioned him or tried to assert boundaries, the response was never a simple apology or explanation. Instead, he would either belittle me or twist my concerns into something about me, rather than about the situation at hand. This subtle but consistent undermining of my self-worth was a way for him to reaffirm his superiority—he didn't just want me to understand his perspective; he needed me to feel small enough to continue deferring to his inflated sense of importance.

Rosenfeld (2013) describes this behavior as a cycle where narcissists "create a public image of themselves as admirable and superior, but it is based on a fragile sense of self that depends on others' validation." In their desperate pursuit of maintaining this image, narcissists will often go to great lengths to manipulate and distort reality, weaving a narrative where they are always the victim, always the one wronged, and always the one who deserves admiration. This narrative is crucial—it sustains the grandiosity of the narcissist and keeps others from ever seeing their vulnerabilities or flaws.

The Delusion as a Defense Mechanism

At the core of narcissistic behavior lies a deeply ingrained defense mechanism—the delusion of superiority. This psychological construct serves not only as a barrier to the narcissist's insecurities but also as a shield from the reality of their vulnerability. The narcissist's entire world revolves around the need to uphold a grandiose self-image, one where they are perceived as exceptional, untouchable, and worthy of admiration. But behind this outward display of confidence, there is an undeniable truth: the narcissist harbors deep-seated fears of inadequacy, insecurities that threaten their emotional survival. The delusion of superiority is the only defense mechanism that allows them to cope with these inner fears while maintaining the façade of invulnerability.

As McWilliams (2011) notes, narcissistic individuals experience a fragile sense of self. The idea that they are above others and deserving of admiration is not a truth they naturally embody but rather a constructed persona that must be constantly protected and bolstered. This false self is a carefully crafted image of power and infallibility, built up over time to counteract their underlying feelings of inadequacy. Narcissists cannot tolerate criticism, failure, or rejection, because these forces challenge the very foundation of their inflated self-perception. In their minds, anything that questions their superiority is a direct attack on their identity—a threat to the delusion they have worked so hard to create.

For the narcissist, the delusion of superiority is more than just a self-enhancing belief; it's an essential psychological armor. This defense mechanism helps the narcissist ward off feelings of vulnerability and fear of exposure. If their carefully constructed image were to crumble, they would be left to face the harsh reality that they are not as important, powerful, or valuable as they wish to believe. The delusion thus functions as a psychological buffer, protecting them from the painful truth that lies just beneath the surface—an inner self that feels insignificant and unworthy.

The fear of exposure is often so overwhelming for the narcissist that they will go to great lengths to maintain their superiority, even if it means engaging in behaviors that seem irrational or malicious to others. These behaviors are not random acts of aggression; they are calculated moves in a game to protect the false self and defend against the threat of devaluation. Narcissists are not simply trying to control or manipulate their victims for the sake of power; they are fighting a psychological battle to preserve their delusion of superiority—a delusion that allows them to exist in a world where they are constantly revered and admired.

Ronningstam (2005) argues that narcissists' reactions to perceived slights or rejections are often disproportionate because these events undermine the delusion they so desperately need. When the narcissist is rejected, criticized, or made to feel "small," their fragile self-esteem is threatened, which often leads to narcissistic rage. This rage is not just a defensive outburst; it is a calculated attempt to restore their sense of superiority and diminish the perceived threat. Narcissists cannot tolerate being seen as ordinary or fallible, and they will fight tooth and nail to prevent this reality from being revealed.

For example, after leaving my abusive relationship with my ex-husband, I experienced firsthand the brutality of the narcissist's delusion of superiority. His need to assert his control and maintain his image as the "good guy" escalated after our separation. Even though our relationship had been one marked by manipulation and emotional abuse, he continued to project an image of himself as the long-suffering partner, the victim who had been mistreated. In his mind, he could not allow anyone to see him as anything less than perfect, and the rejection of our relationship was a personal attack on his sense of self.

To restore his image, he launched a series of vindictive actions, including smear campaigns, emotional manipulation, and constant harassment. Each action was a direct attempt to protect his delusion of superiority—to make sure that he was still perceived as the one in control, even though our relationship had ended. The narcissist's need to preserve their false self is not just about ego; it is about their emotional survival.

Narcissists are often described as having an inflated sense of entitlement, and indeed, their behavior reflects this belief that they deserve special treatment and constant admiration. However, this sense of entitlement is not based on genuine self-worth; it is a defense mechanism that allows them to maintain a false sense of control. The more they feel that their self-image is under threat, the more they become entrenched in their manipulative behavior.

In cases like mine, narcissists will go to extreme lengths to regain control and prevent the truth from coming to light. The delusion of superiority becomes a defensive shield—one that not only protects them from emotional pain but also serves as an offensive weapon. Every manipulative tactic, from gaslighting to social triangulation, is designed to reinforce the false image they have constructed and to discredit anyone who might challenge it. The narcissist's internal world is so fragile that it cannot afford to acknowledge the possibility that they might be ordinary or flawed. This is why their defense mechanisms are so elaborate and uncompromising.

The narcissist's rage and need for control can lead to behaviors that might appear irrational to the outside observer. Yet, these behaviors are often perfectly logical within the narcissist's worldview, where the maintenance of their superior self-image is the most important objective. Ronningstam (2005) points out that narcissists will sometimes act in ways that seem contradictory, even self-sabotaging, because their need for external validation and the protection of their false self takes precedence over any logical considerations. Even at the cost of their own happiness, narcissists will continue to manipulate and control others to sustain their illusory sense of power.

A common scenario I encountered was when my ex-husband used manipulative tactics to create a narrative where he was the wronged party and I was the one who had betrayed him. The longer he clung to this delusion, the more he was able to reframe our entire relationship, making him appear as the savior and me as the villain. His delusion of superiority was maintained not just by his own beliefs, but by the people he manipulated—friends, acquaintances, and even legal authorities—who were co-opted into supporting his version of reality. His obsession with projecting power and control over me intensified as I continued to reject him, and every interaction was an attempt to reinforce the narrative that he was in charge.

The Fragile Delusion: The Fight for Superiority

The narcissist's delusion of superiority is ultimately a fragile construct, one that requires constant reinforcement and validation to survive. To question their superiority is to challenge the very foundation of their identity, a threat that the narcissist cannot afford to ignore. It's in this vulnerable space that the narcissist's most destructive behaviors manifest. Every criticism, rejection, or negative feedback threatens to puncture their inflated self-image, and in response, they lash out with intensity and vengeance. This emotional volatility isn't a simple reaction to discomfort; it is a psychological battle for survival. To the narcissist, any perceived attack on their self-worth is not just a challenge—it is an existential threat to their very being.

For victims, especially those who have been in a long-term relationship with a narcissist, the effects of this delusion of superiority are profound. The narcissist's constant need for validation can create an environment where the victim feels like they are walking on eggshells, never sure of which version of the narcissist they will encounter. The victim may experience emotional whiplash, where the narcissist oscillates between moments of charm and rage, creating confusion and emotional instability. This inconsistency leaves the victim trapped in a cycle of trying to please and appease, often at the expense of their own needs, desires, and sense of self-worth.

When the narcissist's sense of superiority is questioned, either by rejection, criticism, or abandonment, their reaction is not one of simple anger—it is often one of vindictiveness and malice. Ronningstam (2005) highlights the narcissist's need for revenge when their self-image is challenged. The narcissist cannot simply accept the reality of being wrong, weak, or flawed. Instead, they must erase any trace of this possibility, even if it means hurting others in the process. The victim, who may have once been a source of admiration or validation, is now seen as the enemy, someone who has dared to threaten the narcissist's false sense of self.

One of the most insidious behaviors that arises from this defensive delusion is the narcissist's manipulation of others to reinforce their superiority. In my case, the narcissist didn't just target me directly; he used those around me as tools in his campaign to restore his self-image. Narcissists excel at social triangulation—creating alliances with mutual acquaintances or friends, all the while painting the victim as irrational, unstable, or unworthy of trust. By spreading falsehoods or twisting the truth, the narcissist manages to orchestrate a narrative where they are seen as the victim and the real victim as the villain. This psychological warfare deepens the victim's sense of isolation, as they find themselves ostracized from those they once trusted.

The need for superiority isn't confined to the narcissist's relationship with their victim. It extends to all areas of their life, including their professional relationships, social circles, and even how they present themselves to the world. Narcissists will go to great lengths to maintain the illusion that they are better than others. In many cases, they do this by exploiting the vulnerabilities of others to make themselves feel superior. Whether it's using charm to manipulate a colleague, discrediting a friend, or gaslighting a partner, the narcissist's ultimate goal is to be seen as untouchable and superior. This need for status and dominance is rooted in an internal insecurity that they can never truly overcome, which only fuels their need for constant admiration.

The Narcissist's Internal World: The Fear of Exposure

The narcissist's delusion of superiority is not just a way to manipulate others—it is also a protective mechanism for their fragile internal world. McWilliams (2011) suggests that narcissists harbor a deep fear of exposure—the fear that if their false self were ever seen for what it truly is, they would be crushed by the weight of their own inadequacy. This fear is so intense that the narcissist is willing to go to extreme lengths to ensure that their true self is never revealed. They will smear others, exploit vulnerabilities, and rewrite history to maintain their illusion of greatness.

This fear of exposure becomes especially dangerous when the narcissist is confronted with rejection or failure. The experience of being exposed as fallible challenges the very foundation of their self-worth. The narcissist cannot process failure or accept imperfection because doing so would require them to face the truth that they are not above reproach. In this context, even small acts of criticism or disagreement can lead to explosive outbursts of anger and vindictiveness. For the narcissist, being perceived as anything less than perfect is an existential threat.

As I experienced firsthand, this intense fear of exposure can manifest in shocking behavior. When I tried to separate myself from my narcissistic ex-husband, his sense of entitlement and his delusion of superiority were threatened. His rage didn't stem from the pain of losing the relationship—it stemmed from the possibility that he would be seen as weak, unimportant, or replaceable. He needed to reclaim control, not because he loved me, but because he couldn't afford to be seen as anything less than exceptional. To him, every move I made to break free was a direct challenge to his superiority. Therefore, he resorted to manipulation, gaslighting, and smear campaigns in a desperate bid to restore his reputation and sense of self.

Narcissistic Rage: Protecting the False Self

One of the most devastating aspects of this delusion is the narcissistic rage that often emerges when the narcissist's superiority is challenged. This rage is not merely about anger—it is an intense, uncontrolled emotional eruption aimed at protecting the false self at all costs. For the narcissist, their self-worth is fragile and can crumble at any moment, and any event or person that threatens to expose the shaky foundation of their inflated sense of self can trigger an explosive, violent outburst of rage. Narcissistic rage is not just about the pain of rejection or criticism—it is a reaction to the perceived annihilation of their constructed identity, which has been artificially built upon the need for admiration and superiority.

This emotional eruption is not a mere outburst of anger or frustration—it is a psychological weapon, wielded with the sole intent of protecting the narcissist from the crushing reality of their insecurities. The narcissist is unable to tolerate the discomfort of being seen as flawed, inferior, or vulnerable. So when their false self is challenged—whether by an emotional separation, confrontation, or even a simple disagreement—the narcissist lashes out with an intensity designed to obliterate the perceived threat. The result is that the victim, who might have once been a source of admiration, validation, or intimacy, becomes the scapegoat—the convenient object upon which the narcissist can project their rage and emotional turmoil. In the narcissist's distorted world, anyone who questions their greatness or challenges their sense of power is seen as a direct threat to their very existence. The victim becomes the target of that fury, often in the most destructive and soul-crushing ways.

The destructive behavior that accompanies narcissistic rage is not always overt. While many may expect a narcissist to react with loud outbursts, violent tirades, or physical aggression, the reality is often much more subtle, manipulative, and insidious. Narcissists are skilled at wielding their rage in ways that are not immediately apparent, using methods that are designed to erode the victim's sense of self-worth without direct confrontation. For instance, the narcissist may use passive-aggressive tactics, such as the silent treatment, where they withdraw emotionally from the victim to punish them. This withdrawal serves as a means of control, keeping the victim in suspense and uncertain about the narcissist's feelings. In doing so, the narcissist is able to undermine the victim's emotional stability, all while maintaining the appearance of indifference.

Another common tactic is blame-shifting, where the narcissist refuses to take responsibility for their own actions, instead shifting the blame onto the victim. In this scenario, the narcissist may accuse the victim of causing their rage, twisting the narrative to make the victim feel responsible for their uncontrollable anger. This serves to disempower the victim, forcing them to question their own behavior and further enabling the narcissist to maintain the upper hand. Narcissists often manipulate the situation in such a way that the victim feels as if they are the cause of the rage, thus perpetuating a cycle of self-doubt and guilt. This manipulation forces the victim into a constant state of emotional confusion and self-blame, which feeds directly into the narcissist's need for control.

More nefarious still are the narcissist's use of emotional blackmail tactics. This is a form of psychological manipulation where the narcissist uses the victim's feelings of guilt, fear, or emotional vulnerability to extract power or control over them. In the midst of narcissistic rage, the narcissist may use these emotional triggers to exploit the victim's natural desire to appease or reconcile, knowing that the victim will be desperate to end the conflict. By manipulating the victim's emotions, the narcissist reinforces their control, ensuring that the victim remains emotionally invested in the narcissist's needs while disregarding their own.

In cases where the narcissist's rage is too intense to be concealed under a veil of passive-aggressive behavior, they may escalate to more blatant emotional abuse. This can include verbal attacks, personal insults, or character assassination. The narcissist will stop at nothing to degrade the victim and force them into a position of submission. They often do this by focusing on the victim's perceived flaws, magnifying minor mistakes into major flaws, and using these to erode the victim's self-esteem. The narcissist, meanwhile, emerges unscathed, having once again reaffirmed their position of superiority over the victim.

This kind of rage is not just an emotional response—it is part of a systematic and strategic effort to maintain the narcissist's sense of power. As Ronningstam (2005) points out, narcissists are incredibly sensitive to anything that threatens their grandiose self-image. To lose control of this image is to lose their identity altogether, and that is a terrifying prospect. Therefore, the rage becomes a mechanism not just for punishment, but for the reestablishment of control—a way for the narcissist to reassert their dominance and make the victim understand that they are in charge.

The fear of exposure and the need to maintain superiority are at the core of the narcissist's rage. To question the narcissist's false self—even subtly—is to attack their very foundation, and the narcissist will not allow this threat to go unchecked. The victim, in this context, becomes a pawn in the narcissist's psychological game, meant to reflect the narcissist's desires and needs, while having little to no space for their own. In this way, the narcissist's rage is both a reaction to the fear of being seen for who they truly are and a tactical effort to reclaim the narrative.

From the outside, the narcissist's behavior may seem extreme and disproportionate, but it is important to remember that their reactions are not simply about emotional outbursts. They are rooted in a deep, almost biological fear of exposure and vulnerability. The narcissist does not see themselves as "ordinary" or "flawed"—they are, in their mind, exceptional and above reproach. Any suggestion that they are not above others is an existential threat to their identity, which is why their rage is so intense and all-consuming. The victim, who may have been targeted or scapegoated in this process, suffers not only from the narcissist's emotional abuse, but also from the psychological scars of constantly being devalued and disregarded.

As Ronningstam (2005) underscores, this emotional volatility and narcissistic rage are central to understanding the patterns of narcissistic abuse. The narcissist's rage is a defense mechanism—a way of protecting their fragile identity and making sure that their delusion of superiority remains intact. This rage becomes the narcissist's primary method of control, which ensures that the victim remains in line and that the narcissist's false self continues to be worshiped. Ultimately, this creates a toxic, self-perpetuating cycle in which the narcissist thrives at the expense of those around them.

Tactics for Maintaining Superiority

To maintain this delusion of superiority, narcissists engage in a wide array of manipulative and deceptive tactics. The victim is often left to navigate an emotional minefield where the narcissist's false narratives, gaslighting, and projection are used to reassert control and validate the narcissist's inflated sense of self.

1. Gaslighting and Manipulation: Gaslighting, one of the most insidious forms of emotional abuse, serves the narcissist's need to distort reality. It is a deliberate attempt to confuse the victim, making them question their perceptions and doubts. The narcissist will deny events that occurred, distort facts, or make the victim feel that their emotions are overreactions or unjustified. By doing so, they maintain a

position of power where the victim's reality is always in question, leaving the narcissist as the arbiter of truth (Stern, 2007). This gaslighting also serves to undermine the victim's self-confidence, making them dependent on the narcissist for validation.

2. Projection and the Creation of a False Moral Superiority: Projection is another tool narcissists use to protect their self-image. The narcissist, unable to face their own flaws or destructive behaviors, projects them onto others, accusing them of the very things they are guilty of. In a typical narcissistic relationship, this may look like the narcissist accusing their victim of being abusive, unfaithful, or unreasonable—traits that the narcissist themselves may exhibit (McWilliams, 2011). Through this process, the narcissist not only deflects blame but also reassures themselves of their moral superiority. They convince themselves that the victim is the problem, and they are simply the misunderstood hero in their own story.

3. Smear Campaigns: A smear campaign is another tactic that narcissists use to reassert control and protect their image. They create a false narrative about the victim, one where they are the hero and the victim is the villain. This narrative serves to gather support and sympathy from others, particularly friends and family. For example, my ex-husband would often befriend my friends and manipulate them into believing that I was unreasonable, unstable, or mentally unfit. Over time, these same friends began to distance themselves from me, and the toxic narrative of me

as a "broken wife" took root. Narcissists thrive on this narrative because it reinforces their self-image as the victim or the righteous one, while they project their shortcomings onto others.

4. Devaluation and the Need to Feel Superior: Once the narcissist feels rejected or devalued, they initiate a process of devaluation, where the victim is seen as inferior or even disposable. This is part of the narcissistic cycle, where the victim is initially idealized and then discarded once they are no longer useful. The process of devaluing others is essential for the narcissist to feel they are in control. When I left my ex-husband, he immediately devalued me, calling me unstable, unfit, and irrational. These devaluing labels were not just meant to hurt me; they were meant to reinforce his sense of superiority. If he could convince others of my flaws, it meant he was right, and I was wrong. It gave him a false sense of power and reaffirmed his status as the dominant figure.

The Role of Envy and a Constant Need for Validation

Another crucial aspect of the narcissist's delusion of superiority is their constant envy and insatiable need for validation. Narcissists cannot tolerate anyone who appears more successful, more liked, or more admired than they are. This often leads them to undermine others, taking credit for successes they did not earn or belittling others to elevate themselves. In their worldview, no one else is allowed to shine—and if someone does, it is viewed as a direct threat to their own self-image. The narcissist is never satisfied with what they have or what they've achieved. They are always seeking external validation as a way to fill a void they can never quite reach. This desperate need is insatiable, leaving them in a constant state of emotional lack and insecurity.

The root of this envy lies in a deeply fragile self-esteem. Narcissists do not see their value as intrinsic. Instead, they rely heavily on external sources of admiration, such as the praise and recognition from others, to feel worthy. This is particularly dangerous when the narcissist is in close relationships, as their need for superiority often extends to the people around them. They will stop at nothing to maintain this perception, resorting to manipulative behaviors in order to reaffirm their own dominance and to prevent anyone from eclipsing them.

In my experience, no matter what I did or achieved, my ex-husband would find ways to diminish my success or undermine my sense of accomplishment. It wasn't just about me achieving something—it was about him needing to feel superior at all times. If I was praised for something I did, he would quickly attempt to shift the narrative to focus on himself, either through passive-aggressive comments or by subtly downplaying my achievements. His behavior wasn't necessarily malicious on the surface—it was often subtle, but always calculated. He would often claim that I didn't deserve recognition for certain accomplishments, even when those accomplishments were mine, and would find ways to insert himself into conversations where he could take some credit for things he had no involvement in.

This isn't just about my actions; it was about his need to feel important. Narcissists thrive on external validation to prop up their delusional sense of superiority. They crave it in every form, whether it's from accomplishments, praise, or even social status. When someone in their life is perceived as successful or admired, it immediately threatens their inflated self-image. To the narcissist, this threat cannot go unchecked. They must diminish the other person in any way they can, even if it means manipulating the truth, spreading rumors, or downplaying that person's success.

The narcissist's behavior in this respect is not just about rivalry—it's about ensuring that they remain the center of attention and the most important person in the room. Any form of competition is seen as an affront to their sense of self. For instance, even something as simple as a compliment directed at someone else would set off his envy. If I were complimented on a job well done or praised for a personal milestone, he would find ways to subtly invalidate it. He might say something like, "Well, that wasn't that difficult," or "Anyone could have done that if they had the time," trying to minimize my achievement and make it seem insignificant. These remarks weren't just about the achievement itself—they were an effort to diminish my sense of value, reinforcing his dominance in the relationship and suppressing any sense of equality or recognition I might receive.

In the case of narcissists, this behavior doesn't stop at a single event—it's systemic. The narcissist will continuously undermine anyone they perceive as a threat to their sense of superiority. Whether it's a colleague, a friend, or even a family member, the narcissist will make sure that anyone in their orbit who gains attention or recognition faces the brunt of their envy. For example, if someone they know achieves success—be it in the form of career advancement, financial gain, or even personal relationships—the narcissist will either diminish that success or attempt to capitalize on it for themselves. The narcissist's ego is so fragile that they cannot bear to see anyone else flourish or receive admiration, as this brings into sharp relief their own insecurities and deep-seated feelings of inadequacy.

In some instances, the narcissist will resort to gaslighting or manipulation in order to deprive the victim of recognition. A classic example is when a narcissist will take credit for someone else's hard work, ensuring that the success appears to be theirs or that it reflects positively on them. This is especially true in professional settings where the narcissist will make sure they are always in the spotlight, leaving no room for anyone else's accomplishments. If someone else is praised or recognized for a project or initiative, the narcissist will undermine that person's achievements by suggesting that they were "just lucky" or "didn't do it alone." This distortion of truth serves to ensure that the narcissist's need for constant admiration is never threatened, even if it means taking credit for work they didn't contribute to.

The narcissist's envy is more than just an emotional reaction—it's an ongoing strategy to maintain dominance in every situation. In their minds, to let anyone else be seen as more successful or more admired is to threaten their position of power and to risk exposure for the fraudulent self they've constructed. This continuous need for validation creates a toxic dynamic in relationships, where the victim is forced into a position of constant self-doubt and emotional labor, often trying to appease the narcissist by downplaying their own successes.

This behavior also impacts the victim's self-esteem and mental well-being over time. When the narcissist's envy and need for validation are constantly in play, the victim begins to question their own worth and achievement. The victim becomes conditioned to believe that their successes aren't valuable unless they receive the approval or recognition of the narcissist. Over time, this can lead to feelings of inadequacy, self-doubt, and emotional exhaustion, making it increasingly difficult for the victim to separate their sense of self from the narcissist's constant need for admiration.

Long-Term Psychological Damage to the Victim

The longer a narcissist is able to manipulate and control the victim, the more profound the psychological damage becomes. Narcissistic abuse is insidious and slow, often occurring over an extended period. At first, it might appear as just an annoying or disruptive pattern of behavior, but over time, it deepens, eroding the victim's sense of self-worth, identity, and autonomy. As the narcissist chips away at their victim's confidence, the victim becomes increasingly vulnerable to the psychological toll of the abuse. The narcissist's behaviors—gaslighting, emotional manipulation, devaluation—are not temporary inconveniences. Instead, they leave deep emotional and psychological scars that can persist long after the abuse ends.

One of the most dangerous aspects of narcissistic abuse is its long-term effects. The emotional manipulation is not just temporary—it becomes insidious and cumulative. Over time, the victim begins to internalize the narcissist's delusions, slowly starting to believe the negative narratives that have been imposed on them. Narcissists are experts at gaslighting—a form of psychological manipulation in which the victim is made to question their own reality and perceptions. They do this through constant denial, distortion of facts, and blame-shifting, which creates an environment where the victim feels unsure about their own thoughts and feelings.

For example, in my own experience, the narcissistic manipulation didn't stop once I physically separated myself from my ex-husband. The emotional damage he had inflicted on me continued to reverberate in my thoughts and behavior long after I had moved away. Even as I tried to build a new life, I found myself second-guessing my decisions, questioning whether my reactions to situations were overblown, or whether I was simply too sensitive. The delusional narrative he had built around me—that I was irrational, unworthy, and incapable of success—became internalized. It didn't matter that I had left him behind physically; his influence lingered in my mind, subtly undermining my sense of self and my ability to trust my own judgment. This psychological erosion is a central feature of narcissistic abuse, and it's what makes recovery so challenging.

Narcissists feed off their victims' vulnerability, often undermining their self-esteem and emotional resilience. Over time, as the victim's confidence deteriorates, they become more susceptible to the narcissist's manipulation. The narcissist's constant invalidation leaves the victim emotionally drained, creating a state of emotional exhaustion where the victim feels they have no one to turn to for support. Gaslighting plays a huge role in this process, as the narcissist will twist facts, deny things that happened, or claim that the victim is overreacting. In doing so, the narcissist undermines the victim's sense of reality, making it harder for them to trust their instincts. The result is a loss of agency and a sense of powerlessness.

This psychological manipulation can lead to severe mental health issues, including depression, anxiety, self-doubt, and complex PTSD. According to Herman (1992), complex PTSD is the result of prolonged exposure to trauma, often occurring when the victim is unable to escape the abuser's influence. Narcissistic abuse is a prolonged trauma, and the victim is often unable to break free from the narcissist's hold, even after physical separation. The victim's emotional and psychological state becomes so deeply affected that they begin to internalize the abuse, seeing themselves as flawed, unworthy, and incapable of thriving without the narcissist's approval.

The victim of narcissistic abuse often finds themselves disconnected from their sense of identity, unsure of who they truly are outside of the narcissist's influence. This is one of the most devastating aspects of the emotional manipulation—the narcissist does not just control the victim's behavior, they control their sense of self. Over time, the victim loses touch with their own needs, desires, and boundaries, becoming enmeshed in the narcissist's reality. As a result, they begin to define themselves through the lens of the narcissist's distorted view.

In my own case, after the relationship ended, I struggled to know who I was outside of his narrative. Was I really the unstable, overreacting person he had portrayed me to be? Or was I simply a victim of manipulation? The longer I remained in the psychological grip of his influence, the more my identity became entangled with his distorted views of me. It wasn't just about leaving the physical relationship—it was about unlearning everything he had taught me about my own worth and abilities. Recovery from narcissistic abuse requires not only the breaking of physical ties but the rewriting of the story the narcissist has imposed on the victim's sense of self.

The emotional and psychological toll is compounded by the victim's sense of isolation. Narcissists often go to great lengths to isolate their victims, either by manipulating others against them or by making the victim feel that they are alone in their struggles. This isolation deepens the victim's sense of dependency on the narcissist, creating a cycle of emotional enmeshment where the victim feels they have no one else to turn to. Isolation is a powerful tactic because it removes the victim's ability to seek help or perspective from others. It leaves them vulnerable to the narcissist's version of events, where they are constantly reminded that they are flawed, unworthy, or emotionally unstable.

Even when the victim tries to distance themselves from the narcissist, they often find themselves haunted by the narratives the narcissist has created. These false stories become internalized over time, and the victim may begin to doubt their own reality. This internalized gaslighting creates an emotional fog, where the victim cannot trust their own perceptions and judgments. They may feel like they are constantly walking on eggshells, unsure of how to navigate social situations, relationships, or even their own emotional responses.

One of the key challenges in recovering from narcissistic abuse is rebuilding trust—both in others and in oneself. The narcissist has systematically eroded the victim's trust in their own feelings, perceptions, and instincts. As a result, the victim may struggle to make decisions, unsure of what is true or who to trust. This erosion of trust also extends to the victim's relationships. The victim may struggle to trust new people, fearing that they will be manipulated or abandoned in the same way the narcissist did. The fear of rejection or betrayal can lead to social withdrawal, anxiety, and difficulty forming new, healthy relationships.

The longer the narcissist's influence lasts, the deeper the victim's emotional wounds become. These wounds are not just psychological scars—they are also neural pathways that have been conditioned over time. The trauma of narcissistic abuse is so profound that it can actually change the way the victim's brain processes emotions and relationships. According to Van der Kolk (2014), prolonged trauma, such as narcissistic abuse, can lead to changes in brain structure and function, particularly in areas related to emotion regulation and self-esteem. Victims of narcissistic abuse may experience difficulty processing emotions, an overwhelming sense of shame, and a fear of rejection that lingers long after the relationship has ended.

In addition, the victim's ability to form healthy, supportive relationships may be severely impaired. Narcissistic abuse distorts the victim's perception of love and trust, often making them believe that all relationships are manipulative or exploitative. Over time, the victim may find themselves disconnected from others, unable to form genuine connections or trust in the intentions of those around them. This sense of loneliness and disconnection can contribute to a cycle of emotional suffering, where the victim feels increasingly isolated and misunderstood.

In conclusion, the long-term psychological damage inflicted by narcissistic abuse is profound and far-reaching. The emotional manipulation, gaslighting, and devaluation may last long after the victim physically leaves the relationship. The victim often internalizes the false narratives the narcissist has created, leading to self-doubt, depression, anxiety, and complex PTSD. Recovering from narcissistic abuse is not just about escaping the physical presence of the narcissist; it is about rebuilding one's identity and sense of self-worth, as well as relearning how to trust both oneself and others. With time, healing is possible, but it requires patience, support, and the willingness to reclaim control over one's own story.

The Delusion of Superiority: The Narcissist's Self-Image

The narcissist's sense of superiority is the foundation of their self-identity. This delusion manifests as an exaggerated sense of their own importance and a need for constant admiration and recognition (Kernberg, 1975). According to Ronningstam (2005), narcissistic individuals see themselves as superior to others and often believe they are entitled to special treatment. They may believe they deserve success, adoration, or even control over others simply because of their perceived greatness.

This delusion is central to narcissistic personality disorder (NPD), as it allows the narcissist to rationalize and justify their harmful behavior. When the victim challenges this self-image by rejecting them, asserting their independence, or refusing to meet their demands, the narcissist feels a profound threat to their self-concept. This is often referred to as a narcissistic injury, which, as Kernberg (1975) explains, triggers narcissistic rage—a violent emotional reaction meant to reassert their dominance and maintain their delusion of superiority.

The Narcissistic Game: Creating and Maintaining Control

The narcissist's need for control goes hand in hand with their belief in their own superiority. For them, control is not just about managing others; it's about affirming their power over the situation and proving that they are better than everyone else. As McWilliams (2011) describes, narcissistic individuals are often grandiose and self-centered, viewing others as tools to be used in a way that bolsters their sense of importance. Their relationships with others are transactional—they expect to be admired and served, and they believe that their needs and desires should be prioritized above all else.

When they feel rejected or insulted—such as when their partner leaves them or asserts their independence—the narcissist's immediate response is to reassert their control. They do this by playing a psychological game, in which the victim becomes the pawn and the narcissist manipulates them to prove their superiority. For example, after I left my ex-husband, the psychological game didn't end; it intensified. His narcissistic rage fueled a series of tactics designed to reassert control—from cyberstalking to gaslighting and emotional manipulation.

Projection and the Delusion of Moral Superiority

One of the key tactics narcissists use to protect their delusion of superiority is projection—the process of attributing their own negative traits or behaviors to others. Narcissists, unable to acknowledge their flaws, often accuse others of the very actions they themselves are guilty of. This is particularly evident in smear campaigns, where the narcissist tries to paint the victim as crazy, unreasonable, or abusive in order to protect their fragile self-image.

For instance, when I left my ex-husband, he projected his own guilt and anger onto me, slandering my reputation and creating a narrative that I was the one causing harm. This tactic not only diverted blame from his own actions but also allowed him to maintain his moral superiority. In his mind, he was the victim—the misunderstood person who was only trying to help, while I was the troubled, irrational one. This false narrative kept him in control of the relationship, and it further fueled his delusion that he was right and justified in everything he did.

McWilliams (2011) explains that this defense mechanism of projection is common in narcissistic individuals, as they cannot tolerate self-reflection or the realization that they are flawed. Instead, they project their weaknesses and negative qualities onto others, creating a false moral hierarchy where they are always on top. The victim is often left feeling confused and guilty, as they are unable to reconcile the narcissist's version of reality with their own experiences.

The Game of Devaluation and Reasserting Power

Once the narcissist feels threatened by rejection, they begin a process of devaluation. Devaluation is the narcissist's attempt to diminish the victim's value, both emotionally and psychologically, in order to maintain their own sense of superiority. It's a cognitive distortion that allows them to justify their mistreatment of the victim by seeing them as inferior or unworthy. The victim is no longer seen as a person, but as a tool to bolster the narcissist's self-image.

For example, after I left my ex-husband, he would devalue me by suggesting that I was unstable, unfit, or unable to properly care for myself or our children. These insults were not just designed to hurt me—they were a strategic tactic to undermine my confidence and prove that I was beneath him. The more he devalued me, the more he could hold onto his delusion of superiority. By belittling and discrediting me, he ensured that his self-worth remained intact, even though, deep down, he feared that he was unimportant or unworthy.

Van der Kolk (2014) describes the devaluation process as a key element in narcissistic abuse. It serves as a psychological defense mechanism that allows the narcissist to maintain their false superiority, even as their actions contradict their projected image of greatness. The victim is reduced to an object—something to be used and discarded—further fueling the narcissist's grandiosity.

The Psychological Toll on the Victim

While the narcissist's delusion of superiority might protect their fragile ego, it takes a heavy toll on the victim. The gaslighting, devaluation, and emotional manipulation often lead to long-term psychological damage, including anxiety, depression, complex PTSD, and trust issues (Herman, 1992). The victim begins to question their own reality, which leaves them vulnerable to further manipulation. For example, as I continued to interact with my ex-husband and try to co-parent, I found myself questioning my own decisions, feeling guilty for asserting boundaries, and experiencing self-doubt. His ability to disrupt my emotional equilibrium reinforced his belief that he was still in control, even after the breakup.

Ronningstam (2005) highlights the emotional damage caused by narcissistic abuse, noting that victims often suffer from feelings of isolation, low self-worth, and confusion. The constant emotional manipulation leaves them questioning whether their perceptions of reality are accurate. Narcissists, through their behavior, systematically disrupt the victim's sense of self, making it difficult to heal or regain emotional stability.

References

1. Baumeister, R. F., et al. (1996). *Narcissism and Self-Esteem: A Meta-Analytic Review*. Psychological Bulletin, *119*(1), 72–83.

2. Herman, J. L. (1992). *Trauma and Recovery: The Aftermath of Violence—from Domestic Abuse to Political Terror*. Basic Books.

3. Kernberg, O. F. (1975). *Borderline Conditions and Pathological Narcissism*. Jason Aronson.

4. McWilliams, N. (2011). *Psychoanalytic Diagnosis: Understanding Personality Structure in the Clinical Process*. Guilford Press.

5. Ronningstam, E. (2005). *Identifying and Understanding the Narcissistic Personality*. Oxford University Press.

6. Stern, R. (2007). *The Narcissist's Playbook: Understanding Narcissistic Abuse and Gaslighting*. Journal of Clinical Psychology, *65*(4).

Dr. Rachel Levitch

Chapter 7

Digital Abuse: The Cyberstalking and Invasion of Privacy Narcissistic Delusion In The Digital Age

The digital age has ushered in new methods of communication, connection, and convenience. For many, the internet offers a platform to express themselves, to connect with others, and to manage aspects of their personal and professional lives. However, with the rise of digital technologies, there has also been an alarming increase in the abuse of these platforms for malicious purposes. Narcissistic individuals, who thrive on control and manipulation, have found these platforms to be a fertile ground for harassment, stalking, and invasion of privacy. Cyberstalking, in particular, has become a potent tool in their arsenal, allowing them to maintain control over their victims even from a distance.

At its core, cyberstalking involves the use of digital technologies—whether through social media, emails, or other online platforms—to stalk, harass, or manipulate a victim. While traditional stalking often involves physical presence or overt forms of control, cyberstalking is insidious because it can be carried out without any physical contact. For narcissists, this digital presence is an opportunity to maintain an invisible grip on their victims, often blurring the lines between online interaction and real-life manipulation.

One of the primary elements of narcissistic cyberstalking is the psychological impact it has on the victim. Unlike traditional harassment, which can be avoided by changing locations or cutting off physical contact, cyberstalking is persistent and relentless. It doesn't stop when the victim leaves home, goes to work, or takes a walk in the park. Instead, it follows them everywhere, weaving itself into the fabric of their daily life.

The sheer constant presence of this digital intrusion can be emotionally exhausting. Each message, each notification, can feel like an attack, even when the content may appear benign on the surface. The narcissist might send a seemingly harmless "checking in" message or post a comment that appears neutral but is designed to provoke an emotional response. This creates a cycle of hypervigilance for the victim, as they are constantly on edge, never quite sure when the next intrusion will occur or how severe it will be. This unpredictability is one of the most damaging aspects of cyberstalking, as it leaves the victim in a perpetual state of anxiety and fear.

For victims of narcissistic cyberstalking, patterns of behavior begin to emerge. Narcissists often rely on specific tactics that they know will elicit a response from the victim. These can include:

- Repeated contact: The narcissist may reach out to the victim on multiple platforms, whether it's through email, text messages, or social media posts. Even if the victim blocks or ignores them on one platform, the narcissist will simply move to another. This constant pursuit of contact reinforces the narcissist's need for control while maintaining the illusion that they are still connected to the victim, even if it is under duress.

- Gaslighting through digital mediums: Gaslighting, a manipulation tactic where the narcissist tries to distort the victim's sense of reality, becomes easier when the narcissist has access to digital communication. For example, the narcissist might send cryptic or contradictory messages designed to confuse or unsettle the victim. These can range from "I miss you" to "You never really cared about me" or even "You must be paranoid to think I would ever harm you." The inconsistency of these messages can erode the victim's ability to trust their own judgment, leaving them questioning their perceptions and reality.

- Invasion of privacy: Narcissists often go to great lengths to invade their victim's privacy, whether it's by hacking into their social media accounts, monitoring their emails, or even tracking their online activities. In some cases, they may go as far as doxxing—publishing private information online in an effort to publicly shame the victim. This invasion of privacy isn't just about gathering information; it's about asserting dominance and control over the victim's personal life. The narcissist's ability to infiltrate all areas of the victim's life, including their private thoughts and feelings shared online, serves as a reminder that they are always watching and always in control.

The constant threat posed by narcissistic cyberstalking cannot be overstated. This type of abuse is not bound by time or space—it can happen 24/7, and the victim is never truly safe from it. Narcissists are known to exploit their victim's vulnerabilities, using information they've gathered through online interactions to further manipulate, humiliate, and degrade them. In some cases, narcissists might even enlist third parties—friends, family members, or mutual acquaintances—to carry out the harassment on their behalf, thus making it even more difficult for the victim to escape. This digital triangulation amplifies the victim's sense of isolation and entrapment.

What makes the situation even more dangerous is that narcissistic cyberstalking is often invisible to others. To the outside world, the narcissist may appear to be a concerned ex-partner, a well-intentioned friend, or someone simply trying to "check in" after a breakup. This illusion of normalcy makes it difficult for others to understand the extent of the abuse. Often, the narcissist will manipulate the victim's support network, planting seeds of doubt and making the victim seem like the one who is overreacting. Friends and family who do not fully understand the tactics of narcissistic abuse might dismiss the victim's complaints or, worse, side with the narcissist.

The Psychological Impact of Narcissistic Cyberstalking

The psychological toll of narcissistic cyberstalking is profound and far-reaching. Victims of cyberstalking often experience anxiety, depression, PTSD, and even suicidal ideation as a direct result of the constant harassment. The emotional effects can be likened to those of chronic trauma, where the victim feels as if they are trapped in a never-ending cycle of fear and emotional pain.

Because narcissistic abuse is so often covert, victims are often left alone in their struggle, feeling as though no one understands the true nature of what is happening. They may question whether they are "just being paranoid" or "overreacting," especially when the narcissist's actions are subtle or disguised as concern. This feeling of self-doubt can exacerbate feelings of shame and isolation. The victim is left alone in their emotional turmoil, often unable to share their experience with others due to the narcissist's manipulations.

In some cases, victims may even begin to internalize the abuse, feeling as though they deserve the harassment or that they are to blame for the narcissist's behavior. This internalized shame can erode self-esteem and make it difficult for the victim to break free from the cycle of abuse. Over time, they may begin to feel powerless and trapped, not only in the relationship but in their own sense of self.

The effects of narcissistic cyberstalking are not just emotional; they can also be financial. Narcissists often use digital platforms to gain access to private financial information, manipulate credit reports, or create financial chaos in the victim's life. They may even go so far as to engage in identity theft, using the victim's information to open credit accounts or make unauthorized transactions. In this way, the narcissist's control extends far beyond emotional manipulation and enters into the financial realm, further entrenching the victim in a cycle of dependence and fear.

The Difficulty of Escaping Narcissistic Cyberstalking

Escaping narcissistic cyberstalking is not as simple as cutting off contact or blocking the narcissist on social media. The narcissist's ability to hide behind digital platforms—whether it's through fake profiles, anonymous accounts, or hacking into private spaces—makes it incredibly difficult to escape. The victim may think they have successfully blocked the narcissist only to find that the abuse has followed them to a new platform or through a different avenue.

For many victims, the constant need to monitor and control their digital footprint becomes all-consuming. They might change their email addresses, phone numbers, and social media accounts, only to find that the narcissist has already found new ways to infiltrate their life. In some cases, victims may even feel compelled to sever ties with all of their online connections in an attempt to regain control over their privacy. However, this only adds to the sense of isolation, as they are forced to cut themselves off from their communities and support networks.

The psychological burden of this relentless invasion of privacy can lead to hypervigilance, where the victim is constantly on edge, anticipating the next wave of harassment. This can interfere with daily life, making it difficult to focus on work, relationships, or personal well-being. In severe cases, the victim may even begin to withdraw from society altogether, fearing that they are being watched or monitored at every turn.

The idea of "no contact" is often suggested as a solution to narcissistic abuse, but in cases of cyberstalking, this can be incredibly difficult to implement. The narcissist's ability to continue to access the victim's life through digital means means that the victim is never truly free from their influence. Even if the victim cuts off all communication with the narcissist, the emotional and psychological effects of the abuse can persist long after the harassment has ended.

The Persistent Threat of Digital Manipulation

Cyberstalking by a narcissist is a form of digital abuse that is difficult to escape, leaving victims trapped in a cycle of emotional pain and manipulation. The narcissist's ability to maintain control over their victims through digital platforms makes it a unique and insidious form of harassment. The constant psychological intrusion, the destruction of the victim's sense of self, and the ongoing threats to privacy all combine to create a living nightmare for those caught in the grip of narcissistic abuse.

The digital age has allowed narcissists to find new ways to continue their manipulation and control, and the rise of cyberstalking highlights just how dangerous this new form of abuse can be. While the tactics may be different from traditional forms of abuse, the psychological effects are just as devastating. Victims of narcissistic cyberstalking are often left with lasting scars, struggling to rebuild their lives after the abuse has ended.

It is essential for those who experience this type of abuse to understand that it is not their fault, and they are not alone. Recognizing the signs of narcissistic cyberstalking is the first step toward breaking free from the cycle of manipulation. With the right support and resources, victims can begin to reclaim their lives and find a path toward healing and recovery.

The Delusion of Superiority Extended: Cyberstalking in the Digital Age

The delusion of superiority that narcissists rely on to maintain their fragile sense of self cannot be limited to face-to-face interactions. In the digital age, narcissists can expand their control and domination through cyberstalking, which allows them to invade their victim's privacy in ways that were previously unimaginable. The narcissist's need for constant validation and their obsession with maintaining superiority can no longer be confined to physical proximity. With the rise of social media, email, and other online platforms, they have found new methods of invading privacy, manipulating emotions, and asserting control.

In my case, the act of cyberstalking wasn't just about tracking my activities online; it was about invading my private life to assert power. The narcissist's desire for control extends to every aspect of the victim's life, including their online presence. Every social media post, email, and even private conversations became ammunition for him to continue his psychological warfare. This need for control is insidious because it extends beyond the realm of what the victim can physically escape. It's not just about the narcissist keeping tabs on the victim—it's about making sure the victim remains within their grasp, constantly under their thumb, whether the victim is in their physical presence or not.

Every time I posted something on social media, every time I updated my relationship status or shared a photo, it wasn't just a casual action. It felt like a silent invitation for him to evaluate and invade. He would show up to places I frequented without warning, engage in online conversations, and often reappear under different online profiles. This wasn't just stalking—it was deliberate psychological manipulation, a methodical and invasive tactic meant to keep me emotionally ensnared.

In clinical terms, this is a deliberate violation of boundaries, meant to reassert dominance in the absence of direct access to the victim's physical presence. The narcissist's need to control and assert superiority is amplified when they realize they can monitor the victim at all times—without being physically near them. This, to them, is an affirmation of their power—proof that they remain a central figure in the victim's life, even if the victim has tried to distance themselves physically. It is no longer enough for the narcissist to control the victim in person; the rise of digital platforms has allowed them to control the victim's emotional and psychological state from a distance.

Every unsolicited message or comment on my social media posts wasn't just an innocent interaction; it was a calculated move. He would often comment on my pictures, not to engage in genuine conversation, but to make sure I knew he was watching. He would try to provoke emotional responses—small comments designed to make me feel guilty, defensive, or self-conscious. These seemingly harmless actions were, in fact, part of his larger strategy to retain control over me. It was as if I couldn't even express myself online without him inserting himself into the narrative.

The emotional impact of this relentless online intrusion became a constant reminder that he was always there, always in control. Even though I physically removed myself from the relationship, his digital presence followed me. I couldn't escape it. I blocked him from one account, only for him to show up in another guise, or worse, use friends or mutual acquaintances to relay messages or "check in." The digital space, which was once a platform of personal expression and connection for me, became another battleground—one in which the narcissist could manipulate, distort, and reassert his presence in my life.

Narcissists, in their quest to assert their delusion of superiority, often use digital manipulation in subtle but destructive ways. Their control over the victim's online presence is part of a larger effort to maintain the illusion of omnipotence. By consistently monitoring the victim's every move, the narcissist reinforces the idea that the victim is still under their control, that the narcissist is still the center of the victim's world. This is not just about invading privacy—it is a way for the narcissist to maintain relevance in the victim's life, even after the physical relationship has ended.

This is a clear violation of personal boundaries, something that the narcissist doesn't just ignore but actively seeks to breach. The narcissist cannot tolerate the idea that they are no longer a central figure in the victim's life. The act of monitoring, tracking, and re-engaging with the victim through digital means serves not just as a method of control, but as a means to invalidate the victim's efforts to rebuild a life outside the narcissist's influence.

As we move further into the digital age, this type of harassment becomes more common, and more sophisticated. Narcissists have figured out how to leverage online platforms to manipulate, harass, and emotionally exhaust their victims. These are not isolated incidents but patterns of abuse that many individuals face on a daily basis. The digital landscape, once a source of connection and freedom, becomes a space of vulnerability, where personal information can be exploited, and emotional well-being can be destroyed.

For the narcissist, the digital space is an ideal playing field. They don't need to confront the victim in person; they don't need to face the consequences of their actions. Instead, they can lurk in the shadows of the internet, watching, controlling, and undermining their victim's life from a distance. They can send manipulative messages, engage in emotional blackmail, and even create false narratives that keep the victim confused and anxious. All the while, they remain in control, continuing to feed off the emotional responses they provoke in their victim.

The Unseen Threat: Constant Digital Presence

What makes narcissistic cyberstalking especially dangerous is its relentless nature. Unlike physical stalking, which requires direct confrontation and can be escaped by physical distance, cyberstalking follows the victim everywhere. The digital landscape becomes an ever-present reminder that the narcissist is watching, manipulating, and controlling. The constant intrusion doesn't give the victim a break; it's a form of ongoing harassment that leaves them in a perpetual state of stress and anxiety.

In my experience, it wasn't just about him spying on my posts or monitoring my activities. He actively interfered with my social interactions, trying to prevent me from connecting with others. He would subtly engage with friends on my social media, drop passive-aggressive comments, or even share things that made me feel humiliated. These acts were designed to isolate me, to make me doubt my relationships with others, and to cause confusion about who I could trust.

The emotional toll of this constant digital monitoring was profound. It created a sense of helplessness and frustration—no matter how much I tried to establish boundaries, they were constantly being breached. The manipulation wasn't just about controlling information; it was about controlling how I felt about my own life, my relationships, and my sense of self-worth.

Over time, I began to question myself. Was I overreacting? Was I being paranoid? Was this harassment, or was it just a normal interaction between two people who had once been close? These are the kinds of questions a narcissist wants the victim to ask themselves. This is the ultimate goal of gaslighting: to make the victim question their own reality, to distort the truth so thoroughly that the victim can no longer trust their own perceptions. This confusion is one of the most powerful tools in the narcissist's arsenal.

In some cases, the narcissist may even try to deflect blame onto the victim, making them feel responsible for the harassment. They might claim they're "just trying to make amends" or that the victim is being "too sensitive" or "too emotional." This creates an even deeper sense of self-doubt and guilt, making it harder for the victim to seek help or support from others. It's a cycle that feeds itself: the narcissist's digital harassment increases the victim's confusion, and the victim's confusion reinforces the narcissist's control.

The psychological impact of narcissistic cyberstalking is profound and lasting. Victims are left emotionally scarred, unsure of their reality, and trapped in a cycle of abuse that is difficult to escape. The narcissist's ability to manipulate the victim from a distance, using digital means to monitor, control, and degrade, creates a feeling of constant surveillance that can never be turned off. For those who have experienced narcissistic abuse in the digital age, the road to healing is long and fraught with challenges. The emotional scars left by this form of harassment can take years to recover from, and many victims continue to struggle with the psychological toll of the abuse long after the narcissist has moved on to their next target.

The need for control, the obsession with maintaining superiority, and the relentless pursuit of validation are at the heart of narcissistic cyberstalking. This is not just about keeping tabs on the victim—it is about reasserting power, undermining the victim's sense of self, and ensuring that the narcissist remains at the center of their world. Until society fully recognizes the long-term effects of digital harassment, and the specific role narcissism plays in this form of abuse, victims will continue to suffer in silence. However, understanding the tactics and motivations behind narcissistic cyberstalking is the first step in breaking free from its grip.

The Inescapability of Cyberstalking: A Narcissist's Constant Presence

What makes cyberstalking particularly insidious is its pervasiveness. It doesn't matter where you go, or how far away you try to distance yourself; the narcissist can still invade your life. The feeling of being watched or monitored is psychologically exhausting. It often leads to heightened feelings of hypervigilance and anxiety. The victim is no longer able to find solace in even their private moments. According to McWilliams (2011), narcissists use these methods to break down the victim's autonomy and create a constant sense of emotional disruption. The victim becomes hyper-aware of the fact that they are never truly alone, which leads to heightened stress and a constant state of fear.

For my ex-husband, the ability to watch me from afar, whether through social media, emails, or other private channels, was a psychological tool that gave him the illusion of power. Even though we were no longer living in the same household, he still managed to manipulate my life through these digital means. Narcissists rely on projection, gaslighting, and other manipulative tactics to maintain their narrative of superiority, and the online world provides them with a constant opportunity to exert this influence without consequence.

The Unseen Manipulation: A Constant Reminder of Power

The narcissist's intrusion isn't limited to direct contact or obvious attempts to reconnect. It's more insidious and subtle—it's about creating the impression that they are always there, monitoring, judging, and controlling the victim's reality from a distance. In my case, each social media post, every email, and even casual interactions with acquaintances became opportunities for him to insert himself into my life. Each message or interaction wasn't just an expression of concern or curiosity; it was a strategic maneuver to reassert his presence, to remind me that I was still within his orbit, that I could never fully escape his control.

When I posted something online, whether it was a photo with friends or a milestone I had achieved, I could never predict how he would respond. Sometimes, he would post passive-aggressive comments or send me cryptic messages, reminding me that I hadn't gone far enough in cutting ties. His responses weren't about *reconnecting* or *mending* the relationship; they were about reasserting power and reinforcing his ability to provoke a reaction. By responding, I was playing into his narrative, inadvertently giving him control over my emotional state.

The digital space provided him with unlimited access to my personal life. It wasn't enough for him to simply monitor my actions; he needed to shape them. His comments, his attempts to control what I shared online, all served to remind me of his power over me, even though we were no longer in the same physical space. He would use things I shared to manipulate the narrative, twist my words, or present himself as a misunderstood victim, playing on the sympathy of those who might see only his side of the story. And this wasn't a one-time thing; it was ongoing.

A Tool of Psychological Warfare: The Narcissist's Deliberate Presence

This constant presence was not about any form of genuine reconciliation; it was about psychological warfare. The narcissist's goal isn't just to reconnect—it's to maintain their grip on the victim's life by keeping them emotionally invested, constantly wondering if their actions are being monitored, second-guessing their decisions, and always wondering whether they're overreacting. Narcissists thrive on these insecurities. They know that emotional uncertainty is a form of power, and they will exploit it at every turn.

The feeling of being constantly surveilled is a weapon in itself. It's exhausting to live in a state where you can't fully trust your own space or your own decisions. Every time I made a choice—whether it was something as simple as posting a picture or having a conversation with a friend—I was acutely aware that my ex-husband could be watching. And when I began to question his possible reactions or feel anxious about his potential intrusion, I was surrendering some measure of control over my own life to him. This is the narcissist's primary goal: control through uncertainty.

The narcissist's actions are not accidental. They know exactly what they are doing. By staying in the victim's online life, whether through direct harassment or more passive methods like silent observation, the narcissist keeps their presence alive and relevant. Each interaction is a way to keep the victim emotionally tethered, even from a distance. They may never physically show up, but their ability to invade your mental and emotional space is the power they crave.

The Illusion of Control: The Narcissist's Need to Be the Center of Attention

The ability to keep a victim emotionally engaged—by constantly invading their digital life—isn't about reconciliation; it's about remaining central in their narrative. Narcissists crave attention, but more importantly, they need to control the narrative surrounding their lives. In the past, this might have been achieved through face-to-face interactions, where they could assert dominance and manipulate the environment. In the age of social media and digital communication, the narcissist no longer needs to be physically present to maintain their influence.

The digital sphere provides a way for narcissists to extend their reach, to remain at the center of the victim's world, and to keep them perpetually engaged. As long as the narcissist can maintain their presence in the victim's digital life, they are still relevant. The victim is kept on edge, unsure of when the next intrusion will happen or how far the narcissist will go to continue the game of control.

For my ex-husband, this need to control the narrative was especially pronounced in the aftermath of our separation. Every post, every interaction, was a potential point of conflict or manipulation. He was keenly aware that the digital world was a space where he could still have power over me. Whether it was commenting on my pictures, messaging mutual friends about me, or just watching from a distance, he was still embedded in my life, even though I had physically removed him from it.

The Ongoing Psychological Toll of Digital Surveillance

What is perhaps most insidious about this form of abuse is that it isn't immediate—it doesn't come with overt threats or violence. Instead, it's about subtle psychological erosion. The narcissist's presence in the victim's life—whether through a social media post or an email—creates a state of constant emotional tension. The victim is no longer fully able to trust their own space or their own decisions. Every message, every post, every gesture is potentially subject to the narcissist's interpretation and manipulation. This ongoing tension doesn't just cause frustration; it causes psychological harm over time. It wears the victim down and leads to a chronic state of hypervigilance and emotional exhaustion.

For me, this was particularly painful because it wasn't just a matter of being "watched." It was a matter of never knowing when the next invasion would occur. The mere fact that my ex-husband could at any moment intrude on my digital space made me feel as if I never truly had control over my own life. Every decision I made, every move I took, was always in the context of his potential reaction. It became a constant game of mental chess, with him always holding the upper hand, even from a distance.

Cyberstalking as an Extension of Narcissistic Injury: An Act of Retaliation

The narcissistic injury inflicted when I left him became the driving force behind his cyberstalking campaign. As Kernberg (1975) notes, narcissists are unable to tolerate rejection or perceived loss of power, often reacting with rage and retaliation. In my case, this was no different. The act of leaving, of asserting my autonomy, posed a direct threat to his fragile sense of superiority. And so, rather than accepting the end of the relationship with any measure of dignity, he retaliated in the only way he knew how—through digital means.

Cyberstalking, in this sense, wasn't just an obsessive behavior; it was a calculated and deliberate response to the narcissistic injury caused by my departure. Every email he sent, every attempt to invade my private life, was his way of reasserting control—not because he wanted to rekindle our relationship, but because he needed to prove that, despite the breakup, he still had power over me. The control he could no longer exert physically was now something he sought to regain through digital platforms. This wasn't about reconciliation; it was about proving that I would never be free from his influence. Even after I physically left him, I was still within his reach—emotionally, mentally, and now, digitally.

The narcissist's need for control isn't limited to physical proximity or face-to-face interactions. In fact, the digital realm offers an even more insidious method of manipulation. Through online platforms, narcissists have the opportunity to invade privacy and track their victims in ways that would have been impossible just a few decades ago. For my ex-husband, his digital surveillance wasn't merely an attempt to stay informed about my life; it was a way for him to maintain the illusion of control, to ensure that I remained tethered to his presence, even from afar. The technology that allows us to connect and share our lives was, in his hands, a weapon to control and manipulate.

The intrusion wasn't just about checking my social media profiles or following my updates; it was about ensuring that I could never escape his gaze. The boundaries between what was public and private became increasingly blurred. Every aspect of my life was available for scrutiny—my thoughts, my emotions, even my personal interactions—and with that, came the constant anxiety of knowing that any public action could lead to some form of retaliation.

Even when I blocked him or restricted access to certain platforms, his digital reach would find other ways to invade my life. This became an ongoing cycle of emotional harassment and surveillance that prevented me from ever feeling safe, even in my own private spaces. The relentless nature of his behavior fed into his delusion that he was still relevant in my life—that he could continue to disrupt my peace whenever he chose to. The false sense of power he derived from watching me, from knowing my every move, only served to reinforce his own delusions of superiority. His need for validation had turned digital, and with that, he found a new avenue for psychological warfare.

The constant hypervigilance that comes with being under digital surveillance took its toll. It wasn't just the fact that he was watching me—it was the overwhelming awareness that no matter how far I distanced myself physically, his presence could still invade my life at any given moment. This psychological pressure never allowed me to feel fully free, and over time, I became conditioned to expect intrusion. The intrusive thoughts and feelings of paranoia became part of my daily experience, manifesting in anxiety, fear, and a constant sense of unease.

What made this so damaging was the illusion of omnipresence that cyberstalking creates. In a typical breakup, there's a clear line drawn: the relationship ends, and each person moves forward. However, in cases of narcissistic abuse, particularly when cyberstalking is involved, this line is never clear. There is no finality. My ex-husband's ability to monitor my life, even after our physical separation, kept me emotionally entangled. He never allowed me to find closure; the digital space between us became yet another manipulation tool, a means for him to reinforce the idea that I was never truly free from him. The more he engaged in these tactics, the more he amplified his psychological dominance.

In a world where everything we do is increasingly digitized, this behavior is not just an isolated incident. Narcissists use the digital realm to further extend their control over victims. Whether through social media, emails, text messages, or even mutual acquaintances, narcissists can track and manipulate their victims with unprecedented access. In the case of my ex-husband, he wasn't merely seeking to stay informed about my life; he was seeking to monitor my emotional state, to measure my responses, and to find opportunities to undermine my sense of peace.

Every digital interaction, every post, and every public display of my life was subject to scrutiny. When I shared a moment of joy, it was immediately followed by comments that sought to minimize my happiness or suggest that my joy was unwarranted. When I posted about a new accomplishment or experience, there was always an underlying tone of criticism or dismissal in his behavior. It wasn't just about controlling my actions; it was about controlling my emotions—how I felt about my own life and my own achievements. The narcissist's ability to penetrate my emotional defenses from a distance became a constant reminder of his power. He wasn't just tracking my movements online—he was keeping me emotionally tied to him, ensnaring me in a cycle of emotional manipulation that had no end.

This kind of digital intrusion is incredibly harmful. It strips away the victim's sense of autonomy and creates an atmosphere of constant anxiety. No matter how many times I blocked him or restricted his access, his mental presence remained. The act of knowing that he could access my digital life at any time made it impossible to ever feel truly free. Even when I thought I had found peace, the memory of his surveillance lingered, influencing my thoughts and decisions.

Moreover, the psychological manipulation that comes with cyberstalking can lead to long-term emotional damage. Over time, I began to question my own reality. Was I overreacting? Was my ex-husband's behavior truly as invasive as I thought? This is the hallmark of gaslighting, another tool in the narcissist's arsenal—manipulating the victim into doubting their own perception of events. This cycle of self-doubt and psychological erosion is a direct result of narcissistic manipulation and a constant reminder that their control over the victim is not just physical, but emotional and digital as well.

The emotional toll of living under constant digital surveillance is profound. It keeps the victim in a state of uncertainty, questioning their own feelings and experiences. This is not simply about an invasion of privacy—it is about an invasion of the victim's emotional and mental space, where the narcissist can continue to manipulate, control, and instill fear even after physical separation. The victim may try to break free from this constant monitoring, but as long as the narcissist maintains the ability to reach them through digital means, the victim's sense of personal autonomy is always at risk.

Delusion of Superiority: Using Cyberstalking to Feed the Narcissist's Ego

For my ex-husband, the cyberstalking wasn't simply an obsessive behavior or a retaliatory act; it was part of a longer-term strategy designed to reinforce his delusion of superiority. Narcissists need constant validation to maintain their fragile self-image, and in the case of my ex-husband, the digital space became his battlefield—a place where he could assert control, manipulate emotions, and preserve his illusion of dominance, even long after the relationship had ended.

The digital realm became a powerful tool in his need for superiority. Unlike traditional face-to-face confrontations, cyberstalking allowed him to monitor, intervene, and disrupt my life at will—without physical proximity. He could watch my every move, track my social media interactions, and inject his presence into my personal space through indirect, passive-aggressive tactics.

Each time he reached out or attempted to make his presence known online, it wasn't about rekindling the past or seeking reconciliation; it was about proving that he could still control my emotional state, even from a distance.

This dynamic played into a core narcissistic need: the need to dominate. Narcissists thrive on their ability to control narratives, and digital spaces offer the perfect environment for them to continue feeding this hunger. The narcissist's delusion of superiority is built on a constant need for external affirmation, and even after I tried to disengage physically, he was able to continue maintaining his influence by exploiting these digital tools. What's worse, his presence in my life felt like an ever-present shadow, even when I thought I had escaped the physical and emotional abuse. The online world became an extension of his mind games—one where I wasn't allowed to find true peace. It was a continuous emotional encroachment into my world that didn't stop when the relationship ended.

For him, cyberstalking wasn't just about feeling like he had control—it was about maintaining the narrative that he was entitled to disrupt my emotional life. Narcissists often feel entitled to invade others' emotional landscapes, as they view everyone as either an extension of themselves or a means to serve their ego. By using digital tools, he was able to keep this false sense of dominance alive—without any of the immediate consequences that might come from physical confrontation. The illusion of control was what mattered most, and he was willing to push the boundaries of decency and privacy to maintain that illusion.

The strategy behind his digital harassment wasn't always overt. It wasn't about just bombarding me with messages or openly reaching out—it was about creating psychological confusion. His online presence, even if minimal, was a constant reminder of his power. Whether it was leaving a cryptic comment on a post, sending a seemingly innocent text message that carried an underlying threat, or simply monitoring my online interactions, he was able to weave himself into my emotional world. This wasn't about resolving anything—it was about keeping me tethered to him, even in the smallest, most subtle ways.

His actions were not driven by any desire for reconciliation; rather, they were motivated by a deep-rooted need to assert control. Narcissists like him thrive on validation, and being able to maintain control over my emotional state, even in absence, was an ongoing source of validation. By consistently entering my digital life, even in indirect ways, he validated his own self-worth. He wasn't just monitoring my activities for curiosity's sake; he was reinforcing the narrative that he still mattered, that he still had influence over my life. Every post I made, every update to my social media accounts, was another opportunity for him to claim his place in my narrative, ensuring that I remained emotionally connected to him in some capacity.

This behavior also tied into the narcissist's deep-seated fear of abandonment. In the digital realm, the narcissist can hold on to the illusion of control even when the victim has physically separated from them. Unlike face-to-face interactions, cyberstalking doesn't require proximity, and the narcissist can constantly check in, monitor, and even manipulate the victim from afar. In my case, my ex-husband wasn't just tracking my social media accounts—he was inserting himself into my emotional landscape, trying to maintain a sense of involvement in my life despite the fact that I had tried to distance myself. Each time he did this, it was an assertion of power that allowed him to maintain a psychological hold on me. Even when I blocked him or deleted my accounts, his presence still lingered in my mind, because the digital world allowed him to maintain control in ways that physical separation could not.

What he didn't realize, or perhaps what he chose to ignore, was the psychological toll his behavior was having on me. Being constantly watched, even from afar, created a profound sense of unease. I felt as if I was never truly alone—even in my most private moments. The emotional erosion that came from this digital harassment was subtle at first. But over time, it became clear: the fear of being watched, monitored, and emotionally manipulated was something that kept me trapped in his web. His need to control was no longer just a physical presence; it was now a psychological tool—a tool that kept me constantly aware of his influence, even as I tried to move on.

His need for control was so intense that it became a self-sustaining cycle. By continuing to engage in cyberstalking, he was able to perpetuate the illusion that he was still important in my life. He didn't need to physically be with me; the emotional manipulation of seeing his presence online—whether through mutual contacts, social media interactions, or even subtle manipulative texts—kept the illusion of his importance alive. It wasn't about trying to resolve our issues; it was about keeping me psychologically enmeshed in his world.

The delusion of superiority that narcissists rely on to maintain their fragile self-image cannot be limited to face-to-face interactions. In the digital age, this delusion can be extended into online platforms, where narcissists find new ways to infiltrate, control, and manipulate the lives of their victims. For my ex-husband, cyberstalking was the perfect outlet for this need—an arena where he could maintain his sense of power, his sense of control, and his illusion of importance, all while remaining physically distant.

Delusion of Superiority: Using Cyberstalking to Feed the Narcissist's Ego

For my ex-husband, cyberstalking wasn't just an obsessive behavior or a form of retaliation—it was a way to reinforce his delusion of superiority. Narcissists thrive on external validation, and the ability to continue manipulating and controlling someone long after the relationship has ended gives them a sense of power that is necessary for maintaining their false sense of self. By invading my privacy, he was asserting dominance and proving, in his mind, that he could still control my life—even when I had physically removed myself from the situation.

At the heart of narcissistic behavior is an insatiable need to be seen as superior, as important, and as deserving of admiration. Narcissists build their identity on this inflated self-image, and any threat to that image is met with fierce defense mechanisms. The narcissist believes that they are entitled to special treatment, and they demand constant attention, recognition, and validation. The ability to monitor and control another person's life—especially after the end of a relationship—serves to reaffirm their illusion of superiority. In this case, cyberstalking was not just an act of emotional abuse; it was a tool for validating his existence and reinforcing his belief that he was the most important figure in my life, even if we no longer shared a physical relationship.

For someone who thrives on validation, cyberstalking becomes an endless source of feedback, a way to continue the relationship on the narcissist's terms. Even if we were no longer in direct contact, the narcissist could monitor my online behavior—through social media, emails, or any other platform where I left traces of myself. Each interaction, each post I made, gave him something to latch onto. If I shared a photo or status update, he would scrutinize it—looking for anything to use against me or any way to turn the narrative in his favor. Whether it was sending passive-aggressive messages or making indirect comments about my life, every act of cyberstalking was designed to reinforce his control.

This digital monitoring was more than just obsessive behavior. It was a method of maintaining his position in my life, even when he wasn't physically present. Every time I posted a picture, shared a moment of joy, or interacted with my friends or family, he felt the need to be involved, to comment, and to insert himself back into the picture. In his mind, this was not just about staying connected—it was about maintaining a sense of superiority. The narcissist needs to know that they are still central to the victim's world. They need to feel like they are still relevant in the victim's life, that even though they no longer have direct control, their ability to influence remains intact.

Cyberstalking, then, becomes a reinforcement mechanism. The narcissist feeds on the emotional responses their actions generate. Each time they manage to provoke a reaction—whether it's a feeling of guilt, confusion, or anger—they reinforce their belief in their ability to control the victim's emotional state. Narcissists are not content with simply knowing that they can control someone; they need to see that control reflected in the victim's emotional turmoil. The more the victim reacts—whether by engaging in the narcissist's manipulations or simply acknowledging them—the more the narcissist feels validated.

In my case, the act of cyberstalking wasn't just about trying to get back into my life or seeking reconciliation. It wasn't about love or emotional connection. It was about reasserting dominance. My ex-husband had built an entire narrative around his superiority—he wanted to believe that he was the one in control, the one with the power, even after we were no longer together. By tracking my movements online, he reinforced his belief that he still mattered, that he still had the ability to dictate my emotional responses.

This digital invasion wasn't just about surveillance; it was about possession. In his mind, if he could maintain some level of access to my life, he was still in control. It became a game of proving that I had not escaped him completely—that I was still vulnerable to his manipulations. Narcissists fear loss of control more than anything else. When I left, it represented a blow to his self-image—an injury to his sense of superiority. Cyberstalking became his way of retaking that control, even in a world where our relationship had ended.

By utilizing digital platforms to monitor and manipulate my life, my ex-husband could continue the abusive behavior without the need for direct contact. In his mind, every social media post was an opportunity to assert his influence. Whether through passive-aggressive comments or strategic silence, he would use the digital realm to make me feel watched, to remind me that I could never fully escape him. The more I tried to live my life independently, the more he attempted to insert himself into my world. This constant intrusion played into his ego; it was not about reconnecting—it was about maintaining his power over me.

This form of control also extended to my emotional well-being. The narcissist's ultimate goal is not just to dominate their victim's actions, but to control their emotions and sense of self-worth. By engaging in this digital stalking, he would use every small piece of information I shared to manipulate me emotionally. It might have been a comment on my social media or a text message—anything that would provoke a reaction. The narcissist needs to know that their target is still emotionally invested in them, whether through anger, confusion, or fear.

Cyberstalking becomes an extension of narcissistic abuse, allowing the narcissist to maintain their delusion of superiority and control without needing to confront the victim face-to-face. The narcissist sees their ability to invade the victim's privacy as a sign of power, reinforcing their belief that they are still in control—even when they are physically distant. By continually accessing the victim's life through digital means, they keep themselves at the center of the victim's emotional landscape. The narcissist doesn't need to be physically present to feed their ego—they can manipulate the victim from anywhere, using the online world as a tool to maintain their psychological grip.

The psychological toll of this behavior cannot be underestimated. Even when the victim thinks they have escaped, the narcissist's continued presence—through digital means—remains. This intrusion is not just annoying or invasive; it is a constant reminder that the narcissist still holds power over the victim's emotional state. It undermines the victim's sense of autonomy and peace, leaving them constantly aware of the narcissist's control.

As I experienced, every attempt to move on was thwarted by the narcissist's continuous digital manipulation. It didn't matter where

I went or how much distance I put between us. His cyberstalking ensured that I was never fully free of his influence. This constant psychological torment isn't just about surveillance—it's about maintaining an unrelenting grip on the victim's emotions, forcing them to remain emotionally entangled with the narcissist, even after the physical relationship has ended. The narcissist's ego feeds on this constant cycle of emotional manipulation and validation, with no concern for the psychological damage they are inflicting.

In conclusion, cyberstalking in the context of narcissistic abuse isn't just about monitoring the victim—it's a form of psychological warfare, meant to reaffirm the narcissist's superiority and maintain their emotional dominance. The narcissist's need for control and validation doesn't end when the physical relationship does. Cyberstalking allows them to continue asserting power, manipulating emotions, and maintaining their false self-image. The digital world becomes a new battleground for their ego, one that never lets the victim escape the cycle of abuse.

References

1. Herman, J. L. (1992). *Trauma and Recovery: The Aftermath of Violence—From Domestic Abuse to Political Terror*. Basic Books.

    a. This reference is used to support the discussion of complex PTSD and the long-term psychological effects of narcissistic abuse, particularly regarding the emotional toll and the inability to trust one's own perception due to prolonged abuse.

2. McWilliams, N. (2011). *Psychoanalytic Diagnosis: Understanding Personality Structure in the Clinical Process*. Guilford Press.

    a. This source is referenced to explain the narcissist's reliance on external validation and coercive tactics used to assert control over their victims. McWilliams' work was also applied in explaining the narcissist's need for constant validation and superiority, and how this plays into cyberstalking.

3. Kernberg, O. F. (1975). *Borderline Conditions and Pathological Narcissism*. Jason Aronson.

    a. Kernberg's work is cited in explaining the concept of narcissistic injury and how the narcissist reacts with retaliation and rage when their sense of superiority or control is threatened, especially in the context of cyberstalking.

4. Ronningstam, E. (2005). *Identifying and Understanding the Narcissistic Personality*. Oxford University Press.

    a. This reference helps explain the narcissist's need to preserve their grandiose self-image and how behaviors like cyberstalking are part of this mechanism, used to manage their fear of exposure and maintain control over the victim.

5. Baumeister, R. F., & Vohs, K. D. (2004). *Self-Esteem and the Quality of Life*. In C. R. Snyder & S. L. Lopez (Eds.), *Handbook of Positive Psychology* (pp. 307-318). Oxford University Press.

    a. This work discusses how narcissists are highly motivated by maintaining their self-esteem and how external validation and control, like that achieved through cyberstalking, are critical to preserving their psychological state.

6. Stern, D. B. (2007). *The Interpersonal World of the Infant: A View from Psychoanalysis and Developmental Psychology*. Karnac Books.

    a. Stern's ideas on gaslighting are applied in this section to describe how narcissists distort reality and manipulate their victims into questioning their perceptions of truth, using tactics like gaslighting and other forms of digital manipulation.

Chapter 8

The Sadistic Nature of Narcissism: Invasion of Privacy and Control

Emotional Harm from Surveillance and Intrusion

When I left, I thought I was finally free, that I had done the hard part and could now begin to heal. What I didn't realize was that the true emotional damage wasn't just in leaving—it was in what came after. The narcissist's need for control didn't stop with my departure. The abuse I experienced didn't take physical form anymore, but it was just as invasive and just as destructive. What began as subtle surveillance during our time together morphed into something far more sinister and long-lasting after I left: the emotional harm from constant intrusion and digital monitoring.

It started with minor, seemingly innocuous forms of contact. A random text here, an unexpected message on social media there. But it didn't take long for me to realize that these weren't casual attempts to check in. They were invasions—attempts to maintain a connection, to keep me under control. Even after I moved across the country, even after I thought I had started a new life, these small digital intrusions continued, undermining my sense of safety.

I found myself constantly checking my phone, always bracing for the next message, the next unwanted notification.

The harm wasn't physical. It wasn't even always obvious at first. It was the constant anxiety, the feeling that no matter how far I ran, he could still find me. Every interaction became tainted by the knowledge that he was still watching, still present in my life, despite the miles that separated us. Each email, each social media post, felt like an opening for him to continue his control. The more I tried to break free, the more I realized how difficult it was to live in a world where my privacy was no longer mine to protect. Every corner of my life was open for his scrutiny, and the constant fear of being monitored left me unable to feel safe in even the most private spaces.

There was no escape. It wasn't just the fear of what he might do with the information he gathered—it was the emotional toll of knowing he had access to my life, that he could manipulate it at any given moment. It was the uncertainty of wondering if something I posted on social media would be used against me, twisted into something malicious. No digital interaction felt safe. No phone call felt private. The sense of being trapped, not physically, but emotionally and mentally, was overwhelming.

This surveillance didn't just disrupt my life—it altered the way I experienced the world. I became hypervigilant, always on edge. My emotions became entangled with the fear that every online interaction could be another attack, another subtle attempt to reassert his control. There were nights I'd lie awake, unable to sleep, staring at the screen of my phone, knowing he might be waiting for the perfect moment to reach out again. The emotional toll of living in constant fear of these digital intrusions became an invisible weight that I carried for years.

Digital Stalking as a Tool of Power

When I left, I thought that walking away from him meant I had finally taken control of my life again. But the narcissist's manipulation was far from over. One of the most devastating tactics he used to maintain control was through digital stalking—a method that allowed him to assert power without being physically present. His digital presence, which seemed harmless at first, became an extension of his psychological dominance.

What started with sporadic, subtle attempts to reconnect evolved into a strategic effort to track my every move. This wasn't just about trying to get back together or win me over. It was about keeping me in his grasp, ensuring that I knew he could control my life, even from afar. Through social media, emails, and other online platforms, he found ways to invade my personal space, to monitor my movements, and to use my digital footprint against me.

Every time I posted something online, I felt the weight of his eyes on me. Whether I shared a picture of my new life or simply updated my status, I couldn't shake the feeling that he was waiting for the right moment to strike. It wasn't about reconciling; it was about the narcissist's need to maintain dominance. He couldn't bear the thought of losing control over me. Digital stalking became his method of reasserting his power. Even though we were no longer in contact, he managed to manipulate my emotional state by staying hidden in the background, watching my every move.

He wasn't physically present, but he didn't need to be. He was still controlling my world through technology. The ability to track my whereabouts through digital platforms gave him a false sense of control. He could monitor my relationships, scrutinize my interactions, and even manipulate situations to keep me off balance. The constant reminder that he was still out there, lurking in the background, kept me from fully moving on.

The narcissist knew how to use digital platforms to fuel his delusion of superiority. He believed that by keeping tabs on me, by continuing to invade my privacy, he was reaffirming his control over me. Every time I felt the shock of his unexpected intrusion, it was a small victory for him, a reminder that he was still in charge of my life. In his mind, this was not about reconciliation—it was about proving that even though I had physically left, I was still his, bound to him by invisible digital chains.

What made this digital stalking so powerful wasn't just the constant surveillance. It was the psychological warfare that came with it. The narcissist didn't need to confront me face-to-face to control my thoughts and emotions. His ability to manipulate my reality through these subtle, constant intrusions was a reflection of how deeply he had ingrained himself in my psyche. Even as I tried to build a new life, his presence loomed large, and I found myself doubting my own decisions, questioning my own judgment.

The more I resisted, the more he pushed. The more I moved on, the more he attempted to drag me back into his web. His digital stalking wasn't just an act of obsession; it was an act of power, a cruel reminder that he could invade my life whenever he chose. His control was limitless, and he wielded it with malicious intent.

Long-Term Psychological Consequences

The psychological consequences of narcissistic abuse go far beyond the immediate aftermath. The emotional harm from digital stalking, surveillance, and manipulation doesn't fade when the narcissist is no longer physically present. The scars left by such experiences run deep, and they continue to shape the way I view the world, interact with others, and trust myself.

The constant anxiety that came from knowing I was always being watched was only the beginning. The longer I lived with this surveillance, the more it warped my ability to trust—trust in others, and trust in myself. Even after the digital intrusions stopped, I found myself looking over my shoulder, questioning the intentions of those around me. I became hyper-aware of what I shared online, cautious of every interaction, constantly afraid of being exposed or manipulated again.

The emotional toll of living in a state of perpetual hypervigilance left me feeling drained, emotionally exhausted. I couldn't relax. I couldn't trust. I couldn't feel safe. This became a permanent state of unease—a constant reminder that the world I inhabited wasn't fully mine. Even after moving on, after leaving him behind, I found myself haunted by the invisible chains he had placed around me through digital manipulation.

This state of hypervigilance and distrust didn't just affect my relationships with others; it deeply affected my relationship with myself. I began to question my own judgment—wondering if I could ever make the right decision again. If I was capable of seeing the truth, or if my perceptions were always distorted. The narcissist had gaslit me for so long, manipulated my sense of reality to such an extent, that I began to doubt my own thoughts and feelings. This self-doubt became a permanent companion.

It wasn't just the emotional toll; it was the mental exhaustion of living in a state of perpetual fear and doubt. Even in the absence of direct contact, the narcissist's digital presence continued to exert influence over my life. The consequences of his actions were long-lasting. Every time I checked my phone or opened my inbox, I was reminded of the power he once had over me. And though I had physically escaped, I realized that the emotional scars of digital stalking would take far longer to heal.

As time passed, I found myself struggling with the inability to let go. It was as if his influence had permeated so deeply into my psyche that I couldn't easily erase it. It wasn't just about moving on from the relationship—it was about finding a way to reclaim my own autonomy in a world where I had been constantly controlled.

The long-term psychological impact of narcissistic abuse is often not visible to the outside world. It's not something that can be easily measured or understood. But the scars are real. The emotional weight of living under constant surveillance, under the shadow of someone who wants to control every aspect of your life, is a burden that continues to shape who you are long after the abuse has ended. The consequences of this kind of abuse are deeply ingrained in your psyche, and it takes a long time—sometimes a lifetime—to heal.

Despite this, I learned that healing is possible. It's not easy, and it doesn't happen overnight. But the process of rebuilding yourself after narcissistic abuse is one of the most empowering journeys a person can take. Over time, I have learned to trust myself again, to trust others, and to slowly regain control of the narrative of my life. And while the scars remain, they no longer define me. They are part of my story, yes, but they do not control me.

The longer I lived with this digital invasion, the more I realized just how far-reaching its impact truly was. At first, it seemed like an annoyance—an uncomfortable reality that I'd have to face as I adjusted to my new life, a life in which I thought I had finally broken free from the suffocating grip of my ex-husband's narcissism. But the more I encountered these subtle intrusions, the more it became clear that they weren't random acts—they were calculated, deliberate, and deeply malicious. They were designed to keep me in a perpetual state of uncertainty, anxiety, and emotional distress.

What I didn't know at the time was that these moments of intrusion, these texts and messages, were part of a much larger, far more sinister strategy. They were his way of ensuring that no matter how much physical distance there was between us, there was still emotional proximity. He was trying to force me to relive the power dynamics of our past, constantly reminding me that I wasn't free. I was still under his control, even from miles away. It was like he had access to a secret control panel in my mind, pressing buttons to make me react, to make me doubt myself, and to make me feel small and powerless once again.

This behavior wasn't just about asserting dominance. It was an ongoing attack on my sense of self. Every time he reached out—whether through a passive-aggressive message, a covert attempt to "check in," or even a public post that made me feel exposed—it reinforced a lie: that I was never really in control, that I could never really escape. These were not just emotional blows; they were strategic strikes aimed at breaking me down, ensuring that I would always feel tethered to him in some way, even if he wasn't physically present.

It's hard to explain the emotional toll of this kind of manipulation. It wasn't that I feared his presence. It wasn't even about being "chased" in the traditional sense. It was more insidious than that. It was about the mental drain of living in a constant state of fear and hyperawareness, knowing that my private life, my every thought and interaction, could be scrutinized at any moment. There was no safe space, no sanctuary from his intrusion. Whether I was at home or out with friends, whether I was celebrating a personal achievement or simply going about my day-to-day life, there was always a part of me that remained alert, waiting for the next breach.

This emotional harm—this constant surveillance—became a silent weight that I carried, day in and day out. It wasn't the kind of burden you could just shake off. It was there, gnawing at my peace of mind, slowly eroding my ability to trust not just others, but myself. I became hypervigilant about my surroundings, cautious about every text, every email, every social media interaction. The boundaries I once took for granted, the safe spaces where I could feel free and at ease, were no longer mine to enjoy. Instead, they became battlegrounds—mental wars fought in the quiet spaces of my day, when I should have been able to relax.

Digital Stalking as a Tool of Power

Narcissists are often portrayed as emotionally self-absorbed individuals, but when their need for control goes unchecked, their behavior can evolve into something far more sadistic. This is especially true when they face a narcissistic injury, such as the rejection of their false self—a blow that strikes at the core of their identity. In an attempt to repair their fragile ego, the narcissist will go to great lengths to reassert dominance, and often, this involves inflicting emotional pain on their victims.

What begins as an act of control becomes a systematic and sadistic campaign to break down the victim, reaffirming the narcissist's superiority while simultaneously denying the victim any semblance of peace or freedom. The narcissist's behavior is no longer simply about reasserting power—it's about proving their sense of entitlement and superiority by deliberately hurting the victim. This is what makes narcissistic abuse so damaging: it's not just about controlling someone—it's about breaking them down emotionally, making them feel small, insignificant, and powerless.

This sadistic approach is exactly what I began to experience. After I left, his digital stalking didn't just continue—it escalated. The narcissist was no longer trying to "win me back" or "show concern." No, his efforts became about control, and that control was now firmly rooted in the digital realm.

He would regularly check my social media accounts, keep tabs on my friends, and monitor my movements through the online traces I left behind. At first, I didn't realize the full extent of his surveillance. But slowly, I began to notice patterns—unexpected messages at odd hours, strange comments from people I didn't recognize, or, worse yet, messages that mirrored things I'd just told someone in a private conversation. This was the worst part: the feeling that I couldn't have a private thought, a private life, without him knowing about it. And the worst part of all? He was no longer even physically close. I had left him behind, but he was still there, lurking in the shadows of my digital world, keeping watch over every detail of my life.

What truly made the situation unbearable was the fact that I had no control over it. He was using my own personal information—things I thought I had safely tucked away—to invade my privacy and maintain his grasp over me. Whether it was logging into my accounts or manipulating my online presence to make it seem like I was still "his," he was constantly finding new ways to assert his dominance. The sadism in this behavior wasn't just in the invasion itself, but in how it deliberately targeted my emotional state, ensuring that I could never truly escape. The victim's mind became the battleground, and the narcissist was playing a game where they made all the rules.

This form of digital stalking became a tool of absolute power, and it was effective because it never ceased. Even when I was physically miles away, even when I had blocked him on every possible platform, he would still find ways to watch, to control, and to invade my life. The feeling of being watched was suffocating. It wasn't just the constant intrusion—it was the reminder that no matter how far I went, no matter how many walls I built around myself, he was always present.

It's hard to explain how deeply this digital surveillance affects you, especially when you're left feeling like a puppet in someone else's game. The narcissist wasn't interested in re-establishing a relationship. They didn't want reconciliation or healing. They wanted to see me suffer, to see me broken, to see me mentally shackled by the same force that had controlled me throughout our relationship. The abuse became personal—not just an act of emotional manipulation, but a campaign to reaffirm his superiority and destroy my sense of self-worth.

Long-Term Psychological Consequences

The long-term psychological consequences of living under such intense and unrelenting digital control are profound. The mental and emotional toll of being stalked and manipulated for years doesn't just disappear once the narcissist is no longer physically present. If anything, the trauma deepens, and it can take years, sometimes a lifetime, to truly understand and heal from the damage caused.

For me, the hypervigilance didn't end with the last message or the last breach. The damage from those years of constant surveillance became internalized. It wasn't just about being watched anymore; it was about the mental habits that had been formed, the endless state of fear that had been drilled into me, and the psychological scars that I didn't even realize were there until much later.

At its core, narcissistic abuse, especially in its sadistic form, warps your perception of reality. The narcissist's constant gaslighting and emotional manipulation have a way of making you question everything you know about yourself. Over time, you begin to lose touch with your own boundaries, your own thoughts, and your own sense of safety. The narcissist's power extends far beyond the physical realm—it reaches deep into your mind, where it creates lasting damage.

For me, the most significant consequence of this constant digital surveillance was the loss of trust—not just in others, but in myself.

The more I questioned my reality, the more I struggled to trust my own instincts. I would second-guess my decisions, doubt my judgment, and constantly wonder whether I was truly in control of my life, or if I was still, in some way, trapped by the narcissist's influence.

Even when I tried to rebuild, even when I moved forward with new relationships or started fresh in different environments, the scars of digital abuse remained. They were invisible, but they were always there—haunting me in the back of my mind. Would this person hurt me like he did? Are they watching me? Will they manipulate me? These were the questions I couldn't shake, the fears I couldn't outrun.

The most insidious part of it all was the invisibility of the wounds. Unlike physical scars, you can't see the psychological toll narcissistic abuse leaves behind. It's not something you can point to or touch. But it's there—affecting everything from how you approach new relationships to how you view your own identity. In some ways, the emotional pain caused by narcissistic abuse becomes part of you, woven into the fabric of your being. It's a kind of post-traumatic growth that never feels like growth. Instead, it feels like an ongoing battle—one that never truly ends.

The sadistic nature of narcissism doesn't just leave a mark on your psyche—it leaves a permanent imprint on your worldview. When you've been controlled for so long, the concept of freedom becomes foreign. It's no longer something you can easily grasp. Every attempt to rebuild is met with the silent reminder that control can always return, that power dynamics are never as settled as you'd like to believe.

As you begin to heal, the process is less about forgetting than it is about learning to coexist with your scars. These scars don't make you weak—they make you resilient. You learn, over time, to stand firm, to trust your instincts again, to understand that you are not defined by the pain, the manipulation, or the control you once suffered under.

But what I learned, through all the years of struggle, is that healing is a choice—and it's a choice that requires you to stop letting the narcissist's actions control your emotions, your thoughts, and your life. It's a choice to reclaim your privacy, your independence, and your sense of self.

No matter how long the narcissist's shadow looms, you can, eventually, step out into the light—stronger, wiser, and more aware of your own worth than ever before.

False Sense of Self-Worth: How Cyberstalking Reinforces Delusion

At the core of narcissistic abuse is the narcissist's need to maintain a false sense of self-worth, a fragile construct built on a deep and unwavering delusion of grandeur. Unlike most people whose self-esteem is balanced by both external recognition and internal validation, the narcissist's self-esteem relies almost exclusively on external validation (Ronningstam, 2005). It is this need for validation that drives their cruel and manipulative behavior—each interaction, each form of emotional control, is a means to reinforce their distorted sense of superiority.

In the wake of rejection or abandonment, the narcissist experiences what is known as a narcissistic injury, a deep wound to their fragile ego. The narcissistic injury is not merely emotional hurt; it is an assault on the narcissist's very sense of self. To the narcissist, the loss of admiration, power, or control represents an existential threat, leading to anger, resentment, and often, violent retribution in the form of emotional manipulation, gaslighting, or even cyberstalking. This behavior is driven by an unrelenting need to repair the false self—to remind both the victim and themselves that they are powerful, untouchable, and worthy of admiration.

For someone trapped in this cycle of narcissistic abuse, it is difficult to understand just how deeply these actions are rooted in the narcissist's insecurity. Cyberstalking becomes not only a tool of control but also a twisted mechanism of self-validation. Every time the narcissist invades the victim's life—through monitoring social media accounts, hacking emails, or spreading malicious rumors—it is not merely about exerting dominance. It's about proving that they still hold power over the victim, that the victim still reacts, and that their control is absolute, even when the victim is physically distant.

In this context, cyberstalking ceases to be just a tool of power—it becomes part of the narcissist's narcissistic game. Each message sent, each email intercepted, and each surveillance maneuver is an attempt to restore the narcissist's sense of superiority, to show the world, and themselves, that they are still untouchable, still in control. The narcissist feeds off the emotional reactions of the victim, finding a twisted sense of validation in their suffering.

What makes this particularly sadistic is that, for the narcissist, their self-worth is no longer derived from admiration or affection. Instead, it is reinforced through destruction. Every violation of privacy, every intrusive act, feeds into the cycle of narcissistic abuse. The more discomfort and pain the victim feels, the more the narcissist believes they are in control—and the more their fragile ego is propped up. In their twisted worldview, their ability to manipulate the victim and provoke emotional responses proves their superiority. And, in a strange and sadistic way, the victim's emotional turmoil becomes the narcissist's source of self-esteem.

The narcissist's behavior is addictive—like a drug. The more control they have, the more they crave. With each victory, they feel momentarily satisfied, but it is never enough. The validation they receive from breaking the victim's spirit, from forcing them into emotional submission, becomes insatiable. This dependency on external validation drives the narcissist to double down on their abusive tactics, always seeking more control, more manipulation, more emotional devastation. And, as the narcissist feeds on the victim's misery, the cycle continues, ensuring that the narcissist's false sense of self is constantly reaffirmed—at the expense of the victim's peace, security, and self-worth.

As the narcissist continues to engage in cyberstalking, it becomes clear that the ultimate goal is not merely to possess the victim, but to systematically break them down. The more the narcissist invades their private life—whether through digital surveillance, online harassment, or subtle tactics to instill fear and doubt—the more the victim is emotionally devastated. Identity theft, cyberbullying, and the manipulation of the victim's social circle become part of the narcissist's toolbox. By doing this, they reinforce their control, ensuring that their victim feels trapped, unable to escape. This sadistic behavior, under the guise of control, seeks to strip the victim of their independence and self-respect, turning them into puppets in a game where the narcissist dictates every move.

The narcissist's goal is to ensure that the victim never feels truly free. Even after physical separation, the victim is left feeling constantly under siege, their privacy constantly violated, their peace consistently disrupted. The narcissist thrives on this invisible control, knowing that even when they are not present, they can still cause emotional turmoil and continue to assert their dominance.

Ronningstam (2005) articulates how narcissistic individuals maintain their false self by creating and manipulating external narratives—the narcissist will stop at nothing to maintain the image of superiority. They construct elaborate facades and often use others as tools to reinforce their distorted version of reality. This can extend into cyberstalking or other forms of digital abuse as a way of maintaining a narrative that keeps the victim subjugated and under their control. The victim becomes an object in the narcissist's emotional chess game, a piece to be moved, manipulated, and discarded once their utility is over.

The deeper the narcissist's involvement in cyberstalking, the more they begin to believe the false narrative they have created about themselves. They convince themselves that they are entitled to invade the victim's privacy, to control their thoughts and emotions, and to maintain the illusion of a relationship that no longer exists. This delusion reinforces their need to constantly engage in emotional warfare and destructive behaviors. The narcissist's internal reality is so fragile and fragile that the only way they can validate their self-worth is by breaking others down. It is a perverse reflection of how narcissism manifests: through external validation, even at the cost of others' pain and well-being.

From the perspective of the victim, this constant violation of privacy and emotional manipulation leaves scars that are hard to heal. The effects of cyberstalking are not just about lost privacy but about lost agency. The victim's ability to control their narrative is stripped away, and they are forced to play along in a game where the rules are constantly changing. Cyberstalking becomes a means of erasing the victim's reality and replacing it with the narcissist's narrative, one where the narcissist is always in control, always on top, always superior.

This erosion of the victim's sense of autonomy is one of the most damaging consequences of narcissistic abuse. The narcissist's power lies not only in their ability to invade privacy but in their ability to make the victim question their own reality. In this twisted dance, the narcissist becomes the puppet master, and the victim is left to wonder if they are truly free or if they will always be haunted by the narcissist's shadow.

Ronningstam's (2005) exploration of narcissistic self-esteem challenges the conventional understanding of narcissistic behavior. Rather than seeing narcissists as simply self-centered individuals, we must understand that their delusion of superiority is a defense mechanism—one that drives them to engage in behaviors that are emotionally violent and psychologically intrusive. The narcissist's lack of authentic self-worth leads them to use others as emotional fuel, perpetuating a cycle of abuse that, in many cases, can last for years.

The consequences of this digital abuse are more than just short-term inconveniences. Psychological tolls like complex PTSD, anxiety, and depression become permanent fixtures in the victim's life. Victims of cyberstalking often feel isolated and helpless, knowing that they are still being watched, still being controlled by an invisible force. The scars of digital manipulation are often invisible to the outside world, but the emotional and psychological damage is profound.

Through this sadistic campaign, the narcissist not only reaffirms their false sense of self-worth but also reinforces the victim's perception of themselves as unworthy or inadequate. Each violation of the victim's privacy, whether subtle or overt, serves to confirm the narcissist's perceived superiority and the victim's inferiority. This delusion drives the cycle of narcissistic abuse—each time the victim responds, each time they react emotionally, the narcissist feels validated. This becomes an emotional feedback loop, where the narcissist's need for external validation is endlessly fed by the victim's distress.

The more a narcissist's false self is validated through these sadistic acts of control, the more entrenched their behavior becomes. The victim, caught in this toxic dance, may never fully understand the depths of the narcissist's manipulation until they are far removed from the situation. By then, however, the damage may already be done. The sadistic nature of narcissistic abuse, fueled by the narcissist's desperate need for external validation, leaves a trail of emotional destruction, leaving the victim in a perpetual state of self-doubt, anxiety, and fear.

Case Study: The Silent Destruction of Emma's Life

Emma, a woman in her early thirties, had been in a relationship with a man we'll call James for five years. From the outside, their relationship seemed perfect. James was charming, successful, and appeared to be deeply in love with Emma. However, the reality of their dynamic was far different. Emma, like many victims of narcissistic abuse, didn't realize the extent of the manipulation until years later.

At the beginning of their relationship, Emma was swept off her feet by James's charm and overwhelming attention. He made her feel like the most important person in the world, showering her with affection and constantly telling her how lucky he was to have found someone as beautiful, intelligent, and successful as she was. At the time, these compliments seemed genuine and made Emma feel special. But as time went on, the praise began to shift subtly into control.

James's validation of Emma's worth was contingent on her meeting his expectations. He would praise her for things that aligned with his ideals but would quickly dismiss or criticize her when she didn't conform to his needs. This started as small things—comments about her appearance, her career choices, and her interactions with others—but slowly escalated. The more Emma tried to assert her independence, the more James tried to pull her back into his orbit.

In the beginning, Emma didn't realize that James's actions were part of a larger strategy to maintain control over her. James was not just trying to dominate her emotionally; he was trying to build a false narrative about their relationship where he was the unquestioned source of validation in her life. He made sure to reinforce that Emma's worth was tied to his opinion of her, and if she ever stepped out of line, she would be reminded of her place. It was as if James was trying to create a version of Emma that was in constant need of him for approval.

As the relationship progressed, James's behavior became more subtle but more insidious. He would send passive-aggressive texts, accusing Emma of being selfish or disloyal. He invaded her privacy, frequently checking her phone, reading her emails, and even asking friends and family members for information about her activities. At first, Emma thought his actions were just signs of caring—he was "just concerned" about her well-being. But eventually, it became clear that these actions were not driven by love but by a deep-seated need to control her. James was using surveillance as a tool to reinforce his dominance, knowing that the more Emma was kept in check, the more his false self would be validated.

When Emma finally reached her breaking point and ended the relationship, she believed the worst was behind her. But James's behavior escalated in ways she could never have anticipated. He started to stalk her online, monitoring her social media accounts, tracking her location through her phone, and even creating fake profiles to engage with her friends and family. It was as if, even after the end of their relationship, James needed to ensure Emma's life revolved around his presence. His need for validation had grown so intense that it was no longer enough to control her in the physical world; he needed to control her mentally, digitally, and emotionally.

The sadistic nature of James's actions became apparent when he began spreading false rumors about Emma. He used his access to her personal information to manipulate the people around her, creating a narrative that painted Emma as unstable, irresponsible, and untrustworthy. The more Emma fought back, the more James pushed her into emotional turmoil. Each time she confronted him, he gaslighted her, denying everything and making her question her own sanity.

For Emma, the years following her separation from James were marked by an unrelenting sense of self-doubt. Even when she tried to rebuild her life, the emotional scars of narcissistic abuse held her back. She struggled with feelings of anxiety and fear, knowing that at any moment, James could find a way to invade her privacy again. She couldn't even check her social media without feeling the weight of his constant surveillance looming over her. The feeling that she was never truly free from his control became a part of her daily existence, preventing her from fully healing and moving on.

This cycle of control, surveillance, and emotional manipulation, all fueled by James's need for external validation, left Emma in a state of perpetual emotional distress. She was trapped in a web of lies and deception, unsure of where the manipulation ended and where her own sense of self began. For Emma, the narcissist's false self was a toxic force that, once validated, became an ever-present shadow in her life, constantly reminding her of her vulnerability and inadequacy.

It wasn't until Emma sought therapy and began to understand the mechanisms of narcissistic abuse that she started to regain some sense of control. Slowly, she came to realize that James's actions were not about her—it was never about her at all. His behavior was an effort to feed his delusional sense of superiority. He used Emma as a tool, a means to an end: to prove to himself that he could control, dominate, and ultimately break someone down emotionally in a way that validated his false self. Each act of control, every invasion of privacy, was simply a reinforcement of his delusion that he was more important, more powerful, and more worthy of admiration than anyone else. Emma's pain was his fuel.

The consequences of James's behavior were far-reaching. Even after she had cut ties with him, Emma found herself struggling with feelings of self-doubt, depression, and a sense of loss that she couldn't quite explain. The emotional destruction caused by James's cyberstalking and manipulation was not something she could just "get over." It would take years of therapy, self-reflection, and building new relationships for Emma to start to reclaim her sense of identity and self-worth.

Emma's case is not unique. Many victims of narcissistic abuse face a similar battle—a battle in which their privacy, identity, and emotional stability are systematically destroyed by someone who uses manipulation and control to validate their own false self. For these victims, the scars of narcissistic abuse are not just emotional; they are deep psychological wounds that can take years to heal, if they ever truly do. The narcissist's need for control and external validation leaves an indelible mark on the victim's psyche, often leaving them questioning their own reality long after the abuser has gone.

In Emma's case, the sadistic nature of narcissistic abuse was painfully clear. What began as a relationship built on charm and affection turned into an emotionally draining cycle of manipulation, surveillance, and emotional warfare. By the time Emma left, the damage was done—she had been left questioning her self-worth, struggling with anxiety, and haunted by the specter of a narcissist whose need for control seemed endless.

In the aftermath, Emma realized that her struggle was not just with James but with her own perception of herself. Narcissistic abuse doesn't just leave the victim emotionally scarred; it shatters their sense of self, making them believe they are broken, unworthy, or incapable of finding peace. It's a sadistic cycle that takes time to recognize, understand, and ultimately heal. But for Emma, as for many others, the first step toward recovery was realizing that the narcissist's validation was never meant to heal—it was meant to destroy.

The Narcissist's Psychological Distortion: Inflicting Emotional Pain to Build Their Ego

The sadistic nature of a narcissist is rooted in a psychological distortion that drives them to not just control but actively dehumanize their victims. These individuals do not merely seek power; they derive intense satisfaction from inflicting emotional pain, watching their victims suffer, and feeling the illusion of control that this power provides. Their ability to make others feel small, helpless, and insignificant reinforces their belief in their own superiority—a superiority they desperately need to maintain in order to mask the fragility of their ego.

This behavior doesn't just extend to the narcissist's immediate control over the victim. It becomes something far darker: a sadistic pleasure in watching the victim squirm, their pain acting as a reflection of the narcissist's self-worth. The narcissist's inner world is in constant disarray, and rather than seeking emotional connection or understanding, they actively crush the spirit of others to feel secure in their delusions of grandeur.

Narcissists, driven by their need to maintain a false sense of self, are often obsessed with ensuring that their victims never experience emotional peace. When a narcissist feels threatened—whether through abandonment or any perceived rejection of their false self—they lash out with vengeance. But the torment is not only about reclaiming their power; it's about proving their supremacy, asserting that they can control even the most intimate, private aspects of someone's life. And they do this through both physical and psychological means, turning what was once an abusive relationship into a sadistic game of emotional warfare.

To maintain their distorted self-image, narcissists often resort to digital stalking—a form of control that is invisible, constant, and relentless. Through social media surveillance, monitoring emails, or accessing private information, they invade every aspect of their victim's life, keeping them mentally and emotionally tethered to the narcissist's power. For the narcissist, this is not simply an act of obsession; it is a calculated effort to demonstrate their control over the victim's world, even from a distance. Every invasion into the victim's privacy becomes a calculated effort to reinforce the narcissist's sense of superiority.

In my own life, the psychological torment of my ex-husband's control was not limited to physical threats or emotional manipulation—it extended into my every waking moment. Even when I thought I had escaped, he continued to haunt me from afar. Through digital means, he kept tabs on my life, every post, every interaction, every step I took. But this wasn't just about keeping an eye on me—it was about destroying my peace, destabilizing my emotional state, and proving to himself that he could still control my reality.

Dr. Rachel Levitch

At one point, the abuse went beyond mere manipulation. It became dangerously sadistic, moving from psychological warfare to something far more tangible. One of the clearest examples of his sadistic nature came during a drive home one evening, a moment when his desire for control escalated into a terrifying display of violence that still haunts me.

I remember that day clearly, even though it feels like a foggy, distorted memory now—one I can never seem to fully grasp. It was a family trip, a routine drive, something that should have been innocuous. My daughter, barely five at the time, sat in the back seat, her small voice chirping about something trivial, as children often do. The mood was light until it wasn't. I had noticed my ex-husband's temper starting to flare, a quick change in his demeanor that was always the precursor to something darker. But I never could have anticipated what would happen next.

Without warning, he slammed the car to the side of the road, his knuckles white on the steering wheel. My daughter, who had been playfully tugging at my sleeve, stopped in her tracks, sensing the shift in energy. The air became thick with tension—I could feel my chest tighten, my heart rate quicken. I knew something was wrong, but I couldn't place it. I didn't know how to prepare for what would come.

Before I could speak, he had already turned toward my daughter, rage flashing in his eyes. She hadn't done anything wrong—her innocent chatter was just too much for him in that moment. He grabbed her out of her seat, yanking her forward with a force that made my breath catch. She screamed in shock, unable to comprehend why her father, someone she should have been able to trust, was now inflicting violence on her.

In that moment, I was paralyzed. Frozen. I couldn't move, couldn't speak. I was terrified—not just for her, but for me. For the first time, I realized the full extent of the control he had over my life. His violence wasn't just physical; it was emotional, psychological. I screamed at him, desperate to make him stop. But my voice was small, drowned out by the chaotic noise in my head and the fear that gripped me so tightly, it felt as if my lungs were collapsing under its weight.

He turned to me with a look of contempt, his anger flashing as he yanked me forward by my seatbelt. I remember the force of his grip, how he pulled me toward him with such strength that I thought my body might snap. I was helpless, completely paralyzed by fear and uncertainty. My entire sense of agency evaporated in that moment. I could not protect my daughter. I could not protect myself. I had no power.

And it wasn't just the act itself that left me shattered—it was the fear of what would come next, the sense of being utterly powerless in the hands of someone who could destroy my peace and my safety at any given moment. Even as I screamed for him to stop, even as I begged him to let her go, there was something worse than the violence itself—the reality that he could get away with it. I realized, in that split second, how deeply entrenched his narcissistic control had become. He wasn't simply acting out of anger—he was sending a message. He was reminding me that no matter what I did, no matter how far I ran, he could still control the most intimate moments of my life. He could still make me feel utterly powerless.

When he finally let go, I was left in a numb daze, unsure of what had just happened. My daughter, still sobbing in the back seat, looked at me with wide eyes, confused and hurt. But I couldn't even console her. I was broken. I didn't have the strength to comfort her when I couldn't comfort myself. I was exhausted, emotionally drained. And the worst part was, this incident wasn't the first time, nor would it be the last. It became a pattern: outbursts, threats, emotional manipulation, and then sudden violence—each time more brutal and unpredictable than the last. The aftermath of these moments was always the same—shame, guilt, confusion, and an overwhelming sense of powerlessness.

The narcissist's behavior, fueled by his need to assert dominance, was not just about control. It was about proving his superiority by showing us how easily he could break us. The sadistic pleasure he gained from seeing my daughter's fear, my own helplessness, reinforced the fragile sense of self-worth he so desperately clung to. Each act of cruelty was an attempt to keep us mentally enslaved to his emotional whims, to remind us that no matter how far we ran, how much we tried to reclaim our autonomy, we would always be at his mercy.

This incident left scars that I still carry with me—emotional scars, physical scars, and the residual anxiety that would follow me for years. The fear that we were never safe, that at any moment his rage could erupt again, lingered in every corner of my life. It's taken years of therapy, of processing the trauma, of working through the paralyzing fear that defined my days for so long. But the damage was done. The sadistic nature of his narcissism had left an indelible mark on my soul, a constant reminder of how far a narcissist will go to maintain their sense of superiority, and the emotional destruction they leave in their wake.

What makes the sadistic nature of narcissistic abuse so insidious is that it often isn't immediately obvious. It starts small, with subtle manipulations, emotional outbursts, and calculated attempts to control. But over time, it grows, becoming more blatant, more destructive, until the victim feels trapped in a web of emotional warfare. And the longer the abuse continues, the more the narcissist's distorted sense of self is validated by their ability to cause harm without consequence.

This sadistic cycle perpetuates the narcissist's power, allowing them to continue their manipulations and maintain control of the victim's emotional state. It's not just about control—it's about reaffirming their place above everyone else, and reasserting their superiority over those they seek to dominate. And the victim? They are left struggling to rebuild their sense of self-worth, buried under layers of emotional trauma, fear, and self-doubt, unable to understand just how far the narcissist has gone to break them.

"This is the sadistic nature of narcissistic abuse: a cycle of psychological control, manipulation, and emotional destruction that leaves victims scarred for life."

This statement encapsulates the brutal, unrelenting truth about narcissistic abuse—an ongoing cycle that leaves no room for healing, only for further torment and self-doubt. The narcissist's need for control is not just about power—it is a pathological hunger, an insatiable thirst that grows stronger as they manipulate, gaslight, and break down their victim. What makes this abuse so profoundly damaging is the way it infiltrates the victim's sense of self, their trust in reality, and their ability to feel safe.

Dr. Rachel Levitch

The cycle starts with subtle control—manipulation through guilt, criticism, and covert aggression. But, over time, it escalates into an intense emotional war. Every word, every action, every moment is weaponized against the victim to ensure they remain small, vulnerable, and under the narcissist's thumb. For the narcissist, control isn't just about maintaining order. It's about asserting superiority over someone else, eroding their sense of self-worth, and reminding them that they are not enough—never enough to escape the narcissist's grasp.

This cycle begins subtly but quickly accelerates into more overt acts of control, from emotional manipulation to direct attacks on the victim's sense of identity. Once a narcissist has tasted this power, they crave it—just as an addict craves the next high. And like an addict, they will go to any lengths to keep their victim tethered to them, even if it means subjecting them to extreme emotional turmoil. The narcissist's cruelest tactic is to strip away the victim's ability to trust themselves, leaving them paralyzed in confusion and anxiety.

But the abuse doesn't end once the victim recognizes what is happening. The sadistic nature of narcissistic abuse means that even after leaving the narcissist, the victim is often left with a persistent feeling of being chased, like a ghost they can't escape. The narcissist doesn't just disappear when the relationship ends—they shift tactics. The abuse shifts to cyberstalking, smear campaigns, or continual emotional manipulation—all designed to keep the victim tethered in an invisible way. The victim remains in the cycle, even from a distance, because the narcissist uses whatever tools are at their disposal to continue their control.

For the narcissist, it's not just about having control—it's about proving that they can never be out of control. The delusion of their power must remain intact, and the victim is nothing more than a pawn in their twisted game of self-affirmation. Cyberstalking is often a narcissist's most insidious tool, allowing them to continuously invade the victim's life through digital means. Every email, every social media post, every private conversation becomes an opportunity for the narcissist to remind themselves of their superiority by asserting dominance from a distance.

The Cycle of Emotional Destruction

This emotional destruction is deliberate and it is painful. Victims often find themselves constantly questioning their own emotions, their own reality. Gaslighting is a key tactic in this cycle: the narcissist twists and distorts the truth until the victim can no longer tell what is real. Self-doubt takes root, often causing the victim to feel like they're losing their mind. The narcissist's objective is simple: to break the victim's sense of self—to make them feel powerless and confused. And when the victim questions their own perception, when they ask themselves, "Am I really the problem?" they unknowingly fall into the narcissist's trap.

But this abuse isn't just emotional or psychological—it is physical, too. Fear is a constant companion. The victim begins to live in a state of hypervigilance, always looking over their shoulder, always aware that they are being watched—whether through a pair of eyes or through a digital device. This fear does more than exhaust the victim—it chips away at their mental health, leaving them in a state of perpetual uncertainty.

One of the most shocking things about narcissistic abuse is how seamlessly it can blend into daily life. For a long time, I didn't even recognize the full extent of the manipulation. I thought his controlling behavior was just a result of jealousy or insecurity. But what I didn't understand then was that this control was purposeful—every action, every word, was part of a larger plan. He wanted to make me feel small, inferior, unworthy of peace.

The Physical Toll of Psychological Warfare

Narcissistic abuse leaves its marks everywhere—mentally, emotionally, physically. The sleepless nights, the anxiety, the dread that fills your stomach whenever you have to interact with the narcissist. These physical manifestations of the trauma are often the silent scream of the victim's psyche, trying desperately to escape the emotional prison the narcissist has locked them in. It's a constant battle to reclaim your own space, your own mind.

But the damage isn't just in the moments of active abuse. The long-term impact is even more insidious. Narcissistic abuse often leads to complex PTSD—a form of trauma that develops when someone is repeatedly exposed to emotional manipulation and control over a long period. The victim may struggle with trust, self-worth, and even basic functioning in their day-to-day life. It's hard to even describe how exhausting it is to be trapped in that cycle, where every move you make feels like it might trigger the narcissist into an outburst or manipulation. It's a never-ending mental and emotional treadmill, with no way to step off.

In my case, there were moments when the tension became so unbearable that I could feel my body physically reacting to the emotional strain. I would start shaking for no reason, my breath caught in my chest, as though my body had learned to react to his presence before my mind even registered it. Every minute felt like a battle, a fight to hold onto some semblance of sanity in the face of an enemy who could read me—manipulate me—at will.

The Cycle Never Ends

Perhaps the most devastating part of narcissistic abuse is that it doesn't end when you physically leave. The narcissist will lurk—digitally, emotionally, or even physically—waiting for the next opportunity to regain control. This is what makes their power so pervasive: it doesn't end with a breakup, a physical separation, or even a final "goodbye." The narcissist becomes a shadow in the victim's life, always watching, always waiting, always ready to strike. They know that the victim is never truly free—not mentally, not emotionally, and often not even socially.

I remember vividly the days when I thought I had finally escaped. I had moved away, changed my number, blocked him on social media, and started over. But the reality was that the emotional control didn't end there. I had to face the terrifying truth: he was still controlling my life, even from a distance. Every notification on my phone made my heart race. Every email felt like a message from the abyss. Every new relationship I tried to build was haunted by the fear that he would appear again. And the worst part? The shame I felt, as though it was somehow my fault that I couldn't break free.

But this is the nature of narcissistic abuse. The victim is never truly free because the narcissist has embedded themselves so deeply into their psyche. Even when the narcissist is gone, their influence remains, like a parasite that refuses to die. And this is the essence of the narcissist's sadistic cruelty: they may leave physically, but they never truly let go.

References:

Ronningstam, E. (2005). Identifying and understanding the narcissistic personality. Oxford University Press.

- This book gives an in-depth analysis of Narcissistic Personality Disorder (NPD) and how narcissists rely on external validation to maintain their fragile self-esteem. It can support the concepts around narcissistic injury, the need for control, and the sadistic pleasure they derive from manipulating their victims.

Kernberg, O. F. (1975). Borderline conditions and pathological narcissism. Jason Aronson, Inc.

- Kernberg's work can be used to explain the destructive nature of narcissistic personalities and how they manipulate and control others, aligning with the behavior described in your story.

Vaknin, S. (2015). Malignant self-love: Narcissism revisited. Narcissus Publications.

- This reference is helpful for understanding the sadistic nature of narcissistic abuse and the emotional devastation that victims endure. Vaknin explains how narcissists destroy their victims to maintain their sense of superiority, which would back up the claims of sadistic emotional control.

Millon, T. (2011). Personality disorders in modern life. Wiley.

- Millon's work on personality disorders, including narcissism, can support the claims about the narcissist's compulsive need for control and the deep psychological impact their behavior has on their victims. The victim's psychological breakdown due to the narcissist's control tactics fits well with Millon's theory.

Bancroft, L. (2003). Why does he do that? Inside the minds of angry and controlling men. Berkley Books.

- This book sheds light on the mind of a controlling person, specifically focusing on the behavior of abusers. It touches on narcissistic tendencies, including manipulation, control, and the psychological warfare victims face in relationships. This can support the sections on control, manipulation, and the emotional destruction described.

Walker, L. E. (1979). The battered woman. Harper & Row.

- Although this book focuses more on physical abuse, it also touches on emotional manipulation, which is crucial in understanding how narcissistic abuse operates. It would serve as a reference to describe the trauma victims experience, especially when the victim feels paralyzed by fear or trapped in the cycle.

Freedman, L. (2016). Narcissistic abuse: A survivor's guide to breaking free. The Empowerment Institute.

- This survivor-focused guide can be referenced to show how narcissistic abuse continues to affect victims even after they leave, through tactics like digital stalking and ongoing emotional manipulation.

Malkin, C. (2009). Narcissistic predators: The toxic effect of narcissism on relationships. Alpha Books.

- Malkin's book discusses how narcissistic individuals exploit their relationships for personal gain, including the use of control tactics and emotional abuse. It can be used to support the theme of digital stalking and manipulation for personal validation.

Gabbard, G. O. (2009). Psychodynamic psychiatry in clinical practice. American Psychiatric Publishing.

- This reference provides a clinical perspective on narcissism, specifically focusing on the mechanisms of emotional control and manipulation. It will strengthen the claim that narcissistic abuse is rooted in psychological distortion, where the narcissist derives pleasure from the victim's suffering.

Chesney, E. (2011). Psychological abuse in intimate relationships: The struggle for control. Routledge.

- Chesney's work explores emotional and psychological abuse, particularly in the context of narcissism. This book can support the concept of the narcissist's need to control their victim through fear, manipulation, and emotional dominance.

Chapter 9

Digital Abuse: Narcissistic Superiority and Manipulation During Mental Health Struggles

How narcissists exploit victim vulnerability

The relationship with a narcissist begins to resemble something far beyond simple emotional manipulation. It feels more like an emotional war, a constant battle where your vulnerability is the battleground and their need for power is the ammunition. Narcissists are known for exploiting their victims' weaknesses, and when mental health struggles come into play, it amplifies the narcissist's power. The more vulnerable the victim becomes, the more opportunities the narcissist sees to exert control. Narcissists don't just exploit mental health issues—they weaponize them. This is not just about undermining emotional stability; it's about making sure that the victim never regains control over their own narrative.

As a person who has lived through this kind of manipulation, I can recall how my ex-husband—like so many narcissists—sensed my emotional fragility and capitalized on it in ways I couldn't see until I was far away from the situation. There were moments where I was broken, emotionally drained, and desperate for relief from the mental health issues that plagued me. Instead of offering comfort or understanding, he used my pain to further his own needs, reinforcing his power over me and ensuring that I felt more isolated and helpless than ever before.

One of the most insidious ways narcissists exploit their victims' mental health struggles is by undermining their attempts to seek professional help. This not only amplifies the victim's trauma, but it also reinforces the narcissist's control over the situation. In my experience, this was a constant pattern. Whether it was belittling my need for therapy or undermining my decisions to seek medical help, he always knew how to make me feel like I was wrong for trying to improve myself.

Undermining Your Mental Health Support System

I remember a specific incident when I tried to go to therapy for the first time. At that point, I was beginning to understand that my emotional distress wasn't just something that would go away with time—it was a symptom of years of emotional abuse. But as soon as I expressed my desire to seek professional help, he reacted defensively, almost like I had betrayed him. He dismissed my need for therapy, calling it "selfish" and "overdramatic."

"You don't need therapy," he said, his voice dripping with condescension. "We're fine. I'm here for you. What more do you need?"

In that moment, it felt like a direct attack on my self-worth. How could I trust myself when he was so determined to convince me that I didn't need help? I had always believed that my anxiety, depression, and self-doubt were products of my own weakness.

But here he was, denying the reality of my pain and invalidating my need for support.

It wasn't just that he rejected my desire to seek help; it was the psychological games he played that ultimately made me doubt my own judgment. This is the essence of narcissistic manipulation: making the victim believe that their very perception of reality is flawed. By doing this, the narcissist diminishes the victim's ability to trust their instincts and, in turn, makes it much harder for them to make decisions in their best interest. Narcissists need to keep their victims off-balance, disoriented, and confused. They do this by constantly invalidating the victim's emotions, especially when it comes to mental health struggles.

In retrospect, I realize that his rejection of therapy wasn't just about controlling me—it was about ensuring that I stayed emotionally dependent on him. If I went to therapy, it would force me to confront my own emotions and question the dynamic we shared. It might even help me see that I was in an abusive relationship. And that was something he could never allow. The narcissist cannot bear the thought of losing their grip on their victim.

Amplifying Trauma and Reinforcing Control

Once the narcissist has established control over the victim's mental health journey, they will begin to actively amplify the victim's trauma, reinforcing their dominance in the process. Narcissists are adept at identifying vulnerabilities and exploiting them to intensify emotional suffering, ensuring that the victim's pain is sustained or even worsened.

This amplification is not random. It is carefully calculated and executed at moments when the narcissist senses that the victim may be weakening or questioning their power. The narcissist uses the victim's past trauma against them, often playing on their deepest insecurities to further destabilize their emotional state. It's as if the narcissist sees the victim's pain as a resource—a tool to be used for control rather than something to be healed or acknowledged.

During the period when I was trying to manage my depression and anxiety, he would frequently make comments designed to deepen my suffering. He would bring up my past traumas and use them as weapons. For example, if I was upset or anxious, he would say things like, "Remember when you thought you couldn't get through that? What makes you think you'll handle this any better?"

These words weren't meant to comfort me. They were meant to plant seeds of doubt and insecurity in my mind. By making me question my own strength and resilience, he ensured that I stayed emotionally reliant on him. He wasn't just manipulating me in the moment—he was reinforcing the idea that I needed him, and that without him, I was incapable of managing my own emotions or healing my wounds.

Narcissists are masterful at shifting the blame. Instead of acknowledging their own abusive behavior, they often place the responsibility for the victim's emotional state squarely on the victim's shoulders. When I would bring up my distress or feelings of sadness, he would respond by saying things like, "You're just being dramatic. You make everything worse than it is." These words were devastating because they made me feel like I was the problem, not the abuse I was experiencing. The narcissist's goal is always to shift responsibility and maintain control by making the victim feel responsible for their own suffering.

One of the most damaging aspects of narcissistic abuse is that it isn't just about emotional manipulation—it's about actively destroying the victim's sense of self-worth. By attacking the victim's vulnerabilities, amplifying their trauma, and dismissing their emotional needs, the narcissist ensures that their victim remains trapped in a cycle of self-doubt and emotional chaos. The victim can never fully trust themselves because they've been taught that their emotions are not valid or worthy of support. And this leaves them in a state of constant vulnerability, always searching for validation from the narcissist, even as that validation becomes more elusive and toxic.

Psychological Interplay Between Abuser and Victim

The psychological interplay between the narcissist and the victim is complex and layered, rooted in a dynamic of control that leaves the victim psychologically paralyzed. Narcissists do not just manipulate—they play an elaborate game, using every tool at their disposal to ensure that the victim remains emotionally dependent and isolated.

In the context of mental health struggles, this dynamic becomes even more insidious. The narcissist plays the role of both the abuser and the rescuer, giving the victim just enough attention or reassurance to make them believe that they need the narcissist for survival. This false sense of dependence is critical to maintaining control.

When I tried to reach out to friends or family for support during my mental health struggles, he would often sabotage those attempts, either by making me feel guilty for talking to anyone else or by planting seeds of doubt in my mind about the people I trusted. He would say things like, "I don't know why you talk to them—they don't understand you like I do," or "You're just making yourself a burden to them." Over time, these comments wore me down, causing me to question the genuineness of my relationships and further isolating me from the support system I so desperately needed.

The psychological toll of this isolation was profound. I began to internalize the belief that I could no longer rely on anyone but him. He had successfully framed himself as the sole source of my emotional stability, which, of course, only served to reinforce his control. This constant undermining of my relationships with others played directly into his narcissistic superiority complex. By making me feel that I was emotionally dependent on him, he cemented his position as the ultimate authority in my life, dictating my actions and thoughts.

Through these tactics, the narcissist creates a psychological prison—a maze of emotional confusion and dependency. They make the victim believe that they are broken or incomplete without the narcissist's intervention. This leaves the victim trapped in a perpetual state of fear and vulnerability, unable to escape the narcissist's influence.

Shifting the Dynamic: From Caretaker to Resentment

The Narcissistic Response to Vulnerability

Dr. Rachel Levitch

When I reflect on my time spent in therapy or treatment, what stands out most wasn't the struggle to confront my emotional health—it was the subtle but powerful shift in the dynamic between me and my ex-husband. Whenever I sought help, whether it was through inpatient care or outpatient treatment, his reactions weren't about concern or support. Instead, there was a growing sense of resentment.

What began as occasional moments of disinterest gradually turned into a deeply troubling pattern: he began to resent my need for mental health care. It wasn't that he had concern for my well-being or an understanding of the importance of recovery. No, in his mind, my need for help wasn't a sign of vulnerability—it was an offense against him. It was as if my struggle with mental health threatened his delusional sense of superiority and control. I was no longer just a partner in his eyes; I was a challenge to his perfect image. And for a narcissist, the idea that someone else might be helping me "fix" myself was akin to admitting defeat. He could not, and would not, allow anyone or anything to diminish his power.

He took my need for professional care and twisted it into a reflection of his own inadequacies. My therapy, which was supposed to be an act of self-love and healing, became a direct challenge to his ego. In his eyes, I had chosen to address my problems with outsiders—doctors, therapists, and treatment teams—rather than relying solely on him. This was the root of his anger.

What he failed to see was that his role in my life was not one of support. His focus was never on helping me heal or recover. Instead, it was on maintaining the narrative that he was the strong one, the one who could handle everything on his own, the one who didn't need outside assistance. My seeking care became a betrayal of the role he had carved out for himself: the hero, the savior. He couldn't allow me to be the one to seek external care, because that would mean he wasn't the hero, the "solution" to all my problems. Narcissists, particularly those with a fragile sense of self-worth, cannot tolerate the idea that someone else may be seen as a source of healing. They need to be the one who is needed, the one who can never be replaced.

Narcissistic Projection: Using My Healing as His Weakness

This behavior is not unique to my story. It is part of the narcissist's core defense mechanism—projection. Narcissists often project their own weaknesses onto others, as they cannot bear to confront their own imperfections. In my case, the projection was clear: whenever I sought help, he projected his fear of weakness onto me. The narcissist's vulnerability becomes a threat, and in order to shield themselves from confronting this vulnerability, they turn it outward, accusing others of being weak, helpless, or in need of "fixing."

Whenever I came home from a session, a treatment program, or a visit to a doctor, I could feel the shift. I wasn't "coming back to the safety of my home." I wasn't returning to a partner who was there for me. I was coming back to a man who could now look at me as someone lesser—someone whose struggles required treatment, someone who couldn't manage their own emotions, someone whose vulnerability became fuel for his emotional fire.

In his twisted mind, my mental health issues weren't just personal battles to overcome. They were manifestations of my inability to live up to his standards, and this, in turn, gave him an excuse to reassert dominance. My need for help undermined his false belief in his own infallibility, and he needed to reestablish control to restore his sense of superiority.

Every time I would return from a treatment session, my emotional state felt lighter, more hopeful, and more independent. But he couldn't allow that. He couldn't allow me to heal without him feeling diminished. Therefore, as soon as I returned home, he would subtly (and not-so-subtly) undermine my progress.

Reasserting Control: A Game of Power

When I think back to those days, I remember the eerie calmness with which he would manipulate situations. It was a game—a psychological contest of wills where my very need for healing was seen as a loss for him. After all, if I was healing, then it meant I was gaining strength, independence, and clarity. It meant that I no longer needed him to be my emotional anchor, and that was a direct challenge to his ego.

As my mental health improved and I regained clarity, the shift in his behavior became more overt. It wasn't enough that he'd attempted to destroy my self-worth through emotional manipulation and control. Now, with my recovery, he had to bring the focus back onto him. His narcissistic injury—the wound to his ego from my need for external support—was deepening. It was as though, in his mind, my progress was a personal affront to his sense of dominance.

He couldn't allow me to stand on my own. I couldn't just be the partner who had struggled with mental health and then overcome it. No, in his eyes, I needed to remain the "broken" one—the one who depended on him. So he started acting out, making every situation about his needs and forcing the narrative that I was somehow neglecting him. He would complain about his loneliness, his struggles, or how hard everything was for him. It was his way of ensuring that my healing would never overshadow his need for attention.

One particular example that stands out happened shortly after a week-long inpatient stay I had. I had finally been given the space to breathe and focus on myself, away from the chaotic environment he had fostered. Upon returning home, I was mentally exhausted but more centered than I'd been in a long time. Instead of coming back to a calm environment, I walked into a whirlwind of demands, complaints, and responsibilities.

What became clear to me was that my emotional healing wasn't something he could tolerate. I would find myself jumping back into my caretaking role, even though I had only just returned from seeking help. The moment I was back, I was expected to pick up right where I had left off. He shifted the burden of household responsibilities onto me, and any notion of self-care I had was immediately dismissed as selfishness. My need for rest and recovery became a narrative of neglect in his eyes. In his world, I had abandoned him, and he needed to punish me for it.

This was not just an attempt to regain control of the home or to force me into submission. It was a deliberate shift in how he saw our relationship. I was no longer the person deserving of care or understanding. I was now the one who had become a burden—a person who had "failed" because I needed professional help.

Narcissistic Injury: A Threat to His Superiority

In the narcissist's mind, any challenge to their superiority is perceived as a narcissistic injury. The simple act of me seeking therapy or treatment was, in his eyes, a direct insult to his ability to control the situation. Narcissists often go to great lengths to avoid anything that threatens their image of perfection, and in my case, seeking help was exactly that. It wasn't just about addressing my struggles—it was about reminding him that I could live without him, that I could heal without needing his validation.

It's not just about being "right" or "better"—it's about proving that they are the only solution to the victim's problems. It's about their need to be seen as indispensable. So, when I sought help, it threatened to invalidate everything he had convinced himself of. It threatened to expose the cracks in his perfect facade. Instead of recognizing that I needed help for personal reasons, he took it as a sign that he wasn't enough for me—he wasn't strong enough to be the "hero."

This injury fueled his resentment. He had to remind himself (and everyone else) that he was the one who could hold it all together. He had to reinforce his narrative that he was in control.

Conclusion: The Unseen Damage of Narcissistic Manipulation

The damage caused by these narcissistic dynamics is subtle but profound. The way he manipulated my need for self-care into a source of his resentment was insidious. It wasn't just emotional abuse—it was a psychological battle for power, a game where my vulnerability became his tool to control me.

In the end, the narcissist didn't just undermine my mental health efforts. He distorted my reality, twisted my need for healing, and used it to further tighten his grip on me.

The result? A toxic dynamic where my recovery became an act of defiance against his warped sense of superiority. Every step I took toward healing, every moment I took for myself, was seen as a betrayal. And in his world, I was only allowed to heal if it was done on his terms, according to the narrative he created.

The Narcissistic Need for Control: Exacerbating Vulnerabilities

During my time in mental health treatment, I couldn't shake the eerie sense of his presence even when he wasn't physically there. I remember being in group therapy or in individual sessions, trying to open up and finally confront the darkness I'd been holding inside for so long. But every so often, just as I was beginning to open up or start making real progress, my ex would pop up—often in the most unexpected ways—to assert his control over the situation.

It wasn't about guilt, as much as it was about his need to manipulate the narrative. His presence in these sessions was not about supporting me; it was about him maintaining dominance over the space. If a doctor or a therapist asked me about something personal, he would press for answers, often demanding to know what I had discussed, and if he didn't like my response, he'd insist that I change doctors. His sense of entitlement didn't just extend to the physical world—it reached into the confines of my healing space as well. The idea that I could even think of exploring my feelings or health without him was completely intolerable to him. He couldn't bear the thought that my recovery process might reveal flaws in him, or worse yet, that I might begin to gain the strength to break free from his control.

This intrusion was a calculated maneuver. It wasn't enough for him to know that I was struggling. He needed to control how and when I sought help. He wanted to be the sole source of validation, even in moments when he was not physically present. His behavior was textbook narcissism: the more I sought healing, the more he inserted himself into that process, treating my vulnerability as his opportunity for self-validation.

At times, doctors and therapists would comment that his actions were indicative of narcissistic behavior, even suggesting that I was being manipulated. At the time, I would bristle at this diagnosis. I couldn't face it. I would reject any suggestion that he was anything but the misunderstood partner who was just trying to support me. The idea that I had fallen into an emotionally abusive relationship was too painful to admit. It felt like an attack on my relationship, and by extension, an attack on me. But the truth was staring me in the face. My mental health struggles made me vulnerable, and in that vulnerability, he was finding ways to maintain control. His need for superiority and validation fed on my suffering, and his actions continued to reinforce his belief that he was the one holding everything together, while I was the one who couldn't manage.

When I went for inpatient treatment or would schedule outpatient sessions, there would always be a subtle shift in his behavior. At first, he would tell me how much he wanted me to get better. He'd act like the supportive spouse, saying all the right things, showing what seemed like empathy. But as soon as the treatment took hold, and as I started to feel the faintest glimmer of relief, that act would drop. His support turned into resentment.

This, of course, wasn't overt in the beginning. It would start with a slight, passive-aggressive remark: "You're taking so much time for yourself. I hope it's worth it." It would escalate to something like, "Why can't you just handle things like I do? I don't need any help. You need to do better than this." These comments were his way of undermining my efforts, making me question my recovery and my progress. They were crafted to force me into feeling guilty about needing help. His insecure need to remain in control had morphed into a constant reminder that I should never become independent of him. If I was recovering, if I was finding strength on my own terms, that was a threat to his carefully constructed sense of superiority.

Shifting the Dynamic: From Caretaker to Resentment

Once I returned from treatment, I quickly realized that my healing process was not something to be celebrated or supported, but rather something he resented deeply. The dynamic was shifting, but not in the way I had hoped. His "caretaker" role suddenly turned into a role of control, and once again, it felt like I was at the mercy of his moods.

I would return from a week of inpatient care or even after a day of therapy, and what should have been a moment of rest for me was instead met with a barrage of new tasks. It was almost as if he couldn't stand the idea that I was taking time for myself to heal. Instead of offering comfort or helping with the emotional labor of reintegrating, he would inundate me with more responsibilities around the house. Suddenly, I was expected to pick up the slack for everything he had neglected while I was gone—even though I was still recovering. It was as if his needs and his narrative of control had to dominate, even in my most vulnerable moments.

The most disturbing part, however, was that his resentment didn't just end with the responsibilities. He would continue to undermine my healing by commenting on my "weaknesses". If I was still dealing with emotional turbulence, he would claim I wasn't really trying. "You can't keep using that as an excuse," he would say, all while displaying a complete lack of empathy for the fact that I was struggling. If I couldn't manage a household chore because of emotional exhaustion, he would say, "You're just lazy. This is what happens when you go and get 'help'."

His resentment came from a deeply rooted need to maintain control, and my mental health struggles were the perfect catalyst for him to exploit this. The longer I struggled, the more he felt validated in his superiority, and the more he needed to be seen as the hero of our narrative, even though all of his actions only reinforced the dysfunction we were living in. If I was struggling emotionally or physically, he saw it as an opportunity to show me, and anyone else, that he was the stronger one, even though his behavior was incredibly toxic.

The more I fought for my own healing and independence, the more he fought to keep me in a position of subjugation. My recovery was not just a personal journey—it was an existential threat to his position in our relationship. His narcissistic injury—the fear of being rendered irrelevant or less important—manifested in his attempts to sabotage my progress. He saw my recovery as something that would inevitably diminish his place in the world, and that was something he could not tolerate.

The Narcissistic Projection and Reassertion of Control

During my time in therapy, I was continually haunted by a familiar sensation—the feeling that I was constantly under surveillance, even when I was supposed to be receiving the help I so desperately needed. It wasn't just my physical presence that he sought to control; it was my emotional state, my relationships, and the very process of healing itself. This was no longer just a passive desire to dominate; it was a calculated, deliberate effort to reassert control, particularly during the most vulnerable moments of my life. The narcissist thrives on the fear of losing control—and my mental health struggles became a battleground on which his need for power played out.

I remember walking into my therapy sessions, knowing that, despite my best efforts to process my emotions and gain clarity, his influence still loomed over me. He would sometimes pop up unannounced, showing up at my therapist's office under the pretense of being supportive, though it was clear he was not there to help me at all. His real purpose was far darker: he needed to ensure that everything was going according to his narrative, and that I was still playing the role he had assigned me—one of subjugation and dependence. His need to control was so strong that it transcended the boundaries of normal partnership or care.

There was a disturbing sense of entitlement in his behavior, a belief that my mental health recovery wasn't really about *me* at all; it was about him. Whenever I sought help, whether through inpatient care or outpatient therapy, it became, in his eyes, an opportunity for him to reinforce his superiority. He couldn't bear the idea of me becoming emotionally independent, and my dependence on outside support—whether from a therapist, a doctor, or the treatment facility itself—became something he had to subjugate.

Instead of understanding my need for professional care as a sign of strength or self-awareness, he twisted it into a personal offense against him. He saw my seeking help as a rejection of his authority. And the more I tried to reclaim my sense of self-worth, the more his behavior shifted. It wasn't just that he couldn't allow me to heal; he actively undermined my healing process. For him, my recovery wasn't a journey of growth or renewal—it was an attack on his fragile ego, a signal that he was not enough to fix me, and worse, that I could survive without him.

The moments when I began to gain some distance from his influence—whether through inpatient stays, therapy, or simply taking time for myself—were the moments when he felt most threatened. He would react by projecting his own insecurities onto me. If I was making progress, he would attempt to make me feel guilty for it. If I was finding my voice, he would shut me down. He would insist that I was selfish for taking the time to heal, that my focus on "getting better" somehow hurt him, that I was abandoning him, and therefore I was unworthy of the care I sought. The very notion that I might prioritize my own healing became, in his eyes, a betrayal.

This projection wasn't confined to a few isolated incidents. It was an ongoing pattern, one that I learned to dread. As I worked through my trauma, I came to realize that every time I sought help, it wasn't just about addressing my issues. It was about navigating his emotional needs, his desire for control, and his need to maintain dominance in our relationship. He would manipulate the situation to reassert himself as the focal point of the dynamic, regardless of what I needed.

Shifting the Narrative: I Became the Problem

His reassertion of control didn't stop at mere guilt-tripping. He was an expert in shifting the narrative, creating a scenario in which I became the problem. The more I healed, the more he pushed to ensure that his needs remained front and center. As I worked with my therapist to understand my own emotional landscape, he would actively try to insert himself into the process, often taking on the role of the "savior" or hero in our relationship. The moment I sought help or began making progress, he would need to remind me how much he had done for me—even though his actions were the very reason I needed help in the first place.

If I tried to talk about my own experiences, he would quickly reframe the conversation. Suddenly, my mental health struggles were about him—how hard it was for him to deal with my mental health issues, how I was putting him in a difficult position. It was an insidious manipulation that made me feel like I was being unfair or selfish for addressing my own problems. His projection of his emotional needs onto me created a warped reality in which I felt responsible for his well-being, while I simultaneously struggled to maintain any sense of self.

At times, when I shared my experiences in therapy, he would ask pointed questions that made it clear he was fishing for information—not because he was genuinely concerned, but because he needed to control the narrative. If I shared something that didn't align with his image of the perfect family or relationship, he would demand to know why I hadn't kept certain details from him, or why I had disclosed certain emotions. His ego couldn't handle the idea that my internal world was separate from his control.

The worst part was how often he played the victim. If I sought therapy or counseling, his immediate response was to ask, "What's wrong with you? Why can't you just be happy?" As though my mental health challenges were a personal failure, and my inability to simply "handle everything like he did" was a moral failing. But in reality, he didn't want me to be well—he wanted me to be dependent on him. He wanted me to need him, but only on his terms.

Isolation and Reinforcement of the Control Dynamic

The longer I was in treatment, the more I came to understand the role of isolation in his narcissistic control. When I started getting better, he would subtly isolate me from the very people and support systems that were crucial to my recovery. Therapy sessions, where I began to gain insight into my emotional landscape, became a battleground for control. There were times when he refused to show up for the sessions that directly addressed his behavior or impact on me. If the conversation started to shift in a direction that made him uncomfortable, he would disappear from the discussions, leaving me to fend for myself. His absence was calculated—if he wasn't there, he couldn't be held accountable.

At the same time, he would undermine any progress I made by creating chaos at home. He would engage in arguments or subtle emotional abuse, making it impossible for me to focus on my recovery. He didn't just refuse to support me; he actively sabotaged the space I needed to heal. This was his way of reasserting control over the situation. By maintaining the illusion that he was the one who held everything together, he could continue feeding his delusional need for superiority.

His unpredictable behavior—at times cold and dismissive, at other times intensely controlling—kept me off balance. It was as if he needed to constantly test my boundaries, see how far he could push me before I would break. His need for control was so pervasive that it didn't matter if I was physically or emotionally absent; he wanted to control the situation, control my healing, and make sure that I was always dependent on him.

Narcissism and the Emotional Toll of Manipulation

The abuse wasn't just about emotional neglect—it was an intentional, strategic way to ensure that I could never fully recover. Every effort I made to seek professional care was weaponized against me, creating a dynamic where I was guilted and manipulated back into a role that left me feeling like I was never allowed to heal. It is exhausting for victims to navigate the emotional manipulation of a narcissist, especially when their abuser shifts the narrative and guilt-trips them back into their control.

In summary, his tactics during these vulnerable times were an expression of his deeper sadistic nature. He could not allow me the space or freedom to recover. Instead, he took advantage of my vulnerability to reassert control and validate his own superiority. It was an incredibly damaging cycle, one that made it impossible to break free, because every time I began to heal, he would find a way to make me question myself, my progress, and my place in the world.

References

- Ronningstam, E. (2005). Identifying and Understanding the Narcissistic Personality. Oxford University Press.

- McWilliams, N. (2011). Psychoanalytic Diagnosis: Understanding Personality Structure in the Clinical Process. The Guilford Press.

- Kernberg, O. F. (1975). Borderline Conditions and Pathological Narcissism. Jason Aronson, Inc.

Dr. Rachel Levitch

Chapter 10

Foundation First: The Link Between Rejection and Narcissistic Rage

In today's interconnected world, technology has become a tool for both connection and manipulation. While it has brought immense benefits, it has also opened new avenues for toxic behaviors, particularly for individuals with narcissistic tendencies. Cyberstalking, as part of narcissistic abuse, takes advantage of the anonymity and distance provided by digital platforms. It allows the abuser to maintain a level of control and manipulation without having to physically interact with their victim.

One of the most dangerous aspects of cyberstalking is the ease with which the abuser can hide behind the digital veil of technology. This disconnection between the abuser and the victim allows for a sense of power and control that would be more difficult to achieve in a face-to-face interaction. The narcissist can send emails, texts, or social media messages, each carefully crafted to provoke, manipulate, or confuse. The victim, on the other hand, is left to contend with the psychological toll of being constantly surveilled and emotionally attacked, often with no clear way to escape.

The digital nature of cyberstalking makes it particularly insidious. Unlike traditional forms of abuse, where there are clear physical boundaries, cyberstalking allows for an intrusion into every aspect of the victim's life—without any need for the narcissist to physically be present. The abuser can access personal information, create fake social media profiles, or even manipulate technology in ways that are not immediately apparent to the victim.

As the digital age evolves, it's becoming increasingly clear that the impact of narcissistic abuse is not confined to the real world—it spills over into the online space as well. Narcissists use digital platforms to extend their dominance, and cyberstalking becomes a tool for them to assert control, even when they are physically removed from the victim's life. Whether through social media posts, fake profiles, or other forms of online manipulation, the narcissist continues to weave their toxic web, leaving the victim with the persistent sense that they are always being watched, always being controlled, and never truly free.

Different Types of Narcissism: How They Manifest in Cyberstalking

It is important to understand that narcissism is not a one-size-fits-all disorder. Within the broader framework of Narcissistic Personality Disorder (NPD), there are several subtypes, each with its own manifestation of narcissistic behaviors. These subtypes of narcissism can have distinct impacts on how the abuser manipulates and controls their victim, especially when it comes to cyberstalking. Let's explore a few of these subtypes and how they relate to cyberstalking.

Grandiose Narcissist: The Open Display of Superiority

The grandiose narcissist is the most outwardly obvious and easily recognized form of narcissism. These individuals are often confident, assertive, and charming, and they thrive on external validation. At their core, they believe they are superior to others—entitled to admiration and praise, and deserving of special treatment. The grandiose narcissist sees the world through a lens of entitlement, where their needs and desires should always come first. They often use their outward persona to manipulate others into seeing them as perfect, powerful, and untouchable.

In the age of the internet, the grandiose narcissist finds fertile ground to expand their influence and continue their manipulation. Cyberstalking, in their case, often becomes a tool to assert control and maintain their dominance over the victim. They may use social media and other online platforms to create an idealized image of themselves—carefully curated to show them as successful, beloved, and powerful, while simultaneously undermining and attacking the victim in private. The need for validation is constant; they require others to see them as exceptional, and anything that challenges their image is seen as a threat.

In my experience, my ex-husband embodied many of the traits of a grandiose narcissist. After our separation, he started using social media as a platform to project an image of confidence and success. He posted frequent updates on his achievements, showing off new relationships, career milestones, and extravagant vacations—painting a picture of someone who had moved on and was thriving. However, behind the scenes, he engaged in a very different form of control. He used private messages and online comments to send veiled threats and passive-aggressive remarks meant to provoke emotional responses from me. His goal was not simply to hurt me, but to remind me that he was still the one in control—regardless of how his public image portrayed him.

Grandiose narcissists often use their online presence to create a false narrative, casting themselves as victims or heroes while deflecting blame onto the real victim. For example, after our breakup, my ex frequently used public forums like Facebook to post statements that distorted the truth about our relationship. His posts often painted me as emotionally unstable, irrational, and the source of our problems, despite the fact that he was the one who had consistently manipulated, controlled, and belittled me. These posts were not just casual updates—they were calculated moves designed to ensure that anyone who followed him, anyone who saw his posts, would believe that he was the innocent party and that I was the one who was unworthy of sympathy or understanding.

It wasn't just about controlling me—it was about controlling how others saw him, and more importantly, how others saw us as a couple. In his mind, my leaving him was a threat to his sense of superiority and his carefully constructed image. To reassert control, he used social media not only to paint himself as the better partner and father but also to subtly manipulate my emotions. His messages, though seemingly benign or even supportive, were often filled with passive-aggressive undertones meant to provoke feelings of guilt, doubt, and inferiority.

Public Image vs. Private Manipulation: Grandiose narcissists often use social media to broadcast their success and project a persona of happiness or fulfillment. These broadcasts are often in direct contrast to their private behavior, which is aimed at controlling the victim. They may send private messages meant to gaslight the victim into believing that they are the cause of the problem. For example, my ex often reached out under the guise of offering support, but the underlying messages would be filled with subtle accusations about my parenting or personal decisions.

The Narcissist's Need for Validation

For a grandiose narcissist, being seen as successful or admired is a non-negotiable need. This obsession with validation is not just about appearing perfect—it is an emotional necessity. Without this validation, they risk feeling empty, unimportant, and powerless. For instance, whenever my ex experienced a minor setback or rejection, he would flood social media with updates that reflected his triumphs and the adoring support of others. It wasn't just about boasting; it was about compensating for the feelings of inadequacy he felt in private. The need to project power was essential for him to feel whole.

In many ways, this type of narcissism and cyberstalking behavior is deeply interwoven with their fragile self-esteem. The more they engage in these behaviors, the more they reinforce the delusion that they are in control of their environment—and most importantly, the people around them. As Kernberg (1975) explains, narcissistic behavior is rooted in a fragile ego that is constantly at risk of being punctured by rejection, exposure, or devaluation. For someone with grandiose narcissism, even something as seemingly minor as a breakup can feel like an existential threat to their sense of self-worth.

As a victim, it becomes exhausting to live in a world where your ex is constantly reasserting their control in such an insidious way. The emotional exhaustion of dealing with his cyberstalking, manipulation, and distortion of reality left me feeling both trapped and isolated. Each time I tried to move forward, the carefully crafted images of him, broadcasted online, were enough to pull me back into his sphere of influence. The key here is that the narcissist's use of cyberstalking is not a spontaneous reaction—it is a calculated strategy to protect their sense of superiority and to keep the victim in a perpetual state of emotional dependence.

Ultimately, for the grandiose narcissist, cyberstalking isn't just about gaining power over the victim—it's about creating a false narrative that justifies their sense of entitlement, and superiority. It's a mechanism that allows them to maintain control not only over the victim's emotions but also over how the world perceives them.

Vulnerable Narcissist: Cyber Harassment as Emotional Manipulation

The vulnerable narcissist often flies under the radar because they don't outwardly display the grandiose traits associated with narcissism, such as arrogance or self-importance. Instead, they present themselves as sensitive, emotionally fragile, or even self-deprecating. However, at their core, they harbor a deep need for validation and external affirmation. They may struggle with feelings of inferiority and self-doubt, but these feelings are often masked by a defensive need to be seen as important or worthy of sympathy (Miller et al., 2011).

When faced with rejection or devaluation, the vulnerable narcissist can experience an intense narcissistic injury, leading to rage, resentment, and feelings of being personally attacked. Unlike the grandiose narcissist, who responds to rejection with overt aggression or public acts of superiority, the vulnerable narcissist is more likely to engage in covert forms of emotional manipulation. The digital realm, with its anonymity and ease of creating false personas, becomes the perfect platform for this form of covert cyber harassment.

Covert Cyber Harassment and Manipulation

One of the key characteristics of the vulnerable narcissist is their tendency to play the role of the victim, often in an exaggerated or fabricated way (Kernberg, 1975). They might use cyber harassment as a way to garner sympathy, portray themselves as the wronged party, and manipulate others into aligning with them. This can manifest in several harmful behaviors:

Passive-Aggressive Messages: Vulnerable narcissists may send cryptic or passive-aggressive messages to the victim. These messages often evoke pity or guilt, suggesting that the victim is causing emotional harm without directly accusing them. For example, a vulnerable narcissist might send a message that reads: "I never thought you'd do this to me. I've been nothing but good to you, but I guess I don't matter to you at all." This type of communication creates emotional confusion for the victim and fosters feelings of guilt or self-doubt (McWilliams, 2011).

Manipulating Others to Attack the Victim: A particularly insidious tactic employed by vulnerable narcissists is their ability to manipulate others into attacking the victim on their behalf. They may reach out to friends, family members, or even strangers to spread their false narrative of victimhood. The narcissist might say things like, "I'm just so hurt, I don't know what to do. Can you please check on me? I'm just trying to keep it together." In doing so, they create a web of sympathy, making the victim appear cruel or heartless by comparison (Stinson & Johnson, 2012).

Creating Fake Social Media Accounts: Vulnerable narcissists are highly skilled at using social media to manipulate public perception and gain the validation they crave. They may create fake accounts or anonymous profiles to post misleading or distorted information about the victim. This allows them to present themselves as the innocent party, all the while continuing to manipulate the victim's emotions. In this way, the digital space becomes an extension of the narcissist's false narrative of victimhood, allowing them to influence others while escaping direct accountability for their actions (Stern, 2007).

Crafting a False Narrative of Being "Misunderstood": A central characteristic of the vulnerable narcissist is their sense of being misunderstood. This narrative often plays out in digital spaces where the narcissist can post subtle references to their suffering. They may write posts like, "Some people just don't understand how hard it is to feel alone," or, "Sometimes the people closest to us hurt us the most." These posts might seem benign to outsiders, but they are calculated moves to solicit empathy and garner support from a broader audience. By doing so, the narcissist reinforces their sense of self-worth and keeps the victim in a constant state of emotional vulnerability (Miller et al., 2011).

Cyberstalking as a Tool of Control

For the vulnerable narcissist, cyber harassment is not merely about revenge or retaliation—it is about maintaining control over the narrative and, ultimately, over the victim's emotional state. This form of cyberstalking is far more subtle than other forms, but it is no less harmful. Each text message, social media post, or passive-aggressive comment is part of a larger strategy to manipulate the victim's emotions and thoughts, causing them to question their own perceptions of reality (Stinson & Johnson, 2012).

The narcissist's primary goal in cyber harassment is to disrupt the victim's sense of self. Every act of online manipulation—whether it's sending anonymous messages, creating false narratives, or manipulating mutual acquaintances—serves to destabilize the victim's emotional equilibrium. The result is a state of hypervigilance and self-doubt, where the victim begins to question their own boundaries, reality, and sanity (Herman, 1992). Over time, this erodes the victim's mental and emotional health, contributing to anxiety, depression, and post-traumatic stress (PTSD) symptoms.

The Digital Age and the Empowerment of Vulnerable Narcissists

The digital space has provided a platform for vulnerable narcissists to amplify their emotional abuse. Through the use of social media, email, text messages, and anonymous accounts, they can continue to monitor, harass, and manipulate the victim from a distance. The ease with which they can create false identities or distort the truth in online spaces makes it incredibly difficult for the victim to escape the abuse. In this way, the digital world becomes a weapon that allows the narcissist to extend their psychological control into the victim's daily life, unseen and inexorable (Stern, 2007).

The Trauma of Cyber Harassment for Victims

The emotional toll of being subjected to cyber harassment by a vulnerable narcissist is profound. The victim is left constantly on edge, unsure when the next covert attack will come. This creates a heightened state of hypervigilance, where the victim is always anticipating the next message, the next manipulation, or the next attempt to undermine their sense of self.

As research by Dutton & Painter (1993) indicates, such emotional manipulation can lead to long-term psychological damage, often resulting in Complex PTSD (C-PTSD). Victims of cyberstalking experience intense emotional distress, including feelings of helplessness, anxiety, and a loss of trust in their own perception of reality. This is compounded by the fact that, in the case of vulnerable narcissists, the harassment is often indirect and subtle, making it difficult to pinpoint the abuse. The victim may feel isolated, constantly doubting their own feelings and perceptions, making it even more challenging to break free from the narcissist's grip (Herman, 1992).

Malignant Narcissist: Cyberstalking and the Weaponization of Control

The malignant narcissist is the most dangerous and toxic variant of narcissism, blending traits of Narcissistic Personality Disorder (NPD) with those of a sociopath or psychopath. Unlike other forms of narcissism, the malignant narcissist's sense of entitlement and superiority is not just a mask for fragile self-esteem—it is a deeply ingrained belief that they are entitled to control, dominate, and harm others without remorse (Kernberg, 1975). These individuals are marked by a total lack of empathy and often engage in behavior that is not merely self-serving but actively destructive to others.

Their actions are often deliberate, calculated, and sadistic. Malignant narcissists derive pleasure from the emotional or psychological pain of others and will go to extreme lengths to inflict harm. The digital space, with its ability to provide anonymity, offers the malignant narcissist a powerful tool to continue their abusive behaviors long after the victim has left their physical presence. In the case of my ex-husband, his digital stalking and cyber harassment went far beyond simple obsession—it was an active psychological warfare designed to break me down emotionally, assert dominance, and destabilize my mental health.

The Malignant Narcissist's Use of Cyberstalking as Psychological Terrorism

For the malignant narcissist, cyberstalking is not just a means of monitoring or keeping tabs on the victim—it is an extension of their need to assert control and inflict emotional suffering. They view the digital space as a battlefield where they can engage in psychological terrorism. Unlike the vulnerable narcissist, whose goal is primarily to seek sympathy and manipulate the victim emotionally, the malignant narcissist thrives on causing harm and asserting power over their victim.

Tactics of a Malignant Narcissist in Cyber Harassment

Malicious Rumors and Public Smear Campaigns: One of the most common tools used by a malignant narcissist is the creation and dissemination of false narratives. They might spread malicious rumors about the victim online, telling friends, family, or even strangers fabricated stories to discredit and undermine the victim's character. These lies are designed to destroy the victim's reputation and leave them isolated and vulnerable. For example, the narcissist may claim that the victim is mentally unstable, unfit to parent, or even abusive, all the while portraying themselves as the victim of the victim's irrationality.

Identity Theft and Doxxing

A malignant narcissist may escalate their cyberstalking tactics to identity theft or doxxing, a form of cyber harassment where private, personal information (such as phone numbers, addresses, or employment details) is publicly shared with the intent to cause harm. This tactic is especially dangerous as it not only invades the victim's privacy but also exposes them to further attacks, such as threats, harassment from third parties, and financial damage. In some cases, these narcissists may even use their victim's personal information to damage their career or create chaos in their personal life (Dutton & Painter, 1993).

Gaslighting Through Digital Manipulation

The malignant narcissist's use of gaslighting—the manipulation of facts and events to make the victim doubt their reality—is intensified through cyberstalking. They may reframe past incidents, post distorted versions of conversations, or even hack into the victim's private messages to twist their words. This creates a reality where the victim feels as if they are losing their grasp on the truth. For instance, they may attempt to manipulate mutual friends or family members into believing that the victim is paranoid or delusional for even suspecting that they are being stalked online. By doing this, the malignant narcissist further isolates the victim and undermines their credibility (Stinson & Johnson, 2012).

Hacking and Digital Surveillance

A particularly insidious aspect of cyber harassment from a malignant narcissist is their tendency to engage in digital surveillance. This can include hacking into emails, monitoring social media accounts, or using spyware to track the victim's online behavior. This surveillance is not only about maintaining control; it is about instilling fear and hopelessness. Every email read, every text message intercepted, and every private conversation accessed serves to remind the victim that they are never truly free from the narcissist's grip. This constant invasion of privacy leaves the victim in a perpetual state of hypervigilance, making it almost impossible to feel safe or secure in their own life (Miller et al., 2011).

Narcissistic Injury and the Malignant Need for Retaliation

The malignant narcissist's motivation for these harmful behaviors is rooted in what is known as a narcissistic injury—a severe blow to their self-esteem and sense of superiority caused by the victim's rejection or attempt to break free. Narcissists, particularly the malignant type, cannot tolerate any challenge to their delusion of grandeur, and a rejection feels like a personal attack on their very identity (Kernberg, 1975). This narcissistic rage manifests as a relentless desire to exact revenge and prove that they still control the victim's life, even from a distance.

For example, in my own experience, my ex-husband's obsession with maintaining control was exacerbated after I left. His deep sense of betrayal and humiliation led him to engage in cyber harassment and online manipulation, trying to force me back into a place of subjugation. By constantly bombarding me with messages, emails, and false accusations, he not only sought to harm me emotionally but also to reassert control over me. His need to be seen as superior and in control was so overwhelming that he resorted to digital abuse as a form of psychological torture, hoping to break my spirit and force me back into submission (Dutton & Painter, 1993).

The Psychological Impact: Trauma and Isolation

The psychological toll of such cyber harassment is profound. Narcissistic abuse—whether in person or online—leaves long-lasting scars on the victim's mental health. In my case, the constant emotional manipulation and psychological warfare left me with Post-Traumatic Stress Disorder (PTSD) and Complex PTSD (C-PTSD) symptoms, including anxiety, hypervigilance, and intrusive thoughts about my abuser's ongoing tactics. Every email notification, every social media alert, brought with it a surge of fear and dread—because I knew it meant the narcissist was still trying to control and destroy me from afar.

As Herman (1992) notes, the nature of narcissistic abuse often leaves victims in a state of emotional paralysis, unable to function fully in their personal and professional lives. The combination of gaslighting, identity theft, and doxxing only deepens the victim's sense of helplessness, as they are forced to live in fear of further invasions of privacy and manipulations. These tactics reinforce the narcissist's false narrative that they are superior and untouchable, while systematically eroding the victim's sense of self-worth and sanity.

The Path Forward: Recovery from Malignant Narcissistic Abuse

Recovery from cyberstalking and the trauma of malignant narcissistic abuse is a long and difficult process. As McWilliams (2011) describes, victims must work to rebuild their self-esteem and sense of autonomy, often with the help of professional therapy and support networks. It is crucial for victims to set boundaries, limit contact, and regain control over their digital spaces in order to heal from the emotional devastation caused by such intense manipulation. But even with these tools, the road to recovery is often fraught with setbacks as the malignant narcissist may continue their psychological warfare from afar, testing the victim's strength and resolve.

Covert Narcissist: Subtle Manipulation and Cyberstalking

The covert narcissist operates under the radar of overt recognition. Unlike the grandiose narcissist, who thrives on admiration and external validation, the covert narcissist's narcissism is often hidden behind a facade of insecurity or fragility. They don't seek attention in the traditional sense; instead, they expect others to cater to their emotional needs without question. Their narcissistic injury, which occurs when their sense of entitlement is challenged, often manifests in subtle, insidious behaviors. These individuals are masters at playing the victim while quietly orchestrating emotional manipulation.

In the case of my ex-husband, I observed a gradual shift from his grandiose narcissism, where he overtly sought validation and control, to a more covert approach. After our separation, his behavior became less about seeking physical presence and more about emotional manipulation—through a series of indirect and passive-aggressive tactics that served to control, confuse, and manipulate me. He no longer relied on direct confrontation but instead used covert cyberstalking and emotional abuse to maintain his dominance over me.

The Covert Narcissist's Subtle Cyberstalking Tactics

Unlike malignant narcissists, who might engage in overt harassment, covert narcissists prefer subtlety and deniability in their actions. Their cyberstalking is less about public campaigns of defamation and more about private, calculated emotional invasions.

Cryptic Messages and Passive-Aggressive Comments: One of the most common tactics used by covert narcissists is sending cryptic messages or passive-aggressive comments. These messages often contain thinly veiled insults or suggestions, making the victim question their own behavior or sanity. In my case, after the separation, I received vague texts and social media comments from my ex-husband, often disguised as concern for my wellbeing. However, these messages were laced with guilt-tripping and passive accusations, subtly undermining my self-worth. He would send texts saying things like, "It's hard to be the one who's always there for everyone, but I guess that's what people expect," or "I don't know how you do it, but I just hope the kids are okay." These comments, though not explicitly hurtful, carried an underlying emotional weight meant to manipulate and provoke feelings of guilt or self-doubt.

Dr. Rachel Levitch

Triangulation and Enlisting Third Parties: A hallmark of the covert narcissist's manipulation is triangulation—the act of enlisting third parties, such as friends or family, to turn against the victim and create a false narrative about them. The covert narcissist will often approach others with subtle, indirect messages, telling them that they are the true victim, and using third-party validation to undermine the victim's credibility. My ex-husband did exactly this, reaching out to mutual friends and even my own family members, speaking poorly of me without directly confronting me. He would craft stories about how I was the one who was unreasonable or emotionally unstable, playing the role of the concerned but hurt ex-husband, while secretly manipulating the situation to make me seem like the problem. This triangulation not only made me feel isolated but also sowed discord in my relationships with people I trusted, further deepening my sense of confusion and emotional turmoil.

Sympathy Manipulation and False Vulnerability

The covert narcissist also frequently uses sympathy manipulation as a tool. They may pretend to be vulnerable or play the role of the suffering party, casting themselves as the one who has been wronged or abandoned. This tactic is designed to make the victim feel sorry for them, even when they are the true abuser. After I left, my ex-husband's behavior shifted to a more sympathetic and fragile persona. He began to frame himself as someone who was now helpless, struggling to manage the children and life without me. He would frequently tell mutual friends how he was doing everything alone and how he was having a hard time without me.

In doing so, he created a false narrative that positioned me as the villain—the one who abandoned him—while he played the role of the sacrificial, misunderstood ex-husband. He would use social media to post sad, cryptic messages about "overcoming adversity" and "doing it all alone," in an effort to manipulate others into feeling pity for him. These emotional displays were designed to distract from his abusive behavior and reinforce the illusion that he was the one who was suffering the most.

Undermining the Victim's Sense of Reality: The covert narcissist often seeks to distort the victim's perception of reality. They will deny past events, manipulate conversations, and use gaslighting tactics to make the victim question their memories or interpretation of events. This subtle manipulation is especially dangerous because it is difficult for the victim to pinpoint what is happening, leaving them feeling disoriented and unsure of their own judgment. For me, this was one of the most insidious aspects of my ex-husband's behavior. After I left, he would constantly deny things he had said or done, rewriting history to make himself seem like the innocent party. I would remember conversations where he openly admitted certain manipulative behaviors, but then he would flatly deny them, making me feel as though I had imagined the entire interaction. This constant reframing of reality made me feel like I was losing my sanity.

Covert Narcissism in the Digital Age

A Continuous Cycle of Control

The rise of social media and digital communication has provided covert narcissists with an ideal platform to continue their abuse from afar. The anonymity of the online world allows them to engage in covert harassment without immediate consequences, creating a false sense of safety as they carry out their emotional manipulations. For my ex-husband, his use of Facebook, Instagram, and WhatsApp allowed him to send cryptic messages, use mutual acquaintances to spread rumors, and create a narrative that painted him as the one wronged by me. His ability to stalk my social media profiles, comment on posts, and even create fake profiles to interact with me or my friends infiltrated my sense of privacy, leaving me feeling constantly exposed and vulnerable.

The digital space allowed him to manipulate the narrative and project a false image of himself as a victim, while hiding his manipulative behavior behind smiling photos and charming updates. This continual cyber harassment not only continued his control over my life but also made me question the authenticity of my online relationships, knowing that his covert narcissistic tactics could extend into these spaces as well.

The Hidden Impact of Covert Narcissism: A Legacy of Emotional Damage

As McWilliams (2011) explains, the covert narcissist's abuse is often more subtle but equally destructive. The lack of outward aggression allows them to avoid accountability while still carrying out a deeply harmful psychological agenda. Over time, the victim becomes emotionally exhausted, confused, and isolated—leaving them with long-lasting trauma that can be hard to identify. The covert narcissist's ability to disguise their abuse as concern, vulnerability, or confusion often leaves the victim feeling gaslighted, questioning their own perceptions, and doubting their self-worth.

For me, the covert abuse I endured left me emotionally exhausted. It wasn't just about the harassment—it was the constant undermining of my confidence, my relationships, and my own sense of reality. Every time I tried to regain control, my ex-husband's subtle manipulations made me feel as though I was always at fault, always the one who was misunderstood or irrational.

The Link: How Cyberstalking Connects to Narcissistic Abuse

The rise of the digital age has significantly changed the dynamics of narcissistic abuse, providing narcissists with a powerful platform for manipulation and control. No longer confined to physical spaces, narcissistic abuse now extends into cyberspace, where perpetrators can stalk, harass, and emotionally torment their victims without ever having to leave the comfort of their own home. Cyberstalking, as a form of narcissistic abuse, is not just a nuisance—it is an invasive, insidious form of psychological warfare.

Narcissists have always relied on manipulation, devaluation, and control to assert their dominance over others, but the digital world has made it easier for them to operate from the shadows. Whether they use social media, emails, or anonymous online accounts, cyberstalking allows them to continue their abusive behaviors while maintaining a sense of detachment and distance from the victim. What is truly harmful is not just the persistent contact but the psychological weaponry employed through these digital means. Narcissists can create a false narrative, gaslight the victim, or send calculated threats—all while hiding behind the safety of a screen.

The constant intrusion of the victim's private space through cyberstalking exacerbates the trauma already caused by narcissistic abuse. What begins as an isolated instance of harassment gradually escalates into a pervasive, constant presence in the victim's life. The narcissist's need for control, validation, and superiority remains unquenched, even after physical separation. They cannot accept that they no longer have a hold on their victim's life, and cyberstalking becomes a means to continue the emotional control they once held in person.

Different Types of Narcissism: How Cyberstalking Manifests in Various Forms

Understanding the different types of narcissism can help explain how each kind uses cyberstalking in distinct ways, tailored to their particular needs for validation and control. While all forms of narcissism share a core desire to maintain superiority, the tactics they employ can vary greatly. The digital space allows these narcissists to manipulate the narrative and shape perceptions, all while maintaining an aura of plausible deniability.

1. Grandiose Narcissist: The Narcissist as the Victimizer and Hero

The grandiose narcissist is typically the most visible and overt in their manipulation tactics. They crave admiration and view themselves as superior to everyone around them. These individuals often have a larger-than-life persona, marked by arrogance, entitlement, and an unshakeable belief that they are above reproach. They believe the world should cater to their every whim and are prone to using exploitation to ensure that their needs are always met.

In the context of cyberstalking, grandiose narcissists often use digital platforms to reshape the narrative and publicly humiliate their victim. They will often create false social media posts, send threatening emails, or even reach out to friends, family, and colleagues to manipulate them into taking their side. They often cast themselves as the victim in these narratives, portraying the victim as unstable or abusive. This victimization serves to bolster their fragile ego and restore their sense of superiority, as they continue to receive sympathy and support from their followers.

The cyberstalking becomes a form of public spectacle, with the grandiose narcissist using the digital world to keep their audience engaged in the drama. The purpose of this digital dominance is to ensure that the victim is never truly free. As the grandiose narcissist targets the victim online, they maintain a sense of self-validation by receiving praise or attention from others, while simultaneously reasserting their control over the victim's narrative.

2. Vulnerable Narcissist: The Manipulator in Disguise

The vulnerable narcissist is more subtle and introverted in their approach. Unlike the grandiose narcissist, they are not overtly confident or brash, but they still demand validation and recognition. Their self-esteem is fragile, and they often rely on external sources to prop up their sense of worth. While they may appear emotionally sensitive or fragile, their need for control and validation is just as deep as that of the grandiose narcissist.

In cyberstalking, the vulnerable narcissist uses passive-aggressive and covert methods of manipulation. Rather than being overtly threatening or arrogant, they tend to play the role of the victim, casting themselves as the one who has been wronged. They will often cry for sympathy, claiming that the victim is causing them emotional harm, or they will use social media to depict a false narrative of their own suffering. These individuals will often subtly manipulate others by creating an image of themselves as misunderstood or emotionally fragile.

For the vulnerable narcissist, cyberstalking becomes a means to keep the victim emotionally entangled and uncertain. They will send cryptic messages or engage in indirect harassment, often using third parties to harass the victim indirectly. The goal is not only to keep the victim confused and emotionally unstable, but also to create an environment where the victim feels guilty for causing harm to the narcissist.

3. Malignant Narcissist: A Dangerous, Sadistic Force

The malignant narcissist is perhaps the most dangerous type, exhibiting traits of narcissistic personality disorder (NPD) alongside antisocial behavior. These individuals often have a lack of empathy, a desire for cruelty, and an enjoyment of the suffering of others. They may even take sadistic pleasure in the pain and emotional turmoil they cause, using their power to create chaos in the victim's life.

In cyberstalking, malignant narcissists use their digital reach as a form of psychological terrorism. They are methodical and calculated in their approach, often engaging in doxxing (revealing private information), identity theft, or other means of destroying the victim's life. The ultimate goal is to degrade the victim, exerting as much psychological control and pain as possible through the use of technology. Their digital attacks are typically direct, savage, and designed to cause long-lasting emotional scars.

These narcissists often want to control the narrative online, and may go as far as creating fake accounts to pose as others, impersonate the victim, or spread false rumors. Every action is done with the intent to destroy the victim's sense of self-worth and public reputation, ensuring that the victim is never able to escape the clutches of the narcissist's control.

4. Covert Narcissist: The Hidden Manipulator

The covert narcissist is typically more subtle and underhanded in their abuse. These individuals often present as emotionally sensitive or self-deprecating, yet they still believe themselves to be entitled to special treatment and validation. Their narcissism is often masked by a facade of humility or victimhood, which makes their manipulation harder to spot.

In cyberstalking, covert narcissists are experts at gaslighting and subtle emotional manipulation. They may send indirect messages or use third parties to create confusion in the victim's life. Their messages might seem benign or neutral, but they are designed to distort reality and make the victim feel guilty or unreasonable. Covert narcissists thrive on creating situations where the victim second-guesses their own judgment or feels as though they are being irrational.

The covert narcissist is particularly adept at creating a false narrative about the victim's behavior. They may spread rumors, play the role of the victim, and enlist others to reinforce their narrative. In the digital space, they hide behind anonymity and indirect communication to maintain a level of plausible deniability, making it difficult for the victim to prove the abuse.

Dr. Rachel Levitch

The Shared Goal of Cyberstalking—Control and Superiority

Regardless of the specific subtype, all narcissists share a fundamental need for control, validation, and a sense of superiority. In the digital age, cyberstalking has become an efficient, low-risk way for narcissists to manipulate and control their victims without facing direct consequences. Whether they engage in overt harassment or covert manipulation, the goal is always the same: to maintain a position of dominance and to ensure the victim remains under their control.

While the tactics may differ depending on the narcissist's personality, the impact on the victim remains devastating. Cyberstalking, at its core, is a form of psychological abuse, and its effects can be long-lasting and deeply traumatic. The victim is left to contend not only with the physical scars of the abuse but also with the emotional and psychological toll of constant surveillance, manipulation, and harassment.

Index.

A

Addiction
Narcissists are addicted to supply (see 'supply') and will stop at nothing to find it. Survivors can also become addicted to their abuser through trauma bonding (see 'trauma bonding').

Admiration
The extreme and unrelenting need narcissists have to be liked, respected, adored, and placed on a pedestal above all others. Without constant admiration, narcissists feel lost and often become dangerous as they confront the reality of their low self-worth.

Apology
Narcissists don't apologize. They "faux apologize" simply to get what they want when they want. Narcissists don't believe they do anything wrong, so apologies are rare at best and never genuine at worst.

Aggrandizement
The narcissist's need to exaggerate their achievements, talents, or importance in order to appear superior. They inflate their abilities and accomplishments to gain admiration and reinforce their inflated self-image.

Antagonism
Narcissists often use antagonistic behavior to provoke and control others, creating conflict to assert dominance. They may mock, belittle, or undermine others to maintain a sense of power and superiority.

Attention-Seeking
Narcissists constantly seek attention and will do whatever it takes to stay the center of focus. They crave being noticed, whether through drama, self-promotion, or creating situations where they can be admired.

Anger
Narcissists may react with extreme anger when their ego is threatened or when they are not receiving the validation they demand. This anger can be unpredictable and disproportionate to the situation.

B

Blame Shifting
Narcissists avoid taking responsibility by shifting the blame onto others. They never accept fault for their actions, preferring to project their mistakes or shortcomings onto you, making you feel responsible for their wrongdoings.

Boundaries
Narcissists have no respect for boundaries. They ignore, violate, or trample over limits, personal space, and emotional needs. Your comfort, safety, or desires don't matter to them—they operate on the principle that their wants come first, no matter the cost to you.

Breadcrumbing
A manipulation tactic where narcissists offer sporadic affection or attention to keep someone hooked. The intermittent "nice" moments create false hope and emotional dependency, trapping the

victim in a cycle of highs and lows, never knowing when they'll be validated again.

Bluffing

Narcissists often bluff or exaggerate their power, knowledge, or intentions to intimidate others. By creating a false sense of superiority or control, they manipulate people into believing they are more influential or capable than they actually are.

Baiting

Narcissists use baiting to provoke a response from others, often by using insults, jabs, or emotionally charged remarks to trigger an emotional reaction. This is a form of control that gives the narcissist power over the victim's emotions and behavior.

Blocking

When a narcissist doesn't want to be confronted or held accountable, they may block people emotionally, mentally, or even on social media. It's a way of controlling who gets access to them and when, often leaving their victims confused or isolated.

Battering

In some cases, narcissists may resort to emotional or physical abuse, also known as "battering." This is a tactic to break down the victim's resistance, diminishing their self-worth and increasing their dependency on the abuser.

Black-and-White Thinking

Narcissists view the world in extremes—either you're with them, or you're against them. There's no room for nuance, compromise, or understanding. They see people as either "good" or "bad," and once you're labeled as "bad," they'll devalue and discard you without hesitation.

C

Charming
Narcissists use charm to win people over, disarm their targets, and get what they want before revealing their true, manipulative nature.

Childhood
Narcissism is often rooted in childhood experiences, such as over-parenting or neglect, shaping the narcissist's sense of entitlement and superiority.

Coercive Control
A form of manipulation that isolates, monitors, and dominates a victim, often through emotional, psychological, and physical control tactics.

Confabulation
Narcissists create false memories to protect their ego, blending truth with fabrication to distort reality and manipulate others.

Control
Narcissists seek total control over their victims, dictating every aspect of their lives—emotions, actions, and even relationships—to feel powerful.

Covert Narcissism
A more subtle form of narcissism where the individual appears

outwardly normal, often playing the victim while manipulating in the background.

Criticism
Narcissists are hypersensitive to any form of criticism, reacting with anger, defensiveness, or rage when their fragile ego is threatened.

Collapse
Narcissistic collapse happens when their inflated self-image is challenged, leading to extreme emotional reactions like rage, panic, or even sudden discard.

Charm Offensive
A calculated, all-out effort to seduce or manipulate people using excessive charm, often to achieve a hidden agenda or gain power over others.

Comparison
Narcissists constantly compare themselves to others, either elevating their own status or tearing others down to maintain a false sense of superiority.

Compartmentalization
Narcissists separate their emotions and actions into different "compartments," allowing them to treat people or situations inconsistently without feeling guilty.

Cognitive Dissonance
They create confusion by contradicting their actions with their words, leaving their victims questioning reality and doubting their perceptions.

Circling
A tactic where the narcissist avoids answering questions directly and instead goes around in circles, deflecting attention and confusing the victim.

Coldness
Narcissists use emotional detachment or coldness as a form of punishment, withdrawing affection to maintain control or create dependency.

Conditional Love
Love, affection, or approval from a narcissist is only given if their needs are met, creating an environment of uncertainty and emotional manipulation.

Crisis Creation
Narcissists may provoke or exaggerate crises to redirect attention to themselves, create drama, or manipulate others into solving problems they've caused.

Cunning
Narcissists are often highly skilled at manipulation, using cunning strategies to achieve their goals, deceive others, and cover up their true intentions.

Chameleon Effect
Narcissists adjust their behavior, personality, and values to mirror those of others, gaining trust and blending in to manipulate or control the situation.

D

Destabilising

This one is essentially about rocking the boat... making you lose your emotional balance. Narcissists enjoy using all sorts of manipulation tactics that result in you doubting your own memory, being confused, and ending up overwhelmed.

Devalue and Discard

Heard of the narcissistic cycle of abuse... idealise, devalue, and discard? Devaluing is the act of abusing you and watching you fall to your own personal rock bottom. Discarding is not something done by all narcissists but essentially it is the part of the cycle where they throw you out or dump you. The part where they toss you aside because they have no further need for you.

Diversion Tactics

Narcissists are like magicians, they have bags of tricks to use in their efforts to exploit and abuse you. Think gaslighting, blame shifting, word salad, threats, projections, triangulation, and love-bombing (all these terms are explained within this article). All these tricks are used to deflect responsibility.

E

Ego

Think I, my, mine. Self-interest and self-centredness abound with narcissists. You do not matter. Never have. Never will. It's a sad reality really.

Emotional Abuse

Name calling, shaming, threatening, swearing, belittling, and humiliating you, gaslighting, constantly rejecting you. Consider this psychological warfare.

Emotionally Unavailable

Don't expect a warm, close, loving connection with a narcissist. These people lack the most basic of social skills that allow for genuine human connection. Narcissists are unable to form close emotional bonds.

Empathy

Narcissists feel sorry for, or relate to, no one but themselves. As for you… suffering? They don't care. Exhausted? They don't care. Pregnant? They don't care. Sick? They don't care. Crying? They don't care.

Enabler

An enabler is someone who allows the abuser's behaviours to continue. Someone who encourages the narcissist to continue their toxic behaviour. Someone who knows exactly what the narcissist is up to but does nothing to stop it.

Entitlement

What's mine is mine *and* what is yours… is also mine! This is how a narcissist thinks. They will have what they want when they want and why they want. At any and all costs. They fundamentally believe they are entitled to everything they desire.

Exploitation

Narcissists are programmed to take. And they are good at it. Whether it's money, sex, friendships, assets, love… whatever it is… they will use you for their own personal gain. No holds barred.

F

Fantasy

Many narcissists have fantasies of grandiosity and heroism. Dreams where they save the day. Dreams where they are superior to all. They live in a made-up land of power and influence where they play out everything they long for in reality.

Flying Monkey

Flying monkeys are people who have also been manipulated by the abusive narcissist but who join their team and take sides with them, for countless reasons.

G

Gaslighting

A form of emotional abuse, gaslighting involves extreme psychological manipulation that results in you doubting your own memory or recollection of events. For example… "I never said that!" or "You're the crazy one, that never happened".

Grandiose

Many narcissists have an exaggerated sense of self, believe they are far more powerful than they are and have delusions of immortality. Grandiose behaviours are all about domination.

Grey Rock

Grey rocking is a technique you want to use if you are being abused by a narcissist. Essentially grey rocking is a de-escalation strategy you can use to help slow or stop abuse at any particular moment. It involves being unresponsive, or passive, in your interactions with the narcissist. It could be shrugging, nodding, avoiding the use of eye contact or even using silence to stop the narcissist raging. Also known as the 'extinction effect'.

H

Haughtiness

Haughtiness is about being arrogant, superior, and unreasonably and unabashedly proud of oneself... and that is exactly what a narcissist is.

Hoovering

This is the act of sucking you back into the relationship, either after a fight or at the point you decide to leave the relationship. Hoovering can involve promises of changed behaviours, excessive gift giving or things like intense make-up sex where finally they focus on you. All these behaviours are attempts to lure you back in so they can continue exploiting you for their own gain.

I

Idealise

As mentioned earlier, idealise is part of the narcissistic cycle of abuse where the abuser idealises you before devaluing and discarding you. The idealise stage is where they work the hardest to lock you into the relationship. This is the time you see their best behaviour... where they make the most effort to make the relationship work. Where they put you on a pedestal and tell you all the things they know you'll love to hear.

Insecure

Narcissists are deeply insecure, despite their external behaviours. Remember, misery enjoys company so they love the challenge of turning a happy, confident person into a shell of their former selves. Everything you see is a mask that hides their true self-loathing.

Injury

Similar to narcissistic collapse mentioned earlier, narcissistic injury is experienced by the narcissist when they encounter real or even *perceived* criticism, judgement, loss or abandonment etc. Instead of engaging in healthy reflection and learning something about themselves in these moments, the narcissist often feels humiliation or rejection due to their fragile sense of self. These feelings can result in all sorts of defensive behaviour including use of the silent treatment, stonewalling, making unfounded accusations, gaslighting and physical violence, amongst other things. Narcissists who have experienced a narcissist injury can, and often do, hold grudges for a long time.

J

Jealousy

Narcissists are jealous people. Let me say that again. Narcissists are jealous people. They're jealous of your strengths because they remind them of their weaknesses. They want you, because they *want* you, and because if they can't have you then no one can have you. This is why many women end up murdered at the hands of their current or ex-partner.

Joke

Everything is a joke to a narcissist. Everything. One of their classic lines … "You're so serious, *relax*, it was just a joke!". Narcissists use jokes to justify personal attacks on you that they expect you to just laugh off.

K-

L

Love Bombing

This usually happens at the start of a relationship with a narcissist, during the exciting honeymoon/idealise period where connection is buzzing and romance is blooming. Love bombing involves showering the victim with praise, gifts, loving words and otherwise excessive or over the top acts aimed at getting you to take the 'bait'… hook, line, and sinker.

Lying

Narcissists love lying like kids love lollies. Lying is their go-to for seeking admiration, blame shifting, hiding their flaws, and protecting their insecure sense of self.

M

Malignant

These narcissists are reckless, unhinged, and dangerous. Malignant narcissists often have strong anti-social traits or a diagnosis of anti-social personality disorder and are usually the type to kill their current or former partner.

Manipulation

Narcissists use a variety of tactics to manipulate and wear down their victim. Whatever their strategy, the aim of manipulation is to see them get what they want and it is driven by complete and utter selfishness and desire for power and control.

Mirroring

Mirroring involves copying and acting. As narcissists lack social and relational skills, they offer study their victims and look for clues for how to act 'normally', how to fit in. This is how many narcissists go un-noticed by those other than their partner. Mirroring allows them to behave in ways that 'most' people do so they can avoid being called out for their behaviour.

N

Narcissistic Mask

Narcissists have several versions of themselves who show differently depending on who they are with. This is why their friends see one version of them, but you see a totally different version of them behind closed doors. Narcissists know exactly which 'mask' to wear based on who they are with and what they are wanting.

No Contact

No contact is a form of self-preservation. A way to set and enforce strict boundaries in order to begin healing from prolonged, and often extreme, abuse. No contact means not allowing any contact with the narcissist. This may involve deleting and blocking their phone number, blocking emails from them, and refusing to respond to any contact attempts.

O

Overt

Overt narcissists are loud and proud, the ones who openly and unashamedly use and abuse others in order to advance themselves. These narcissists are aggressive and in your face.

P

Projection

Projection involves the narcissist shifting their unwanted 'stuff' onto you. It may be the traits they don't like about themselves, their shortcomings, or their flaws. Their aim is to blame you for everything they hate about themselves because they haven't got the capacity to sit in their own 'stuff'. So, if they are lying, they'll call you a liar. If they say you're jealous of them, it's because they are jealous of you.

Q-

R

Rage

Narcissistic rage is scary and unpredictable. They fly off the handle, often at the smallest of things which wouldn't worry most people. Perceived threats to their self-worth or self-esteem often trigger their rage which can intensify when they are called out for their behaviour or when the victim tries to leave and expose them for who they really are. Narcissists rage like toddlers throw tantrums... because they haven't the skills to process their emotions.

Reactive Abuse

In my opinion, not enough is known about reactive abuse, yet another weapon used by the narcissist to isolate and shame victims. Reactive abuse is when you have a normal reaction to an abnormal situation, likely after years and years of prolonged abuse. For example, when you finally let go and lash out or scream at an abuser, they use your reaction as proof that *you* are in fact the abuser. Reactive abuse is understandable, but it can see people get in serious trouble with the law if the narcissist reports it to police, so therefore it is essential that you remove yourself from the relationship asap.

Red Flags

Usually only seen in retrospect, once the abuse has already commenced, seeing the red flags is often the saddest moment of awakening for a victim. Red flags are warning signs that something isn't right, and often even if they are seen early in the relationship, people downplay them because they believe they are in love. Classic narcissistic red flags are an overinflated sense of self, lying, saying 'I love you' way to soon, love bombing, arrogance, lack of respect for your boundaries etc.

S

Smear Campaign

Smear campaigns launched by a narcissist can be devastating. They will lie about you and spread all sorts of untruths about you to those closest to you in order to have people see their version of events or their 'truth'. This behaviour is used to essentially 'get in first' so they can shut people up and convince them that you are a terrible person who is in fact abusing *them*. Smear campaigns are intentional and systematic.

Stonewalling

Stonewalling can happen in any relationship but in a relationship with a narcissist it takes on a whole new meaning. In simple terms, stonewalling is the act of refusing to communicate with someone. It could be the use of the silent treatment, using excuses to avoid communicating or outright refusing to listen to your needs.

Supply

Narcissists thrive on supply. Without it they cannot function. Supply is about the constant need for admiration and attention but also relates to the victim and what they can 'offer' the abuser in more general terms. When you supply a narcissist with praise, love, sex, money, a place to live or other such things, they happily drain you of these things to advance themselves.

Silent Treatment

The go-to manipulation tactic of many covert narcissists, the silent treatment is a passive-agressive behaviour intended to coerce or control you. This type of behaviour is not your typical 'going quiet for a while to sort out your emotions'… it is a deliberately used tactic and is considered a form of emotional abuse in the context of narcissism.

T

Threats

Just one tool in the narcissist's toolbox. Threats can be made to further manipulate and control the victim to have them do what the narcissist wants them to do. Often used to have a victim obey or agree with them, threats can be made against the victim, their pets, their family etc.

Trauma Bond

Ever wondered why you feel 'addicted' to your ex? Or why it takes so many times to leave them? Trauma bonding is why! Trauma bonding is a complex concept to explain in just a few short sentences but it involves the bio-chemical bonds developed during traumatic experiences shared with others. In the context of narcissistic abuse, trauma bonding explains the victim making excuses for the abuser, downplaying their abuse, or covering their tracks. Trauma bonding involves dopamine, serotonin, adrenaline, oxytocin and cortisol, all of which play their part in keeping a victim connected to their abuser.

Triangulation

Triangulation involves the use of a third person to further destabilise you. Narcissists love comparing you to others to make you do something they want you to do or to remain in control of you. A classic example of triangulation is comparing you to their ex to make you feel inferior and ultimately to have you do the thing you are reluctant to do or uncomfortable about doing.

U-

V

Validation

Narcissists have an obsessive need for validation and admiration that they use to strengthen their fragile sense of sense. Narcissists never validate the suffering of their victims because, in their eyes, everything is your problem, not theirs.

Vampire

When it comes to 'emotional vampires', people who suck you dry of everything you once were, narcissists take the cake. These people will twist and turn and manipulate and screw you for everything before leaving you to wallow in what they would call self-pity.

Victim Card

Abuse you as they do, narcissists love playing the victim card. This is because they cannot tolerate the reality of the truth you confront them with as it misaligns with their fantasy land version of themselves. Narcissists use this strategy to gain sympathy from those who they can play and manipulate later on.

Violence

Narcissists can, and often do, use violence against their victim to further isolate them and bring on shame and other heavy feelings that keep people stuck in toxic relationships. Many narcissists are paranoid and extremely vindictive, so their violence comes from a place of protecting their ego and sense of entitlement.

Vulnerabilities

You know those deep, private secrets or insecurities you shared with a narc at the start of your relationship when you were falling in love? Prepare for them all to be used against you later. Narcissists have no shame in throwing your normal, human insecurities in your face in efforts to make you hate yourself so that you are further destabilised and therefore, settle for further abuse. This is an ultimate power trip for a narcissist.

W

Word Salad

Ever walked away from a conversation with a narcissist thinking "What the heck was THAT?". Conversations with narcissists are never 'normal'. They go round and round, back and forth and in every other direction you can imagine. The 'word salad' concept relates to their ability to take a bit of this and a bit of that before throwing it all together, mixing it up and tossing it at you. You cannot make sense of nonsense and narcissists love watching their victim struggle to understand them as it shifts the focus of the conversation away from them.

X-

References:

Kernberg, O. F. (1975). Borderline Conditions and Pathological Narcissism. Jason Aronson.

Dutton, D. G., & Painter, S. (1993). The Battered Woman Syndrome. Guilford Press.

McWilliams, N. (2011). Psychoanalytic Diagnosis: Understanding Personality Structure in the Clinical Process. Guilford Press.

Herman, J. L. (1992). Trauma and Recovery: The Aftermath of Violence—From Domestic Abuse to Political Terror. Basic Books.

Stinson, P., & Johnson, L. (2012). The Impact of Narcissistic Abuse on Victims: Understanding the Dynamics of Emotional Exploitation. Journal of Clinical Psychology, 68(2), 234-243.

www.ingramcontent.com/pod-product-compliance
Lightning Source LLC
Chambersburg PA
CBHW050527100526
44581CB00009B/159/J